T0135735

Jens Keiner

Fast Polynomial Transforms

Logos Verlag Berlin

λογος

Bibliografische Information der Deutschen Nationalbibliothek

Die Deutsche Nationalbibliothek verzeichnet diese Publikation in der Deutschen Nationalbibliografie; detaillierte bibliografische Daten sind im Internet über http://dnb.d-nb.de abrufbar.

ISBN 978-3-8325-2850-8

Logos Verlag Berlin GmbH
Comeniushof, Gubener Str. 47,
10243 Berlin
Tel.: +49 (0)30 / 42 85 10 90
Fax: +49 (0)30 / 42 85 10 92
http://www.logos-verlag.de

Contents

List of Figures

List of Tables

Preface

Orthogonal polynomials have received, and continue to receive, much attention in the mathematical community. Developed from a study of continued fractions by Chebyshev in the 19th century and fostered by, among others, Markov and Stieltjes, orthogonal polynomials have found applications in many areas of mathematics and physics. More recently, the field of orthogonal polynomials has "mushroomed enormously", as Walter Gautschi has put it in his enjoyable book [23]. Piling up an enormous number of theoretical and practical results, research in this area continues to expand into previously unknown terrain. This text is restricted to the treatment of the well-established classical orthogonal polynomials and their close relatives, the classical associated functions.

The availability of modern day computers and the numerous applications also demand efficient algorithms to handle computational problems involving orthogonal polynomials. Some of these areas have so far been mainly of theoretical interest and prove to be a formidable challenge for the design of efficient algorithms. For example, classical orthogonal polynomials and classical associated functions play an essential role for Fourier analysis on the hyperspheres $\mathbb{S}^d$ and the related rotation groups $\mathrm{SO}(d)$. Both types of manifolds are important, for example, in the case of the sphere $\mathbb{S}^2$ for the analysis of satellite data which are usually recorded in spherical coordinates. The main goal of this text is to develop efficient algorithms related to the various applications of classical orthogonal polynomials and the classical associated functions.

Another efficient algorithm, one that has strongly influenced large areas of research, clearly is the fast Fourier transform (FFT). It is no exaggeration to call it one of the most important algorithms of the 20th century [12]. Since the description of a divide-and-conquer method for the calculation of the discrete Fourier transform by Cooley and Tukey in 1965 [13], the FFT has become one of the most influential algorithms ever. Today, the FFT has undergone many improvements and is an essential part of a large fraction of algorithms used today.

In recent years, a generalization of the FFT to arbitrary point configurations in the time (or space) domain, called the non-equispaced fast Fourier transform (NFFT), has contributed to a number of new numerical methods. A software library is publicly available free of charge [40]. The NFFT offers a more flexible way to approach a number of problems that previously have been difficult to access for Fourier methods; see, for example, [43, 68] and the references therein. As it turns out, the NFFT is also an important tool for the discrete Fourier transform on the sphere $\mathbb{S}^2$, the rotation group $\mathrm{SO}(3)$, and other manifolds, also in higher dimensions, for one simple reason: Typically, it is desirable or even required that the manifold be sampled non-uniformly with respect to the chosen coordinate system. One reason for this is that a uniform sampling usually entails a number of numerical problems. For example, on the sphere $\mathbb{S}^2$, a uniform sampling in spherical coordinates leads to points that cluster near the poles. Therefore, it is rather essential to have an algorithm that works efficiently for arbitrary nodes.

A problem with Fourier analysis on the mentioned manifolds is that one needs to handle expansions that no longer only involve complex exponentials, but also classical orthogonal polynomials and classical associated functions. To enable the application of the NFFT algorithm, an efficient method is needed to modify the expansions at hand so that they attain a suitable form. This must often be considered the hardest part as one routinely faces numerical challenges. In this thesis, we provide a complete framework for efficient and, as far as numerical results indicate, numerically robust algorithms to handle these problems. To achieve this, we formulate new theoretical results that characterize the *connection problem* for classical orthogonal polynomials and classical associated functions. This problem is concerned with the linear transformation that converts one expansion in classical orthogonal polynomials (or associated functions) into another one in a different sequence of classical orthogonal polynomials (or associated functions). Based on the theoretical results, we develop new efficient methods that provide the desired transformations. Our numerical examples show that the new methods offer competitive results.

Chapter 1 contains a compilation of mostly basic material on classical orthogonal polynomials and their associated functions. While many results can be found in standard references, the material has been adopted into a consistent form that is suitable for our needs. Most importantly, a number of explicit expressions for the connection coefficients is given. These numbers describe how a sequence of orthogonal polynomials may be represented through the members of another. We also observe a number of special cases that are important for algorithmic considerations in later chapters. Moreover, we give a complete definition of classical associated functions. To my best knowledge and despite the fact that most of these functions are well-known (at least in certain circles), reference literature on this topic seems to be scarce. This made it necessary to provide a concise study of this matter. Our approach is based on a modification to the Rodrigues formula for classical orthogonal polynomials and should be consistent with the literature, at least up to conventional scaling.

Chapter 2 is divided into two parts. In each, we develop a method for the efficient conversion between different expansions in classical orthogonal polynomials and in the closely related associated functions, respectively. We start with some basic material on certain classes of structured matrices that are important for our investigations. The single most important type of structured matrices that we will encounter are the *semiseparable* matrices. Interestingly, these have been investigated independently in different fields, see [77], and have recently received a growing amount of attention. The main goal of this chapter is to establish a link between the connection problem for classical orthogonal polynomials and their associated functions, and the class of semiseparable matrices. Our exploration of this topic was inspired by earlier work on the subject by Rokhlin and Tygert [70] who obtained such connection for the associated Legendre functions. They combined this with an algorithm developed by Chandrasekaran and Gu [9] to efficiently calculate the eigendecomposition of symmetric diagonal plus semiseparable matrices. As will be shown, this can be exploited for the desired purposes.

In the first part of Chapter 2, we describe an algorithm for the efficient calculation of the eigendecomposition of diagonal plus triangular generator representable semiseparable matrices. This algorithm was first published by the author in [38] and is based on an idea similar to that of Chandrasekaran and Gu in [9]. In addition, we provide also a simple extension to higher semiseparability ranks. It is then shown that the connection matrices between different sequences of classical orthogonal polynomials are properly scaled eigenvector matrices of certain diagonal plus triangular generator representable semiseparable matrices. The entries of these matrices are calculated explicitly for all important cases. This substantially expands upon the results obtained in [38] which covered only the Gegenbauer polynomials. For the first time, we provide results for all classical orthogonal polynomials. We show how the described algorithm that obtains the eigendecomposition of the calculated matrices can be exploited to obtain an entirely new method to convert between expansions in different classical orthogonal polynomials.

In the second part of Chapter 2, we briefly review the algorithm from [9] for the efficient calculation of the eigendecomposition of symmetric diagonal plus generator representable semiseparable matrices. An application, albeit unrelated to the content of this text, was explored in [46]. In addition to the original algorithm, we also give a simple extension to higher semiseparability ranks. Then we show that the connection matrices between different sequences of classical associated functions contain properly scaled eigenvectors of certain symmetric diagonal plus generator representable semiseparable matrices. The entries of these matrices are calculated explicitly for all relevant cases. This work substantially expands upon the findings in [70] where a similar result for the associated Legendre functions was obtained. For the first time, we provide results for all classical associated functions, including the generalized associated Jacobi functions.

In Chapter 3 an alternative method is developed for the conversion between different sequences of classical orthogonal polynomials. As of now, it is unfortunately not applicable to the classical associated functions. The method is based on the observation that the connection coefficients have a certain smoothness. This can be exploited by simple, yet powerful approximation techniques. More precisely, a number of results is given that state the exact convergence rate with which interpolants at the Chebyshev points converge to the desired coefficients. This enables the use of a variation of the well-known fast multipole method [28] to accelerate the calculation of the transformation. The work in this area was motivated by a similar result by Alpert and Rokhlin [1] for the connection between Chebyshev and Legendre polynomials, albeit with weaker theoretical results. Their findings were later generalized by the author in [39] to the Gegenbauer polynomials, proving for the first time the exact convergence rate. In this text, we also extend the technique to all classical orthogonal polynomials.

To demonstrate a few applications of the methods that we have developed, two nonequispaced fast Fourier transform algorithms are described in Chapter 4, one for the sphere $\mathbb{S}^2$ and the other for the rotation group SO(3). A method, similar to the first algorithm, was published by Rokhlin and Tygert in [70]. However, we combine this method, which is based on the findings in Chapter 2, with the NFFT for the first time. This allows for arbitrarily placed nodes on the sphere $\mathbb{S}^2$. Discrete Fourier transforms on the sphere are important in many applications. Related work includes [41, 42, 44, 45]. The second algorithm can be seen as a generalization of the previous concepts to the rotation group SO(3). While the methods are very similar, we use, for the first time, the results from the second part of Chapter 2 to establish the technique on the rotation group SO(3).

Since this thesis repeatedly makes use of a number of properties of classical orthogonal polynomials, an extensive formula reference is provided in the Appendix A. There, all relevant expressions can also be found for different normalizations, whereas the results in the text are usually only given for one particular normalization. To the reader this should serve as a reference for all expressions that are needed to implement the described numerical methods for classical orthogonal polynomials and classical associated functions.

The material has been arranged such that it can be read sequentially. General knowledge about orthogonal polynomials and structured matrices is not mandatory, but highly recommended. Chapters 2 and 3 contain a large number of tables and figures for illustration of numerical results. The reader may skip this material safely.

Feedback in any form that can help to improve the presentation of the material or fix errors is highly appreciated and may be directed to the following email address:

`fast.polynomial.transforms@googlemail.com`

Nevertheless, any infelicities in this work are entirely my fault.

The work was conducted at the Universität zu Lübeck between 2005 and 2009 and at the University of New South Wales in 2007. The support of the German Academic Exchange Service (DAAD) which made possible the research period in Australia is gratefully acknowledged. I would like to thank Prof. Daniel Potts, Prof. Jürgen Prestin, and numerous other colleagues for countless fruitful discussions and for sharing their insight with me.

Lübeck
09.09.2009

Jens Keiner

1 – Orthogonal polynomials

This introductory chapter is to present basic material on orthogonal polynomials with a particular focus on the classical orthogonal polynomials and their associated functions. Most of the results are found in standard references [2, 10, 23, 63, 64, 65, 73, 74] and have been adapted from there. Proofs are given for a number of other results that are usually not found in the literature.

1.1 Definition and existence

Orthogonal polynomials can be introduced in various ways. A definition that is sufficient for our purposes is given by Gautschi [23, p. 1]: Let $\lambda(x)$ be a non-decreasing function on the real line $\mathbb{R}$ with finite limits for $x \to \pm\infty$ and assume that the induced positive measure $\mathrm{d}\lambda$ has finite moments of all orders

$$\mu_n = \mu_n(\mathrm{d}\lambda) := \int_{\mathbb{R}} x^n \,\mathrm{d}\lambda(x), \qquad n = 0, 1, \ldots, \quad \text{with } \mu_0 > 0. \tag{1.1}$$

Let $\mathbb{P}$ denote the space of real polynomials and let $\mathbb{P}_n \subset \mathbb{P}$ be the restriction to those of degree at most n. Then for two polynomials $p, q \in \mathbb{P}$ one may define an inner product by

$$\langle p, q\rangle = \langle p, q\rangle_{\mathrm{d}\lambda} := \int_{\mathbb{R}} p(x)q(x)\,\mathrm{d}\lambda(x). \tag{1.2}$$

If $\langle p, q\rangle = 0$, then p and q are said to be *orthogonal* to each other. If we take $p = q$, then

$$\|p\| := \sqrt{\langle p, p\rangle} = \left(\int_{\mathbb{R}} \left(p(x)\right)^2 \mathrm{d}\lambda(x)\right)^{1/2}$$

is the induced norm of p. We restrict ourselves to polynomials of a continuous variable, usually denoted x. Most of the theory, however, carries through to polynomials of a discrete variable; see, e.g., [64, p. 106].

Definition 1.1 *A sequence of polynomials $\{p_n\}_{n\in\mathbb{N}_0}$ with degree* $\deg p_n = n$ *is called* orthogonal (with respect to a measure $\mathrm{d}\lambda$) *if it satisfies*

$$\begin{aligned} \langle p_n, p_m\rangle &= 0, \qquad n, m = 0, 1, \ldots, \quad \textit{with } n \neq m, \\ \|p_n\| &> 0, \qquad n = 0, 1, \ldots. \end{aligned}$$

In most cases, the measure $\mathrm{d}\lambda$ uniquely defines a sequence of orthogonal polynomials $\{p_n\}_{n\in\mathbb{N}_0}$ up to a multiplicative constant in each p_n; see Theorem 1.9. Two important cases of such scalings are the *monic* and *orthonormal* variants.

Definition 1.2 *Let $\{p_n\}_{n\in\mathbb{N}_0}$ be a sequence of orthogonal polynomials. Then k_n denotes the leading coefficient of the polynomial $p_n(x) = k_n x^n + \cdots$, and $h_n := \|p_n\|^2$ the square of its norm.*

Definition 1.3 *Let $\{p_n\}_{n\in\mathbb{N}_0}$ be a sequence of orthogonal polynomials. Then $\{\bar{p}_n\}_{n\in\mathbb{N}_0}$ denotes the corresponding sequence of* monic orthogonal polynomials

$$\bar{p}_n(x) := k_n^{-1} p_n(x) = x^n + \cdots,$$

with leading coefficient $\bar{k}_n = 1$ and squared norm $\bar{h}_n := \|\bar{p}_n\|^2$. Likewise, $\{\tilde{p}_n\}_{n\in\mathbb{N}_0}$ denotes the sequence of orthonormal polynomials

$$\tilde{p}_n(x) := h_n^{-1/2} p_n(x),$$

with $\tilde{k}_n$ as leading coefficient of $\tilde{p}_n$ and the squared norm $\tilde{h}_n = 1$.

The following result shows how the squared norm $\bar{h}_n$ and leading coefficient $\tilde{k}_n$ for the monic and orthonormal polynomials, respectively, can be calculated from the known squared norm h_n and the leading coefficient k_n.

Lemma 1.4 *Let $\{p_n\}_{n\in\mathbb{N}_0}$ be a sequence of orthogonal polynomials with leading coefficients k_n and h_n. Then*

$$\bar{h}_n = \frac{h_n}{k_n^2}, \quad \text{and} \quad \tilde{k}_n = \frac{k_n}{\sqrt{h_n}}.$$

PROOF. For the monic polynomial $\bar{p}_n = k_n^{-1} p_n$, we calculate

$$\bar{h}_n = \|\bar{p}_n\|^2 = \frac{\|p_n\|^2}{k_n^2} = \frac{h_n}{k_n^2}.$$

The identity for $\tilde{k}_n$ follows immediately from $\tilde{p}_n = h_n^{-1/2} p_n$. □

Remark 1.5 It should be noted that above definition of normalized orthogonal polynomials carries a slight sense of arbitrariness. For each normalized polynomial $\tilde{p}_n$ also the polynomial $-\tilde{p}_n$ has unit norm. Our definition ensures that the sign of the leading coefficients of normalized and non-normalized variants, $\tilde{p}_n$ and p_n, are the same.

Remark 1.6 Throughout this work, results will usually be given only for one normalization. Appendix A contains an extensive collection of formulae for all three normalizations.

Orthogonal polynomials satisfy $\langle p_n, p_m \rangle = \delta_{n,m} h_n$ with the usual *Kronecker delta*

$$\delta_{n,m} := \begin{cases} 1, & \text{if } n = m, \\ 0, & \text{else.} \end{cases}$$

Owing to the orthogonality relation, orthogonal polynomials provide a basis for polynomial spaces. The following result can be found in [23, p. 2].

Lemma 1.7 *Let $\{p_n\}_{n\in\mathbb{N}_0}$ be a sequence of orthogonal polynomials. Then the set $\{p_j : 0 \leq j \leq n\}$ constitutes a basis for the space $\mathbb{P}_n$.*

PROOF. Clearly, $\mathbb{P}_n$ is an $(n+1)$-dimensional vector space. If

$$\sum_{j=0}^{n} \gamma_j p_j = 0,$$

then, by orthogonality, taking the inner product with the polynomials p_i, $i = 0, 1, \ldots, n$, on both sides reveals that $\gamma_j = 0$ for $j = 0, 1, \ldots, n$. Thus, the polynomials $p_0, p_1, \ldots, p_n$ must be linearly independent and thereby constitute a basis for the space $\mathbb{P}_n$. □

We have defined the inner product in (1.2) in a loose sense since it is generally not guaranteed that it will be positive definite. Since positive definiteness is a sensible and important requirement that is satisfied by most inner products that we may define, let us make the definition precise.

Definition 1.8 *The inner product in* (1.2) *is said to be* positive definite *on $\mathbb{P}$ if it satisfies $\|p\| > 0$ for any $p \in \mathbb{P}$ with $p \neq 0$.*

We are now ready to state a sufficient criterion for the existence and uniqueness of a sequence of orthogonal polynomials, with respect to a measure $\mathrm{d}\lambda$. For this, we must require positive definiteness.

Theorem 1.9 *If the inner product* (1.2) *is positive definite, then there exists a unique infinite sequence of monic polynomials $\{\bar{p}_n\}_{n\in\mathbb{N}_0}$ that is orthogonal with respect to the measure $\mathrm{d}\lambda$.*

PROOF. The polynomials $\bar{p}_n$ can be generated by applying the *Gram-Schmidt process* to the monomials $e_n(x) := x^n$, $n = 0, 1, \ldots$. Positive definiteness of the inner product ensures that the recursively generated polynomials

$$\bar{p}_n = e_n - \sum_{j=0}^{n-1} \frac{\langle e_n, \bar{p}_j \rangle}{\langle \bar{p}_j, \bar{p}_j \rangle} \bar{p}_j, \qquad n = 0, 1, \ldots,$$

are uniquely defined and, by construction, are orthogonal to all polynomials $\bar{p}_j$ with $0 \leq j < n$. □

1.2 Properties

Let in this section a measure $\mathrm{d}\lambda$ be given as before and we assume that it has infinitely many points of increase. We will also assume that the measure $\mathrm{d}\lambda$ is absolutely continuous, whereby $\mathrm{d}\lambda(x) = w(x)\,\mathrm{d}x$ with some non-negative integrable weight function w. Let $\{p_n\}_{n\in\mathbb{N}_0}$ denote a sequence of polynomials, orthogonal with respect to the measure $\mathrm{d}\lambda$, with its monic and orthonormal variants $\{\bar{p}_n\}_{n\in\mathbb{N}_0}$ and $\{\tilde{p}_n\}_{n\in\mathbb{N}_0}$, respectively.

1.2.1 Symmetry.

Definition 1.10 *An absolutely continuous measure* $\mathrm{d}\lambda(x) = w(x)\,\mathrm{d}x$ *is called symmetric with respect to the origin, if its support interval is* $[-a, a]$ *for some* $0 < a \leq \infty$, *and* $w(-x) = w(x)$ *holds for all* $x \in \mathbb{R}$.

Symmetry of the measure $\mathrm{d}\lambda$ implies that each orthogonal polynomial p_n is either an even or an odd function. This depends on the parity of the degree n and is formalized in the following result which can be found in [23, p. 6].

Theorem 1.11 *Let* $\mathrm{d}\lambda$ *be a symmetric measure. Then*

$$p_n(-x) = (-1)^n p_n(x), \qquad n = 0, 1, \ldots. \tag{1.3}$$

PROOF. We define the polynomials $\hat{p}_n(x) := (-1)^n p_n(-x) = (-1)^n k_n \bar{p}_n(-x)$. Then

$$\langle \hat{p}_n, \hat{p}_m \rangle = (-1)^{n+m} k_n k_m \langle \bar{p}_n, \bar{p}_m \rangle = \delta_{n,m} \hat{h}_n,$$

that is, the polynomials $\hat{p}_n(x)$ are orthogonal. Since $k_n^{-1}\hat{p}_n$ is a monic polynomial, we have $\hat{p}_n(x) = p_n(x)$ by uniqueness of the monic orthogonal polynomials; cf. Theorem 1.9. □

1.2.2 Zeros.

The following two results state important facts about the zeros of orthogonal polynomials. They can be found in [23, p. 6].

Theorem 1.12 *For* $n \geq 1$, *all zeros of the polynomial* p_n *are real, simple, and located in the interior of the support interval of the measure* $\mathrm{d}\lambda$.

PROOF. Since

$$\int_{\mathbb{R}} p_n(x)\,\mathrm{d}\lambda(x) = 0, \qquad n = 1, 2, \ldots,$$

the polynomial p_n must change the sign at least once inside the support interval of $\mathrm{d}\lambda$. Let x_j, $j = 1, 2, \ldots, k$, be all such points. If we had $k < n$, then

$$\int_{\mathbb{R}} p_n(x) \prod_{j=1}^{k} (x - x_j)\,\mathrm{d}\lambda(x) = 0$$

by orthogonality. This is impossible since the integrand has constant sign. Thus, we must have $k = n$. □

Theorem 1.13 *For $n \geq 1$, the zeros of the polynomial p_{n+1} alternate with those of the polynomial p_n, that is,*

$$\tau_1^{(n+1)} < \tau_1^{(n)} < \tau_2^{(n+1)} < \tau_2^{(n)} < \cdots < \tau_n^{(n)} < \tau_{n+1}^{(n+1)},$$

where $\tau_i^{(n)}$ and $\tau_i^{(n+1)}$ are, in ascending order, the zeros of p_n and p_{n+1}, respectively.

PROOF. See Remark 1.25 on page 6 for a sketch of a short proof. □

1.2.3 Three-term recurrence. For convenience, it is customary to formally extend a sequence of orthogonal polynomials $\{p_n\}_{n\in\mathbb{N}_0}$ by an element p_{-1}.

Definition 1.14 *Let $\{p_n\}_{n\in\mathbb{N}_0}$ be a sequence of orthogonal polynomials. Then the polynomial p_{-1} is defined by $p_{-1} = 0$. Furthermore, we let $k_{-1} = h_{-1} = 1$.*

Arguably, the single most important property of orthogonal polynomials is that they satisfy a *three-term recurrence.* The following result is found in similar form in [23, p. 10].

Theorem 1.15 *Let $\{p_n\}_{n\in\mathbb{N}_0}$ be a sequence of orthogonal polynomials. Then the polynomials p_n satisfy the three-term recurrence and initial conditions*

$$\begin{aligned} p_{n+1}(x) &= (a_n x - b_n)p_n(x) - c_n p_{n-1}(x), \quad n = 0, 1, \ldots, \\ p_{-1}(x) &= 0, \quad p_0(x) = k_0, \end{aligned} \tag{1.4}$$

with some coefficients $a_n, b_n, c_n, k_0 \in \mathbb{R}$, where $a_n, c_n \neq 0$ for $n = 1, 2, \ldots,$ and $a_0 \neq 0$.

PROOF. Since $x \cdot p_n(x)$ is a polynomial of degree $n+1$ it can be represented as a linear combination of the polynomials p_j for $j = 0, 1, \ldots, n+1$,

$$x \cdot p_n(x) = \sum_{j=0}^{n+1} \gamma_j p_j(x), \quad \text{with } \gamma_j = \frac{\langle x \cdot p_n, p_j \rangle}{\langle p_j, p_j \rangle}.$$

With the shift property $\langle x \cdot, \cdot \rangle = \langle \cdot, x \cdot \rangle$, obviously enjoyed by the inner product (1.2), the identity

$$\frac{\langle x \cdot p_n, p_j \rangle}{\langle p_j, p_j \rangle} = \frac{\langle p_n, x \cdot p_j \rangle}{\langle p_j, p_j \rangle}$$

is obtained. This evaluates to zero when $j < n-1$, owing to orthogonality of the polynomial p_n to those of strictly smaller degree. We have consequently $\gamma_j = 0$ for $j = 0, 1, \ldots, n-2$, whereby

$$x \cdot p_n(x) = \frac{\langle x \cdot p_n, p_{n+1} \rangle}{\langle p_{n+1}, p_{n+1} \rangle} p_{n+1}(x) + \frac{\langle x \cdot p_n, p_n \rangle}{\langle p_n, p_n \rangle} p_n(x) + \frac{\langle x \cdot p_n, p_{n-1} \rangle}{\langle p_{n-1}, p_{n-1} \rangle} p_{n-1}(x).$$

Solving this equation for p_{n+1} yields the desired result. For $n = 0, 1, \ldots,$ the fact that p_{n+1} is a proper polynomial of degree $n+1$ also implies $a_n \neq 0$. For c_n with $n = 1, 2, \ldots,$ we note that

$$c_n = \frac{\langle p_{n+1}, p_{n+1} \rangle}{\langle x \cdot p_n, p_{n+1} \rangle} \frac{\langle x \cdot p_n, p_{n-1} \rangle}{\langle p_{n-1}, p_{n-1} \rangle} = \frac{k_{n-1} k_{n+1}}{k_n^2} \frac{h_n}{h_{n-1}} \neq 0. \tag{1.5}$$

□

Remark 1.16 Although the coefficient c_0 is free of choice (it multiplies the polynomial $p_{-1} = 0$ in (1.4)) it is convenient to define $c_0 = 0$.

If the measure $d\lambda$ is symmetric, the three-term recurrence (1.4) is simplified. This is made precise in the following result.

Lemma 1.17 *Let $\{p_n\}_{n\in\mathbb{N}_0}$ be a sequence of polynomials that is orthogonal with respect to a symmetric measure. Then $b_n = 0$ is satisfied in (1.4).*

PROOF. With Theorem 1.11 (1.4), we can verify for $n = 0, 1, \ldots,$ the identity

$$\begin{aligned}
0 &= p_{n+1}(x) - (-1)^{n+1} p_{n+1}(-x) \\
&= \big(a_n x\, p_n(x) - b_n p_n(x) - c_n p_{n-1}(x)\big) \\
&\quad - (-1)^{n+1}\big(- a_n x\, p_n(-x) - b_n p_n(-x) - c_n p_{n-1}(-x)\big) \\
&= \big(a_n x\, p_n(x) - b_n p_n(x) - c_n p_{n-1}(x)\big) - \big(a_n x\, p_n(x) + b_n p_n(x) - c_n p_{n-1}(x)\big) \\
&= -\, 2 b_n p_n(x).
\end{aligned}$$

This implies $b_n = 0$ because we have $p_n \neq 0$ while x is free of choice. □

It is at times useful to write the three-term recurrence (1.4) in an alternative form as follows.

Corollary 1.18 *The orthogonal polynomials* $\{p_n\}_{n\in\mathbb{N}_0}$ *satisfy the equation*

$$x\, p_n(x) = a'_n p_{n+1}(x) + b'_n p_n(x) + c'_n p_{n-1}(x), \quad n = 0, 1, \ldots, \tag{1.6}$$

with

$$a'_n = \frac{1}{a_n}, \quad b'_n = \frac{b_n}{a_n}, \quad c'_n = \frac{c_n}{a_n}, \qquad n = 0, 1, \ldots.$$

Of course, there are corresponding three-term recurrences for the monic and normalized variants, $\{\bar{p}_n\}_{n\in\mathbb{N}_0}$ and $\{\tilde{p}_n\}_{n\in\mathbb{N}_0}$, respectively. Knowing the three-term recurrence coefficients a_n, b_n, c_n, the leading coefficients k_n, and the squared norms h_n allows to derive these forms.

Lemma 1.19 *The monic orthogonal polynomials* $\{\bar{p}_n\}_{n\in\mathbb{N}_0}$ *satisfy*

$$\begin{aligned}
\bar{p}_{n+1}(x) &= (\bar{a}_n x - \bar{b}_n)\bar{p}_n(x) - \bar{c}_n \bar{p}_{n-1}(x), \quad n = 0, 1, \ldots, \\
\bar{p}_{-1}(x) &= 0, \quad \bar{p}_0(x) = \bar{k}_0 = 1,
\end{aligned} \tag{1.7}$$

with

$$\bar{a}_n = 1, \quad \bar{b}_n = b_n \frac{k_n}{k_{n+1}}, \quad \bar{c}_n = c_n \frac{k_{n-1}}{k_{n+1}}, \qquad n = 0, 1, \ldots.$$

PROOF. Replace $p_j = k_j \bar{p}_j$ for $j = n-1, n, n+1$ in (1.4). □

Corollary 1.20 *The monic orthogonal polynomials* $\{\bar{p}_n\}_{n\in\mathbb{N}_0}$ *satisfy the equation*

$$x\, \bar{p}_n(x) = \bar{a}'_n \bar{p}_{n+1}(x) + \bar{b}'_n \bar{p}_n(x) + \bar{c}'_n \bar{p}_{n-1}(x), \quad n = 0, 1, \ldots,$$

with

$$\bar{a}'_n = 1, \quad \bar{b}'_n = \bar{b}_n, \quad \bar{c}'_n = \bar{c}_n, \qquad n = 0, 1, \ldots.$$

Lemma 1.21 *The orthonormal polynomials* $\{\tilde{p}_n\}_{n\in\mathbb{N}_0}$ *satisfy*

$$\begin{aligned}
\tilde{p}_{n+1}(x) &= (\tilde{a}_n x - \tilde{b}_n)\tilde{p}_n(x) - \tilde{c}_n \tilde{p}_{n-1}(x), \quad n = 0, 1, \ldots, \\
\tilde{p}_{-1}(x) &= 0, \quad \tilde{p}_0(x) = \tilde{k}_0 = h_0^{-1/2},
\end{aligned}$$

with

$$\begin{aligned}
\tilde{a}_n &= a_n \sqrt{\frac{h_n}{h_{n+1}}} = \frac{1}{\sqrt{\bar{c}_{n+1}}}, \\
\tilde{b}_n &= b_n \sqrt{\frac{h_n}{h_{n+1}}} = \frac{\bar{b}_n}{\sqrt{\bar{c}_{n+1}}}, \qquad n = 0, 1, \ldots. \\
\tilde{c}_n &= c_n \sqrt{\frac{h_{n-1}}{h_{n+1}}} = \sqrt{\frac{\bar{c}_n}{\bar{c}_{n+1}}}.
\end{aligned} \tag{1.8}$$

PROOF. Replace $p_j = h_j^{1/2} \tilde{p}_j$ for $j = n-1, n, n+1$ in (1.4). □

Corollary 1.22 *The orthonormal polynomials* $\{\tilde{p}_n\}_{n\in\mathbb{N}_0}$ *satisfy the equation*

$$x\,\tilde{p}_n(x) = \tilde{a}'_n\tilde{p}_{n+1}(x) + \tilde{b}'_n\tilde{p}_n(x) + \tilde{c}'_n\tilde{p}_{n-1}(x), \quad n = 0, 1, \ldots,$$

with

$$\tilde{a}'_n = \frac{1}{\tilde{a}_n}, \quad \tilde{b}'_n = \frac{\tilde{b}_n}{\tilde{a}_n}, \quad \tilde{c}'_n = \frac{\tilde{c}_n}{\tilde{a}_n}, \qquad n = 0, 1, \ldots. \tag{1.9}$$

1.2.4 Jacobi matrix.

Definition 1.23 *Let* $\{p_n\}_{n\in\mathbb{N}_0}$ *be a sequence of orthogonal polynomials. Then the corresponding* Jacobi matrix *is the infinite tridiagonal matrix*

$$\mathbf{J}_\infty = \begin{pmatrix} b'_0 & a'_0 & & 0 \\ c'_1 & b'_1 & a'_1 & \\ & c'_2 & b'_2 & \ddots \\ 0 & & \ddots & \ddots \end{pmatrix}.$$

Its $n \times n$ *principal minor matrix is denoted* $\mathbf{J}_n$*. The Jacobi matrices* $\bar{\mathbf{J}}_\infty$ *and* $\bar{\mathbf{J}}_n$*, as well as* $\tilde{\mathbf{J}}_\infty$ *and* $\tilde{\mathbf{J}}_n$ *are defined accordingly for monic and orthonormal polynomials.*

With the vector $\mathbf{p}(x) = \big(p_0(x), p_1(x), \ldots, p_{n-1}(x)\big)^{\mathrm{T}}$ the recurrence (1.6) may be rewritten in matrix-vector form,

$$x\,\mathbf{p}(x) = \mathbf{J}_n\mathbf{p}(x) + a'_n p_n(x)\mathbf{e}_n, \tag{1.10}$$

where $\mathbf{e}_n = (0, 0, \ldots, 0, 1)^{\mathrm{T}}$ is the nth coordinate vector in $\mathbb{R}^n$. This form reveals that the zeros of p_n are the eigenvalues of the matrix $\mathbf{J}_n$. The following result is found in [23, p. 13].

Theorem 1.24 *The zeros* $\tau_i^{(n)}$*,* $i = 1, 2, \ldots, n$*, of the polynomial* p_n *are the eigenvalues of the Jacobi matrix* $\mathbf{J}_n$*. The corresponding eigenvectors are the vectors* $\mathbf{p}\big(\tau_i^{(n)}\big)$*.*

PROOF. The assertion follows from (1.10) by replacing x with $\tau_i^{(n)}$ for $i = 1, 2, \ldots, n$. Note that $\mathbf{p}\big(\tau_i^{(n)}\big)$ is not the null vector since $p_0\big(\tau_i^{(n)}\big) = k_0 \neq 0$. □

Remark 1.25 The fact that the zeros of the polynomial p_n are the eigenvalues of the matrix $\mathbf{J}_n$ allows for a simple proof of Theorem 1.13. Since $\mathbf{J}_n$ is the first principal minor of $\mathbf{J}_{n+1}$, the eigenvalues of $\mathbf{J}_n$ separate those of $\mathbf{J}_{n+1}$; see [82, p. 103] for a proof of this result.

An interesting property of the Jacobi matrix $\mathbf{J}_\infty$ is that it is symmetrized under normalization of the orthogonal polynomials p_n.

Lemma 1.26 *The Jacobi matrix* $\tilde{\mathbf{J}}_\infty$ *corresponding to the orthonormal polynomials* $\tilde{p}_n$ *is symmetric.*

PROOF. By virtue of (1.8) and (1.9), we obtain

$$\tilde{a}'_n = \sqrt{\bar{c}_{n+1}}, \quad \tilde{b}'_n = \bar{b}_n, \quad \tilde{c}'_n = \sqrt{\bar{c}_n}.$$

Therefore, we have

$$\tilde{\mathbf{J}}_\infty = \begin{pmatrix} \bar{b}_0 & \sqrt{\bar{c}_1} & & 0 \\ \sqrt{\bar{c}_1} & \bar{b}_1 & \sqrt{\bar{c}_2} & \\ & \sqrt{\bar{c}_2} & \bar{b}_2 & \ddots \\ 0 & & \ddots & \ddots \end{pmatrix}.$$

□

1.3 Classical orthogonal polynomials

1.3.1 General theory. According to most definitions, see for example [10, p. 150], [64, p. 21], [74, p. 141], and [23, p. 26], *classical orthogonal polynomials* are those satisfying a linear second-order differential equation of hypergeometric type, defined as follows.

Definition 1.27 *A differential equation of the form*

$$\sigma y'' + \tau y' + \lambda y = 0, \qquad \text{with } \sigma \in \mathbb{P}_2,\ \tau \in \mathbb{P}_1, \text{ and } \lambda \in \mathbb{R} \tag{1.11}$$

is called a differential equation of hypergeometric type *or* hypergeometric differential equation.

According to Chihara [10, p. 150], the equation (1.11) was shown by Bochner [6] to fully characterize the class of classical orthogonal polynomials. Historically, the classical polynomials and many of their common properties, including that they solve a hypergeometric differential equation, had been known before.
A hypergeometric differential equation is a special case of a *Sturm-Liouville* equation [3, p. 497]. These usually have singularities in their solutions unless the parameter λ takes certain values. Finding these values can be thought of as an eigenproblem.

Remark 1.28 Every solution of (1.11) is an eigenfunction to the eigenvalue λ of the differential operator $\mathcal{D} = \mathcal{D}(\sigma, \tau)$ that is defined by

$$\mathcal{D} = -\sigma \frac{\mathrm{d}^2}{\mathrm{d}x^2} - \tau \frac{\mathrm{d}}{\mathrm{d}x}.$$

Given the differential operator $\mathcal{D}$, consider the problem of finding (non-singular) eigenfunctions and their corresponding eigenvalues λ. Problems of this type typically arise in mathematical physics while solving a partial differential equation in non-cartesian coordinates after applying the method of *separation of variables.* Under certain conditions it can be shown that there is a discrete set of eigenvalues λ_n, $n = 0, 1, \ldots$, whose corresponding eigenfunctions form an orthogonal system with respect to a certain inner product. In the particular setting of (1.11), these eigenfunctions turn out to be the classical orthogonal polynomials.
That equation (1.11) indeed admits for polynomial solutions follows from the fact that the differential operator $\mathcal{D}$ carries polynomials into other polynomials of the same degree. Following Nikiforov and Uvarov [64], let us establish an explicit formula for these polynomial solutions: the well-known *Rodrigues formula.*

Lemma 1.29 *Let y be a solution to* (1.11). *Then the nth derivative $y_n := y^{(n)}$, $n \in \mathbb{N}_0$, satisfies the equation*

$$\sigma y_n'' + \tau_n y_n' + \mu_n y_n = 0, \tag{1.12}$$

with $\tau_n = \tau + n\sigma'$, $\mu_n = \lambda + n\tau' + \frac{n(n-1)}{2}\sigma''$.

PROOF. It is straightforward to verify that $y_1 = y^{(1)}$ satisfies the differential equation

$$\sigma y_1'' + \tau_1 y_1' + \mu_1 y_1 = 0,$$

with $\tau_1 = \tau + \sigma'$, $\mu_1 = \lambda + \tau'$. The proof for $n > 1$ follows by induction. □
The converse is also true. It can be shown (see [64, p. 6]) that every solution y_n of (1.12) with $\mu_j \neq 0$, $j = 0, \ldots, n-1$, is of the form $y_n = y^{(n)}$, where y is a solution of (1.11). This observation allows one to construct particular solutions of (1.11) by setting $\mu_n = 0$. Then y_n is a constant solution of (1.12). This means that when

$$\lambda = \lambda_n := -n\tau' - \frac{n(n-1)}{2}\sigma'',$$

equation (1.11) has a particular solution of the form $y = p_n$, that is, a polynomial of degree n. These thoughts give rise to the *Rodrigues formula* presented in the following.

Theorem 1.30 *Let p_n be a polynomial of degree n satisfying* (1.11). *Then*

$$p_n^{(m)} = \frac{A_{n,m}B_n}{w_m}\frac{\mathrm{d}^{n-m}}{\mathrm{d}x^{n-m}}w_n = \frac{A_{n,m}B_n}{\sigma^m w}\frac{\mathrm{d}^{n-m}}{\mathrm{d}x^{n-m}}(\sigma^n w), \tag{1.13}$$

with

$$A_{n,m} := A_m(\lambda_n), \quad A_m(\lambda) := (-1)^m \prod_{j=0}^{m-1} \mu_j(\lambda), \; A_0(\lambda) = 1,$$

a normalizing constant $B_n \in \mathbb{R} \setminus \{0\}$, $w_n(x) := \frac{1}{\sigma(x)}\mathrm{e}^{\int \frac{\tau_n(x)}{\sigma(x)}\,\mathrm{d}x}$, *and* $w := w_0$.

Proof. We multiply equations (1.11) and (1.12) by the functions

$$w(x) = \frac{1}{\sigma(x)}\mathrm{e}^{\int \frac{\tau(x)}{\sigma(x)}\,\mathrm{d}x} \quad \text{and} \quad w_n(x) = \frac{1}{\sigma(x)}\mathrm{e}^{\int \frac{\tau_n(x)}{\sigma(x)}\,\mathrm{d}x} \tag{1.14}$$

to obtain the self-adjoint forms

$$(\sigma w\, y')' + \lambda w y = 0, \qquad (\sigma w_n y_n')' + \mu_n w_n y_n = 0.$$

From (1.14) the identity $w_n = \sigma^n w$ follows up to a positive multiplicative constant. Now, we start with the self-adjoint form of the hypergeometric differential equation for any $0 \le m \le n$, i.e., we take

$$(\sigma w_m y_m')' + \mu_m w_m y_m = 0.$$

Recall that $\sigma w_m = w_{m+1}$ and $y_m' = y_{m+1}$. Then,

$$\begin{aligned} w_m y_m &= -\frac{1}{\mu_m}(w_{m+1}y_{m+1})' \\ &= \frac{1}{\mu_m\mu_{m+1}}(w_{m+2}y_{m+2})'' \\ &= -\frac{1}{\mu_m\mu_{m+1}\mu_{m+2}}(w_{m+3}y_{m+3})''' \\ &= \frac{A_m}{A_n}(w_n y_n)^{(n-m)}. \end{aligned} \tag{1.15}$$

If $y = p_n$ is a polynomial of degree n then $y_n = p_n^{(n)} = B_n$ is a constant with some $B_n \in \mathbb{R} \setminus \{0\}$. Thus, from (1.15) the following explicit expression for the polynomials $y_m = p_n^{(m)}$ is obtained,

$$p_n^{(m)} = \frac{A_{n,m}B_n}{w_m}\frac{\mathrm{d}^{n-m}}{\mathrm{d}x^{n-m}}w_n = \frac{A_{n,m}B_n}{\sigma^m w}\frac{\mathrm{d}^{n-m}}{\mathrm{d}x^{n-m}}(\sigma^n w)\,.$$

□

An important special case is the formula for the polynomial p_n itself.

Corollary 1.31 *For $m = 0$ in* (1.13), *we obtain the formula*

$$p_n = \frac{B_n}{w}\frac{\mathrm{d}^n}{\mathrm{d}x^n}(\sigma^n w)\,, \tag{1.16}$$

and all polynomial solutions of (1.11) *are defined by* (1.16) *up to a constant multiplicative factor. Each polynomial p_n corresponds to the eigenvalue*

$$\lambda_n = -n\tau' - \frac{n(n-1)}{2}\sigma''$$

of the differential operator $\mathcal{D}$; see Remark 1.28.

To derive the different classes of polynomials that arise from (1.11), three different cases for the degree of the polynomial σ can be distinguished. For each, one can write the linear polynomial τ in a general form with two degrees of freedom, say, $\alpha, \beta \in \mathbb{R}$. Furthermore, the function w from (1.14) can be shown to satisfy the differential equation

$$(\sigma w)' = \tau w. \tag{1.17}$$

Solving this equation, one obtains, up to constant factors, all possible forms of the function w,

$$w(x) = \begin{cases} (b-x)^\alpha (x-a)^\beta, & \text{if } \sigma(x) = (b-x)(x-a), \\ (x-a)^\alpha \mathrm{e}^{\beta x}, & \text{if } \sigma(x) = (x-a), \\ \mathrm{e}^{\alpha x^2 + \beta x}, & \text{if } \sigma(x) = 1. \end{cases}$$

Here, a and b are for our purposes real constants. By a linear change of variable, the expressions for σ and w can be standardized into the following forms:

$$w(x) = \begin{cases} (1-x)^\alpha (1+x)^\beta, & \text{if } \sigma(x) = 1 - x^2, \\ x^\alpha \mathrm{e}^{-x}, & \text{if } \sigma(x) = x, \\ \mathrm{e}^{-x^2}, & \text{if } \sigma(x) = 1. \end{cases} \tag{1.18}$$

The last display omits the special case $\sigma(x) = x^2$ which generates the *Bessel polynomials.* These are not orthogonal on a real domain but in the complex plane; see [64, p. 24]. Under the last transformations, equations (1.11) and (1.17) retain their form and the corresponding polynomials p_n remain polynomials in the new variable and are, as before, defined by a Rodrigues formula of the form (1.13).

A consequence of the Rodrigues formula is that the function w, under mild restrictions, can be shown to be the weight function that defines the inner product with respect to which the polynomials p_n are orthogonal. The following result can be found in a slightly different form in [64].

Theorem 1.32 *Let the function w from* (1.14) *satisfy the conditions*

$$\sigma(x) w(x) x^j \Big|_{x=a,b} = 0, \qquad \textit{with } j = 0, 1, \ldots, \tag{1.19}$$

at the endpoints of an interval $[a, b]$ on the extended real line, and let w be positive inside this interval. Then the polynomials p_n corresponding to the eigenvalues λ_n are orthogonal with respect to the measure induced by w on $[a, b]$,

$$\int_a^b p_n(x) p_m(x)\, w(x)\, \mathrm{d}x = \delta_{n,m} h_n, \qquad \textit{with } h_n > 0.$$

Proof. We assume $n \neq m$ and take the differential equations satisfied by p_n and p_m, respectively,

$$\left(\sigma w\, p_n'\right)' + \lambda_n w\, p_n = 0, \qquad \left(\sigma w\, p_m'\right)' + \lambda_m w\, p_m = 0.$$

We multiply the first equation by p_m, the second by p_n, and subtract the first from the second. We now have

$$p_n\, (\sigma w\, p_m)' - p_m\, (\sigma w\, p_n)' = (\lambda_n - \lambda_m)\, w\, p_n\, p_m. \tag{1.20}$$

The left-hand side can be written in the form

$$p_n\, (\sigma w\, p_m)' - p_m\, (\sigma w\, p_n)' = \frac{\mathrm{d}}{\mathrm{d}x} \big(\sigma w\, W(p_n, p_m)\big),$$

where $W(p_n, p_m) := \begin{vmatrix} p_n & p_m \\ p_n' & p_m' \end{vmatrix}$ is the *Wronskian*, that is, a polynomial. We integrate both sides to obtain

$$\int_a^b p_n(x)p_m(x)\, w(x)\, \mathrm{d}x = \frac{1}{\lambda_n - \lambda_m}\Big[\sigma(x)w(x)W(p_n(x), p_m(x))\Big]_a^b,$$

and since $W\big(p_n(x), p_m(x)\big)$ is a polynomial in x, the right hand side vanishes by (1.19). The positivity of the value h_n is ensured by the positivity of the weight function $w(x)$ inside the support interval $[a, b]$. □

1.3.2 Examples. According to the different degrees of the polynomial σ (see (1.18)) we obtain three different types of classical orthogonal polynomials. These are usually normalized in some ad-hoc way and are therefore neither monic nor orthonormal. This section is to give a concise overview over all classical polynomials. A comprehensive reference with the most important properties and related constants and formulae, also for the monic and normalized variants, is given in the Appendix A.

Hermite polynomials

If $\sigma(x) = 1$, then $w(x) = \mathrm{e}^{-x^2}$ and

$$\tau(x) = -2x, \quad \lambda_n = 2n.$$

The corresponding polynomials are the *Hermite polynomials* H_n which are orthogonal over the real line $(-\infty, \infty)$. They satisfy the Hermite differential equation

$$y''(x) - 2xy'(x) + 2ny(x) = 0, \quad \text{with } y = H_n,$$

and have the Rodrigues formula

$$H_n(x) = (-1)^n \mathrm{e}^{x^2} \frac{\mathrm{d}^n}{\mathrm{d}x^n}\Big(\mathrm{e}^{-x^2}\Big).$$

The corresponding three-term recurrence and initial conditions are

$$\begin{aligned} H_{n+1}(x) &= 2xH_n(x) - 2nH_{n-1}(x), \quad n = 0, 1, \ldots, \\ H_{-1}(x) &= 0, \quad H_0(x) = 1. \end{aligned}$$

Hermite polynomials play an important role in probability theory and mathematical physics, e.g., as eigenfunctions of the Fourier transform or in Schrödinger's equation for the harmonic linear oscillator; see, e.g., [2, p. 278], [74, p. 133], and [7, p. 346]. They do not carry any parameter, so there is only a single sequence of Hermite polynomials. Since this work is concerned with methods to convert between different polynomial families within the same class of orthogonal polynomials, they will not play a role for the rest of this text.

Laguerre polynomials

If $\sigma(x) = x$, then $w(x) = x^\alpha \mathrm{e}^{-x}$ and

$$\tau(x) = 1 + \alpha - x, \quad \lambda_n = n.$$

To ensure the existence of the moments μ_n in (1.1), we must require $\alpha > -1$. The corresponding polynomials are the *Laguerre polynomials* $L_n^{(\alpha)}$ normalized by $L_n^{(\alpha)}(1) = \binom{n+\alpha}{n}$. They are orthogonal over the positive real line $[0, \infty)$ and satisfy the Laguerre differential equation

$$xy''(x) + (1 + \alpha - x)y'(x) + ny(x) = 0, \quad \text{with } y = L_n^{(\alpha)}.$$

The Rodrigues formula reads

$$L_n^{(\alpha)}(x) = \frac{1}{\Gamma(n+1)} x^{-\alpha} \mathrm{e}^{x} \frac{\mathrm{d}^n}{\mathrm{d}x^n} \left(x^{\alpha+n} \mathrm{e}^{-x} \right).$$

Laguerre polynomials satisfy the three-term recurrence and initial conditions

$$(n+1)L_{n+1}^{(\alpha)}(x) = (-x + (2n+\alpha+1))L_n^{(\alpha)}(x) - (n+\alpha)L_{n-1}^{(\alpha)}(x), \quad n = 0,1,\ldots,$$
$$L_{-1}^{(\alpha)}(x) = 0, \quad L_0^{(\alpha)}(x) = 1.$$

They have many applications, mainly in quantum mechanics, and bear a close connection to Hermite polynomials; see [2, p. 282] and [7, p. 353]. Sometimes they are called *generalized Laguerre polynomials* or *associated Laguerre polynomials*; see, e.g., [65, p. 48]. This convention is not used in this work as it would introduce confusion with other denominations.

Jacobi polynomials

If $\sigma(x) = 1 - x^2$, then $w(x) = (1-x)^\alpha (1+x)^\beta$ and

$$\tau(x) = -(\alpha+\beta+2)x + \beta - \alpha, \quad \lambda_n = n(n+\alpha+\beta+1).$$

To ensure the existence of the moments μ_n in (1.1), we must require $\alpha, \beta > -1$. The corresponding polynomials are the *Jacobi polynomials* $P_n^{(\alpha,\beta)}$ normalized by $P_n^{(\alpha,\beta)}(1) = \binom{n+\alpha}{n}$. They are orthogonal over the closed interval $[-1,1]$ and are solutions to the Jacobi differential equation

$$(1-x^2)y''(x) - \left((\alpha+\beta+2)x + \alpha - \beta\right)y'(x) + n(n+\alpha+\beta+1)y(x) = 0,$$

where $y = P_n^{(\alpha,\beta)}$. The Rodrigues formula is

$$P_n^{(\alpha,\beta)}(x) = \frac{(-1)^n}{2^n \Gamma(n+1)} (1-x)^{-\alpha}(1+x)^{-\beta} \frac{\mathrm{d}^n}{\mathrm{d}x^n} \left((1-x)^{n+\alpha}(1+x)^{n+\beta} \right).$$

From this, we also obtain the useful identity

$$P_n^{(\alpha,\beta)}(x) = (-1)^n P_n^{(\beta,\alpha)}(-x). \tag{1.21}$$

Notice the change of the roles of α and β on the right hand side. If $\alpha = \beta$, then the weight function w is symmetric and (1.21) is equivalent to (1.3). Jacobi polynomials satisfy the three-term recurrence and initial conditions

$$\begin{aligned} &2(n+1)(n+\alpha+\beta+1)(2n+\alpha+\beta)P_{n+1}^{(\alpha,\beta)}(x) \\ &= \left((2n+\alpha+\beta)_3 x + (2n+\alpha+\beta+1)(\alpha^2-\beta^2)\right)P_n^{(\alpha,\beta)}(x) \\ &\quad - 2(n+\alpha)(n+\beta)(2n+\alpha+\beta+2)P_{n-1}^{(\alpha,\beta)}(x), \qquad n = 0,1,\ldots, \\ &\qquad P_{-1}^{(\alpha,\beta)}(x) = 0, \quad P_0^{(\alpha,\beta)}(x) = 1. \end{aligned} \tag{1.22}$$

Here, $(a)_b := a(a+1)\cdot \cdots \cdot (a+b-1)$ is the rising factorial or *Pochhammer symbol.* Notice that the recurrence is not usable if $\alpha = \beta = -1/2$ and $n = 1$ to obtain the polynomial $P_1^{(-1/2,-1/2)}$. But from the fact that $P_1^{(-1/2,-1/2)}(x)$ is an odd linear polynomial, normalized to $P_1^{(-1/2,-1/2)}(1) = \binom{1/2}{1} = \frac{1}{2}$, we must have

$$P_1^{(-1/2,-1/2)}(x) = \frac{x}{2}.$$

Jacobi polynomials have numerous applications. They arise, for example, in the study of spherical symmetries and rotation groups. A few of these areas are mentioned below.

Legendre polynomials

Legendre polynomials P_n are Jacobi polynomials $P_n^{(\alpha,\beta)}$ with $\alpha = \beta = 0$,

$$P_n(x) = P_n^{(0,0)}(x) = \frac{(-1)^n}{2^n \Gamma(n+1)} \frac{\mathrm{d}^n}{\mathrm{d}x^n}\Big(\big(1-x^2\big)^n \Big),$$

which are orthogonal with respect to the symmetric measure $\mathrm{d}\lambda(x) = \mathrm{d}x$. They are solutions to the Legendre differential equation

$$(1-x^2)P_n''(x) - 2xP_n'(x) + n(n+1)P_n(x) = 0.$$

The recurrence formula (1.22) is simplified to

$$\begin{aligned} (n+1)P_{n+1}(x) &= (2n+1)xP_n(x) - nP_{n-1}(x), \quad n = 0,1,\dots, \\ P_{-1}(x) &= 0, \quad P_0(x) = 1. \end{aligned}$$

Legendre polynomials can be used for series expansions of radial functions on the two-dimensional unit sphere $\mathbb{S}^2$; see, e.g., [62].

Chebyshev polynomials of first and second kind

Chebyshev polynomials of first kind T_n and Chebyshev polynomials of second kind U_n are Jacobi polynomials for $\alpha = \beta = -1/2$ and $\alpha = \beta = 1/2$, respectively, with a different normalization,

$$\begin{aligned} T_n(x) &= \frac{\Gamma(1/2)\Gamma(n+1)}{\Gamma(n+1/2)} P_n^{(-1/2,-1/2)}(x) = \cos(n\theta), \\ U_n(x) &= \frac{\Gamma(3/2)\Gamma(n+2)}{\Gamma(n+3/2)} P_n^{(1/2,1/2)}(x) = \frac{\sin\big((n+1)\theta\big)}{\sin\theta}, \end{aligned} \tag{1.23}$$

where the trigonometric formulae are valid for $x \in [-1,1]$ with $\theta = \arccos x$ (at the borders we need to take limits). Chebyshev polynomials satisfy the Chebyshev differential equations of first and second kind,

$$\begin{aligned} (1-x^2)T_n''(x) - xT_n'(x) + n^2T_n(x) &= 0, \\ (1-x^2)U_n''(x) - 3xU_n'(x) + n(n+2)U_n(x) &= 0. \end{aligned}$$

The three-term recurrences and initial conditions are

$$\begin{aligned} T_{n+1}(x) &= 2xT_n(x) - T_{n-1}(x), \qquad n = 1,2,\dots, \\ T_{-1}(x) &= 0, \quad T_0(x) = 1, \quad T_1(x) = x, \end{aligned}$$

and

$$\begin{aligned} U_{n+1}(x) &= 2xU_n(x) - U_{n-1}(x), \qquad n = 0,1,\dots, \\ U_{-1}(x) &= 0, \quad U_0(x) = 1. \end{aligned}$$

Note that the recurrence formula for the Chebyshev polynomials of first kind T_n starts at $n = 1$. By virtue of (1.23), Chebyshev polynomials are closely related to Fourier series, a fact that underlines their great importance for numerous applications.

Gegenbauer polynomials

Gegenbauer polynomials $C_n^{(\alpha)}$, with $\alpha > -1/2$ and $\alpha \neq 0$, are Jacobi polynomials with a different normalization,

$$C_n^{(\alpha)}(x) = \frac{\Gamma(\alpha+1/2)\Gamma(n+2\alpha)}{\Gamma(n+\alpha+1/2)\Gamma(2\alpha)} P_n^{(\alpha-1/2,\alpha-1/2)}(x).$$

The parameter α must not vanish since this would leave above formula indeterminate. However, the monic and normalized versions $\bar{C}_n^{(\alpha)}(x)$ and $\tilde{C}_n^{(\alpha)}(x)$ are well-defined, even when $\alpha = 0$. The corresponding polynomials are identical, respectively, to the monic and normalized Chebyshev polynomials of first kind. Gegenbauer polynomials are solutions to the Gegenbauer differential equation

$$(1-x^2)y''(x) - (2\alpha+1)xy'(x) + n(n+2\alpha)y(x) = 0, \quad \text{with } y = C_n^{(\alpha)}.$$

The three-term recurrence and initial conditions are

$$(n+1)C_{n+1}^{(\alpha)}(x) = 2(n+\alpha)xC_n^{(\alpha)}(x) - (n+2\alpha-1)C_{n-1}^{(\alpha)}(x), \quad n = 0,1,\ldots,$$

$$C_{-1}^{(\alpha)}(x) = 0, \quad C_0^{(\alpha)}(x) = 1.$$

Gegenbauer polynomials have applications as a generalization of Chebyshev and Legendre polynomials to higher dimensional spheres; see [62].

1.4 Connection coefficients for classical orthogonal polynomials

The *connection coefficients* are those coefficients that allow one to express a sequence of orthogonal polynomials $\{p_n\}_{n\in\mathbb{N}_0}$ in terms of another sequence $\{q_n\}_{n\in\mathbb{N}_0}$ of orthogonal polynomials. Determining these coefficients is an elementary theoretical question. For example, Askey [4] considered this problem among a line of other related ones, like the linearization of products of orthogonal polynomials. However, results about the connection coefficients between classical orthogonal polynomials also have practical impact. This is the topic of Chapters 2 and 3. Therefore, the purpose of this section is not only to survey basic results that are mostly found in the literature, but to state a number of special cases that can be exploited for numerical computations. Usually, most of the latter observations do not appear in the usual references, although they are rather simple consequences of known results.

1.4.1 General theory.

Definition 1.33 *Let $\{p_n\}_{n\in\mathbb{N}_0}$ and $\{q_n\}_{n\in\mathbb{N}_0}$ be two different sequences of orthogonal polynomials. Then every polynomial p_j, $j = 0,1,\ldots,$ can be represented as a linear combination of the polynomials $q_0, q_1, \ldots, q_j$,*

$$p_j = \sum_{i=0}^{j} \kappa_{i,j} q_i.$$

The polynomials $\{p_n\}_{n\in\mathbb{N}_0}$ are called the source polynomials *and the polynomials $\{q_n\}_{n\in\mathbb{N}_0}$ are called the* target polynomials. *The coefficients $\kappa_{i,j}$ are called the* connection coefficients *between $\{p_n\}_{n\in\mathbb{N}_0}$ and $\{q_n\}_{n\in\mathbb{N}_0}$. For $i < 0$, $j < 0$, or $j < i$, the connection coefficients are defined by $\kappa_{i,j} = 0$.*

If $\langle\cdot,\cdot\rangle_{\mathrm{d}\lambda}$ is the inner product with respect to which the target polynomials q_n are orthogonal, then the connection coefficients are clearly given by

$$\kappa_{i,j} = \frac{\langle q_i, p_j\rangle_{\mathrm{d}\lambda}}{\langle q_i, q_i\rangle_{\mathrm{d}\lambda}}. \tag{1.24}$$

Remark 1.34 While connection coefficients $\kappa_{i,j}$ exist for every pair of orthogonal polynomials, only the cases where the sequences $\{p_n\}_{n\in\mathbb{N}_0}$ and $\{q_n\}_{n\in\mathbb{N}_0}$ belong to the same type of classical orthogonal polynomials will be of interest for the rest of this text. For example, we will only consider the connection problem for two different sequences of, say, Jacobi polynomials, but not for a sequence of Hermite and another sequence of Jacobi polynomials.

The following theorem shows that the connection coefficients can be generated by a recurrence that follows directly from the three-term recurrences satisfied by source and target polynomials. This is a slightly modified version of Theorem 1 in [59, p. 295].

Theorem 1.35 *Let $\{p_n\}_{n\in\mathbb{N}_0}$ be a sequence of orthogonal polynomials satisfying the three-term recurrence and initial conditions*

$$\begin{aligned} p_{n+1}(x) &= (a_n x - b_n)p_n(x) - c_n p_{n-1}(x), \quad n = 0, 1, \ldots, \\ p_{-1}(x) &= 0, \quad p_0(x) = k_0, \end{aligned} \tag{1.25}$$

that have leading coefficients k_n. Furthermore, let $\{q_n\}_{n\in\mathbb{N}_0}$ be a sequence of polynomials, orthogonal with respect to the measure $\mathrm{d}\lambda$, that satisfy the three-term recurrence and the initial conditions

$$\begin{aligned} q_{n+1}(x) &= (\hat{a}_n x - \hat{b}_n)q_n(x) - \hat{c}_n q_{n-1}(x), \quad n = 0, 1, \ldots, \\ q_{-1}(x) &= 0, \quad q_0(x) = \hat{k}_0, \end{aligned} \tag{1.26}$$

with $\hat{h}_n = \langle q_n, q_n\rangle_{\mathrm{d}\lambda}$ and leading coefficients $\hat{k}_n$. Then the connection coefficients $\kappa_{i,j}$ in the formula

$$p_j = \sum_{i=0}^{j} \kappa_{i,j} q_i$$

satisfy the recurrence formula

$$\begin{aligned} \kappa_{i,j} = a_{j-1}&\left(\frac{1}{\hat{a}_i}\frac{\hat{h}_{i+1}}{\hat{h}_i}\kappa_{i+1,j-1} + \frac{\hat{b}_i}{\hat{a}_i}\kappa_{i,j-1} + \frac{\hat{c}_i}{\hat{a}_i}\frac{\hat{h}_{i-1}}{\hat{h}_i}\kappa_{i-1,j-1}\right) \\ &- b_{j-1}\kappa_{i,j-1} - c_{j-1}\kappa_{i,j-2}, \end{aligned} \tag{1.27}$$

for $0 \le i \le j$ with the initial conditions

$$\kappa_{i,j} = \begin{cases} \dfrac{k_0}{\hat{k}_0}\left(\dfrac{a_0}{\hat{a}_0}\hat{b}_0 - b_0\right), & \text{if } i = 0 \text{ and } j = 1, \\ \dfrac{k_i}{\hat{k}_i}, & \text{if } i = j. \end{cases} \tag{1.28}$$

Proof. Let us start with the general case. First, we use (1.24) and the three-term recurrence (1.25) to obtain

$$\kappa_{i,j} = \frac{\langle q_i, p_j\rangle_{\mathrm{d}\lambda}}{\langle q_i, q_j\rangle_{\mathrm{d}\lambda}} = \frac{1}{\hat{h}_i}\big(a_{j-1}\langle q_i, x p_{j-1}\rangle_{\mathrm{d}\lambda} - b_{j-1}\langle q_i, p_{j-1}\rangle_{\mathrm{d}\lambda} - c_{j-1}\langle q_i, p_{j-2}\rangle_{\mathrm{d}\lambda}\big). \tag{1.29}$$

The shift property of the inner product, i.e., $\langle\,\cdot\,, x\,\cdot\,\rangle_{\mathrm{d}\lambda} = \langle x\,\cdot\,,\,\cdot\,\rangle_{\mathrm{d}\lambda}$, allows us to apply the three-term recurrence (1.26) to expand the polynomial $x \cdot q_i$; cf. Corollary 1.18. Then,

$$\langle q_i, x\cdot p_{j-1}\rangle_{\mathrm{d}\lambda} = \langle x q_i, p_{j-1}\rangle_{\mathrm{d}\lambda} = \frac{1}{\hat{a}_i}\langle q_{i+1}, p_{j-1}\rangle_{\mathrm{d}\lambda} + \frac{\hat{b}_i}{\hat{a}_i}\langle q_i, p_{j-1}\rangle_{\mathrm{d}\lambda} + \frac{\hat{c}_i}{\hat{a}_i}\langle q_{i-1}, p_{j-1}\rangle_{\mathrm{d}\lambda}.$$

It remains to combine this result with (1.29) and identify the remaining inner products with the connection coefficients, e.g., $\langle q_i, p_{j-1}\rangle_{\mathrm{d}\lambda} = \hat{h}_i\kappa_{i,j-1}$. This completes the proof of the recurrence formula.

For the initial conditions, assume $i = j$ first. Both p_i and q_i are orthogonal to every polynomial of strictly smaller degree, so the following identity is readily obtained,

$$\kappa_{i,i} = \frac{\langle q_i, p_i\rangle_{\mathrm{d}\lambda}}{\hat{h}_i} = \frac{\langle q_i, k_i x^i\rangle_{\mathrm{d}\lambda}}{\hat{h}_i} = \frac{k_i}{\hat{k}_i}\frac{\langle q_i, \hat{k}_i x^i\rangle_{\mathrm{d}\lambda}}{\hat{h}_i} = \frac{k_i}{\hat{k}_i}\frac{\langle q_i, q_i\rangle_{\mathrm{d}\lambda}}{\hat{h}_i} = \frac{k_i}{\hat{k}_i}. \tag{1.30}$$

Finally, for the coefficient $\kappa_{0,1}$, the proof amounts to another direct calculation,

$$\begin{aligned}
\kappa_{0,1} &= \frac{\langle q_0, p_1\rangle_{\mathrm{d}\lambda}}{\hat{h}_0} \\
&= \frac{\langle \hat{k}_0, (a_0 x - b_0)k_0\rangle_{\mathrm{d}\lambda}}{\hat{h}_0} \\
&= \frac{1}{\hat{h}_0}\frac{k_0}{\hat{k}_0}\langle \hat{k}_0, (a_0 x - b_0)\hat{k}_0\rangle_{\mathrm{d}\lambda} \\
&= \frac{1}{\hat{h}_0}\frac{k_0}{\hat{k}_0}\left(\frac{a_0}{\hat{a}_0}\langle \hat{k}_0, (\hat{a}_0 x - \hat{b}_0)\hat{k}_0\rangle_{\mathrm{d}\lambda} + \left(\frac{a_0}{\hat{a}_0}\hat{b}_0 - b_0\right)\langle \hat{k}_0, \hat{k}_0\rangle_{\mathrm{d}\lambda}\right) \\
&= \frac{1}{\hat{h}_0}\frac{k_0}{\hat{k}_0}\left(\frac{a_0}{\hat{a}_0}\langle q_0, q_1\rangle_{\mathrm{d}\lambda} + \left(\frac{a_0}{\hat{a}_0}\hat{b}_0 - b_0\right)\langle q_0, q_0\rangle_{\mathrm{d}\lambda}\right) \\
&= \frac{k_0}{\hat{k}_0}\left(\frac{a_0}{\hat{a}_0}\hat{b}_0 - b_0\right).
\end{aligned}$$

□

Not surprisingly, in the case of symmetric measures, the recurrence is simplified. The result is an immediate consequence of the last theorem.

Corollary 1.36 *Let two polynomial sequences $\{p_n\}_{n\in\mathbb{N}_0}$ and $\{q_n\}_{n\in\mathbb{N}_0}$ be given that satisfy the assumptions of Theorem 1.35. Assume furthermore, that both sequences are orthogonal with respect to symmetric measures. Then the connection coefficients $\kappa_{i,j}$ satisfy the simplified recurrence formula*

$$\kappa_{i,j} = a_{j-1}\left(\frac{1}{\hat{a}_i}\frac{\hat{h}_{i+1}}{\hat{h}_i}\kappa_{i+1,j-1} + \frac{\hat{c}_i}{a_i}\frac{\hat{h}_{i-1}}{\hat{h}_i}\kappa_{i-1,j-1}\right) - c_{j-1}\kappa_{i,j-2}, \tag{1.31}$$

for $0 \le i \le j$ with initial conditions $\kappa_{i,i} = k_i/\hat{k}_i$, $i = 0, 1, \ldots$, and $\kappa_{0,1} = 0$. This implies $\kappa_{i,j} = 0$ if $i + j$ is odd.

PROOF. We take the result from Theorem 1.35 and notice that $\hat{b}_i = b_{j-1} = 0$ by symmetry of the corresponding measures; cf. Lemma 1.17. □

Once the connection coefficients between two sequences of orthogonal polynomials are obtained, they are also available for differently normalized variants.

Lemma 1.37 *Let $\{p_n\}_{n\in\mathbb{N}_0}$ and $\{q_n\}_{n\in\mathbb{N}_0}$ be two sequences of polynomials with leading coefficients k_n, $\hat{k}_n$, and squared norms h_n, $\hat{h}_n$, $n = 0, 1, \ldots$, respectively. Furthermore, let $\{p_n^*\}_{n\in\mathbb{N}_0}$ and $\{q_n^*\}_{n\in\mathbb{N}_0}$ be differently normalized versions of the same polynomials with leading coefficients k_n^*, $\hat{k}_n^*$ and squared norms h_n^*, $\hat{h}_n^*$, $n = 0, 1, \ldots$, respectively. If in the equation*

$$p_j = \sum_{i=0}^{j} \kappa_{i,j} q_i, \qquad j = 0, 1, \ldots, \tag{1.32}$$

the connection coefficients $\kappa_{i,j}$ are known, then in the equation

$$p_j^* = \sum_{i=0}^{j} \kappa_{i,j}^* q_i^*, \qquad j = 0, 1, \ldots,$$

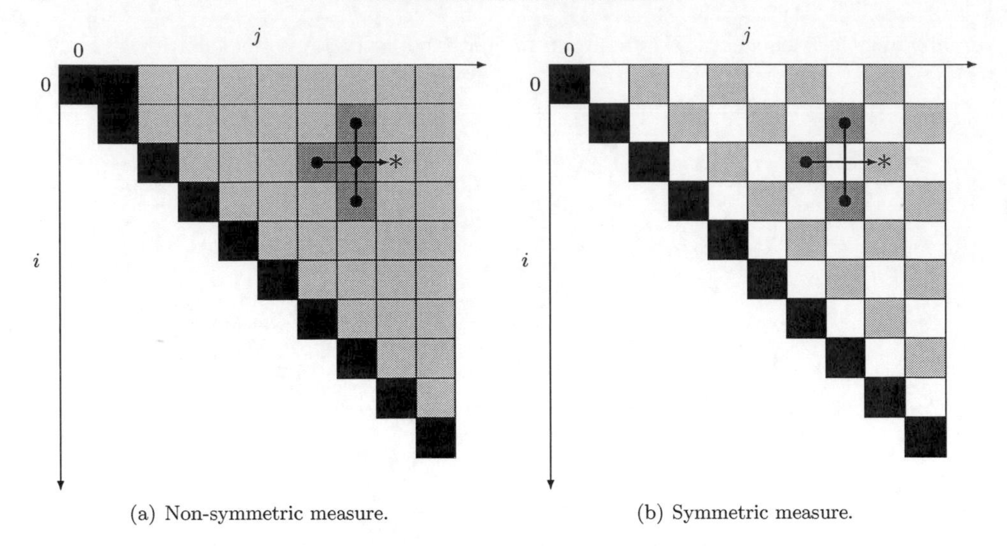

(a) Non-symmetric measure. (b) Symmetric measure.

Figure 1.1: Schematic representation of a triangular connection matrix $\mathbf{K} = (\kappa_{i,j})$ between two sequences of orthogonal polynomials, and the recurrence for the connection coefficients $\kappa_{i,j}$. Non-shaded areas represent coefficients that always vanish. Black squares represent entries given by the initial conditions; these are $\kappa_{i,i}$, $i = 0, 1, \ldots$, and $\kappa_{0,1}$. Gray squares stand for the rest of the coefficients that are determined by the three-term recurrence. For the computation of one of these coefficients $\kappa_{i,j}$ (represented by $*$) the entries $\kappa_{i-1,j-1}$, $\kappa_{i,j-1}$, $\kappa_{i,j-2}$, and $\kappa_{i+1,j-1}$ (represented by $\bullet$) have to be known. In the case of a symmetric measure, the recurrence is simpler since the dependence on $\kappa_{i,j-1}$ is removed. Also, the coefficient $\kappa_{0,1}$ is known to be zero as is the rest corresponding to the non-shaded boxes. The rest of the coefficients is aligned in a checkerboard pattern.

the connection coefficients $\kappa^*_{i,j}$ *are given by*

$$\kappa^*_{i,j} = \frac{k^*_j}{k_j}\frac{\hat{k}_i}{\hat{k}^*_i}\kappa_{i,j}, \qquad i, j = 0, 1, \ldots.$$

If furthermore $k_n k^*_n > 0$ *and* $\hat{k}_n \hat{k}^*_n > 0$*, i.e., the sign of the leading coefficients is the same, then*

$$\kappa^*_{i,j} = \sqrt{\frac{h^*_j}{h_j}\frac{\hat{h}_i}{\hat{h}^*_i}}\kappa_{i,j}, \qquad i, j = 0, 1, \ldots.$$

PROOF. We replace $p_j = (k_j/k^*_j)p^*_j$ and $q_i = (\hat{k}_i/\hat{k}^*_i)q^*_i$ in (1.32). This proves the first assertion. For the second, we replace $p_j = (h_j/h^*_j)^{1/2}p^*_j$ and $q_i = (\hat{h}_i/\hat{h}^*_i)^{1/2}q^*_i$ in the same equation. □

The recurrence for the connection coefficients $\kappa_{i,j}$ is illustrated in Figure 1.1. While, in principle, it allows to numerically compute every connection coefficient $\kappa_{i,j}$, it is only of marginal use in practice. To see this, let us first define the *connection matrix*.

Definition 1.38 *Let* $\{p_n\}_{n\in\mathbb{N}_0}$ *and* $\{q_n\}_{n\in\mathbb{N}_0}$ *be two sequences of orthogonal polynomials, and denote by* $\kappa_{i,j}$ *the connection coefficients between them. Then the upper triangular* *matrix*

$$\mathbf{K} := (\kappa_{i,j})_{i,j=0}^{N} \in \mathbb{R}^{(N+1)\times(N+1)}, \quad N \in \mathbb{N}_0,$$

is called the connection matrix of degree N between $\{p_n\}_{n\in\mathbb{N}_0}$ and $\{q_n\}_{n\in\mathbb{N}_0}$.

This matrix allows for writing the conversion between expansions in different systems of orthogonal polynomials in a more succinct matrix-vector notation.

Lemma 1.39 *Let a function f be represented by a finite expansion in a sequence of orthogonal polynomials $\{p_n\}_{n\in\mathbb{N}_0}$, i.e.,*

$$f = \sum_{j=0}^{N} x_j p_j,$$

with $N \in \mathbb{N}_0$. If $\{q_n\}_{n\in\mathbb{N}_0}$ is another sequence of orthogonal polynomials, then the corresponding expansion coefficients y_j in

$$f = \sum_{j=0}^{N} y_j q_j,$$

and the coefficients x_j in the previous expansion are related by the equation

$$\mathbf{y} = \mathbf{K}\,\mathbf{x}, \quad \text{with } \mathbf{y} = (y_j)_{j=0}^{N},\ \mathbf{x} = (x_j)_{j=0}^{N}, \tag{1.33}$$

where $\mathbf{K}$ is the connection matrix of degree N between $\{p_n\}_{n\in\mathbb{N}_0}$ and $\{q_n\}_{n\in\mathbb{N}_0}$.

For most problems under consideration, one needs an efficient way to apply the connection matrix $\mathbf{K}$. Even though all its entries can be evaluated with a total of $\mathcal{O}(N^2)$ arithmetic operations using the recurrence formula (1.27), this is usually considered too expensive as one is often interested only in applying the matrix $\mathbf{K}$ to a vector like in (1.33). Computing the entries of $\mathbf{K}$ explicitly and then multiplying the usual way is in most cases not the most efficient way.

As we shall see later in Chapters 2 and 3, there exist algorithms that apply $\mathbf{K}$ to a vector at a more favorable cost than $\mathcal{O}(N^2)$. These methods will not be based on the recursive procedure, but will use explicit expressions for the connection coefficients $\kappa_{i,j}$. This compels us to first seek these expressions.

1.4.2 Explicit expressions. A basic problem is to obtain explicit expressions for the connection coefficients $\kappa_{i,j}$ between two sequences of orthogonal polynomials. Another question, although not important here, is to ask whether these coefficients are positive. These and related problems have a long history. Also, many results in this area that have been thought to be new in the first place, have later been re-discovered in older works. As a starting point for the study of results about connection coefficients, reading [4] or [2] is recommended. Both texts contain almost all important results used in this work that are related to the classical orthogonal polynomials. One also finds a lot of background on relevant historical developments there. Other references include [36, 59, 73].

The approach to connection coefficients used in this work is based on [59] using the recurrence that was proven in Theorem 1.35 and Corollary 1.36 for symmetric measures. It should be noted that this is by no means the only valid way to go. In most cases, there are several options how expressions for the connection coefficients can be derived; see [4]. The recurrence formulae (1.27) and (1.31) can help generating explicit expressions for the connection coefficients $\kappa_{i,j}$ by using a procedure that relies on enough sample expressions that are generated by the recurrence. While this approach might not be mathematically as sophisticated as others, it has the advantage that it requires only very little extra knowledge about the orthogonal polynomials at hand. To obtain a concrete expression for

the connection coefficients $\kappa_{i,j}$ the following program from [59, p. 296] can be followed. It consists of two steps.
First, one uses the recursive formulae and initial conditions from Theorem 1.35 or Corollary 1.36 to generate the connection coefficients $\kappa_{i,j}$ for, say, $i, j = 0, \ldots, n$, for some finite and not too large n. It is convenient, if not imperative, to realize this step with a symbolic computation software such as MATHEMATICA[1]. The computed expressions are then, if possible, simplified and rearranged. This should lead to a guess how the connection coefficients $\kappa_{i,j}$ might be defined explicitly.
The second step consists in proving that the guessed formula is indeed correct by showing that the recurrence formula (1.27) and the initial conditions (1.28) are satisfied. It can be implemented with a symbolic computation software to check the respective identities.
In the following, a number of results which yield explicit expressions for the connection coefficients between various families of classical orthogonal polynomials are given. We will observe several special cases where the corresponding connection matrix **K** has a particular structure. The terms used to characterize these matrices are explained in Chapter 2.
Generally, the expressions will involve fractions of gamma functions, some of which can formally become singular in certain special cases. To avoid obscurities which could lead to the impression that the obtained expressions would not be always well defined, let us fix how the results should be interpreted.
For a family of orthogonal polynomials with one or more parameters, e.g., the Gegenbauer polynomials $C_n^{(\alpha)}$ carry a single parameter α, the connection coefficients between two sequences of polynomials obviously depend on these parameters. It can be checked easily that all quantities in the definition of the mentioned classical orthogonal polynomials depend smoothly on the respective parameters. Thus, Theorem 1.35 also asserts that the generated connection coefficients will depend continuously on the involved parameters. In problematic cases it is therefore allowed to take limits in the expressions for the connection coefficients, with respect to the parameters. The following finite limit is then frequently encountered.

Lemma 1.40 *Let $n, m \in \mathbb{N}_0$. Then*

$$\lim_{\nu \to m} \frac{\Gamma(n-\nu)}{\Gamma(-\nu)} = \begin{cases} (-1)^n \dfrac{\Gamma(m+1)}{\Gamma(m-n+1)}, & \text{if } n \leq m, \\ 0, & \text{else.} \end{cases}$$

PROOF. First, assume $n \leq m$. The function $\Gamma(z)$ has isolated poles of order one at $z = n - m$ and $z = -m$ with respective residues

$$\operatorname{Res}(\Gamma, n-m) = \frac{(-1)^{m-n}}{\Gamma(m-n+1)}, \qquad \operatorname{Res}(\Gamma, -m) = \frac{(-1)^m}{\Gamma(m+1)}.$$

Since $\Gamma(z)$ is analytic in the punctured open discs of radius one centered at either pole, we suppose without loss of generality that $0 < |m - \nu| < 1$ and consider the Laurent series expansion around $z = n - m$ and $z = -m$, each multiplied by $(m - \nu)$, i.e.,

$$(m-\nu)\Gamma(n-\nu) = \sum_{j=1}^{\infty} a_j (m-\nu)^j + \frac{(-1)^{m-n}}{\Gamma(m-n+1)},$$

$$(m-\nu)\Gamma(-\nu) = \sum_{j=1}^{\infty} b_j (m-\nu)^j + \frac{(-1)^m}{\Gamma(m+1)},$$

[1] MATHEMATICA is a registered trademark of Wolfram Research, Inc.

with certain $a_j, b_j \in \mathbb{C}$, $j \in \mathbb{N}$. Since the limits

$$\lim_{\nu\to m} (m-\nu)\Gamma(n-\nu) = \frac{(-1)^{m-n}}{\Gamma(m-n+1)}, \qquad \lim_{\nu\to m} (m-\nu)\Gamma(-\nu) = \frac{(-1)^m}{\Gamma(m+1)}$$

exist, we have

$$\lim_{\nu\to m} \frac{\Gamma(n-\nu)}{\Gamma(-\nu)} = (-1)^n \frac{\Gamma(m+1)}{\Gamma(m-n+1)}.$$

For the case $n > m$, we note that while $\Gamma(n-m)$ is finite, the function $1/\Gamma(z)$ has a zero of order one at $z = -m$. Therefore, the limit $\lim_{\nu\to m}\Gamma(n-\nu)/\Gamma(-\nu)$ must vanish. □

The last result should be kept in mind for each explicit expressions for connection coefficients $\kappa_{i,j}$. It is understood that for parameter combinations for which the numerator or denominator (or both) would be undefined due to poles of the gamma function, one has to take the limit with respect to one parameter and consider Lemma 1.40 to obtain a valid expression.

Laguerre polynomials

The Laguerre family of orthogonal polynomials carries a single parameter $\alpha > -1$. An explicit expression for the connection coefficients between these polynomials is given in Askey [4, p. 57], without any reference, but mentioning that it is very old. It can also be obtained by the "guessing" procedure from above, although the result itself is not given in [59]. The expression for the connection coefficients turns out to be reasonably simple, indeed so concise that it serves as a good example for patterns common to all classical orthogonal polynomials.

Definition 1.41 *Let $\{\bar{L}_n^{(\alpha)}\}_{n\in\mathbb{N}_0}$ and $\{\bar{L}_n^{(\beta)}\}_{n\in\mathbb{N}_0}$ with $\alpha, \beta > -1$ be two families of monic Laguerre polynomials. Then the connection coefficients $\bar{\kappa}_{i,j}$ in the formula*

$$\bar{L}_j^{(\alpha)} = \sum_{i=0}^{j} \bar{\kappa}_{i,j} \bar{L}_i^{(\beta)}, \quad j = 0, 1, \dots,$$

are denoted $\bar{\kappa}_{i,j} = \bar{\kappa}_{i,j}^{\mathrm{L},(\alpha)\to(\beta)}$ or $\bar{\kappa}_{i,j}^{(\alpha)\to(\beta)}$ if it is clear that the related polynomials are the Laguerre polynomials. The corresponding connection matrix is denoted $\bar{\mathbf{K}}^{\mathrm{L},(\alpha)\to(\beta)}$ or $\bar{\mathbf{K}}^{(\alpha)\to(\beta)}$ for short.

Theorem 1.42 *Let $\alpha, \beta > -1$. Then the connection coefficients between the sequences of Laguerre polynomials $\{\bar{L}_n^{(\alpha)}\}_{n\in\mathbb{N}_0}$ and $\{\bar{L}_n^{(\beta)}\}_{n\in\mathbb{N}_0}$ are given by*

$$\bar{\kappa}_{i,j}^{(\alpha)\to(\beta)} = \frac{(-1)^{i+j}}{\Gamma(\alpha-\beta)} \frac{\Gamma(j+1)}{\Gamma(i+1)} \frac{\Gamma(j-i+\alpha-\beta)}{\Gamma(j-i+1)}, \qquad \text{with } 0 \le i \le j.$$

Proof. Verify that the recurrence formula (1.27) and initial conditions (1.28) are satisfied. □

Let us look at a special case, namely when the difference between β and α is a positive integer. Using Lemma 1.40, the following simple formula is readily obtained.

Corollary 1.43 *Let $-1 < \alpha < \beta$ such that $\beta-\alpha$ is a positive integer. Then the connection coefficients between the sequences of Laguerre polynomials $\{\bar{L}_n^{(\alpha)}\}_{n\in\mathbb{N}_0}$ and $\{\bar{L}_n^{(\beta)}\}_{n\in\mathbb{N}_0}$ are given by*

$$\bar{\kappa}_{i,j}^{(\alpha)\to(\beta)} = \begin{cases} \binom{\beta-\alpha}{j-i} \frac{\Gamma(j+1)}{\Gamma(i+1)}, & \text{if } 0 \le i \le j \le i+\beta-\alpha, \\ 0, & \text{else.} \end{cases}$$

This implies that the degree-n connection matrix

$$\bar{\mathbf{K}}^{(\alpha)\to(\beta)} = \begin{pmatrix} \overbrace{1 & * & \dots & *}^{\beta-\alpha+1} & & & \\ & 1 & * & & \ddots & & \\ & & 1 & * & & * \\ & & & 1 & \ddots & \vdots \\ & & & & \ddots & * \\ & & & & & 1 \end{pmatrix}$$

is stricly $(0, \beta-\alpha)$-banded; see Definition 2.2 on page 60.

PROOF. Use Theorem 1.42 in conjunction with Lemma 1.40. □

One can then take this theme even a step further and look at the case when $\beta = \alpha + 1$.

Corollary 1.44 *Let $-1 < \alpha$. Then the connection coefficients between the sequences of Laguerre polynomials $\{\bar{L}_n^{(\alpha)}\}_{n\in\mathbb{N}_0}$ and $\{\bar{L}_n^{(\alpha+1)}\}_{n\in\mathbb{N}_0}$ are given by*

$$\bar{\kappa}_{i,j}^{(\alpha)\to(\alpha+1)} = \begin{cases} 1, & \text{if } j = i, \\ i+1, & \text{if } j = i+1, \\ 0, & \text{else.} \end{cases}$$

Thus, the connection matrix

$$\bar{\mathbf{K}}^{(\alpha)\to(\alpha+1)} = \begin{pmatrix} 1 & 1 & & & \\ & 1 & 2 & & \\ & & 1 & \ddots & \\ & & & \ddots & n \\ & & & & 1 \end{pmatrix}$$

is strictly $(0,1)$-banded; see Definition 2.2 on page 60.

PROOF. The proof is a direct consequence of the preceding Corollary. □

The last two results are remarkable because they reveal special cases where the connection matrix is not densely populated. This is usually an appealing property from a computational point of view, since the cost of an $n \times n$ matrix-vector multiplication can then be reduced from $\mathcal{O}(n^2)$ to $\mathcal{O}(n)$.

A similar observation about structuredness of the connection matrix can be made in the dual case, i.e., when $\beta - \alpha$ is a negative integer. But let us start here with the special case $\beta = \alpha - 1$.

Corollary 1.45 *Let $0 < \alpha$. Then the connection coefficients between the sequences of Laguerre polynomials $\{\bar{L}_n^{(\alpha)}\}_{n\in\mathbb{N}_0}$ and $\{\bar{L}_n^{(\alpha-1)}\}_{n\in\mathbb{N}_0}$ are given by*

$$\bar{\kappa}_{i,j}^{(\alpha)\to(\alpha-1)} = \frac{(-1)^j\Gamma(j+1)}{(-1)^i\Gamma(i+1)}, \qquad \text{with } 0 \le i \le j.$$

Thus, the connection matrix

$$\bar{\mathbf{K}}^{(\alpha)\to(\alpha-1)} = \operatorname{triu}(\mathbf{u}\,\mathbf{v}^{\mathrm{T}}),$$

with

$$\mathbf{u} = \left((-1)^i/\Gamma(i+1)\right)_{i=0}^{n}, \quad \mathbf{v} = \left((-1)^j\Gamma(j+1)\right)_{j=0}^{n}$$

is upper (1)-generator representable semiseparable*; see Definition* 2.6 *on page* 62.

PROOF. The result is a direct consequence of Theorem 1.42. □

Now, we are ready to show that larger steps, i.e., $\beta = \alpha - k$, $k \in \mathbb{N}$, result in connection matrices with semiseparability rank equal to k.

Corollary 1.46 *Let $-1 < \beta < \alpha$ such that $\alpha - \beta$ is a positive integer. Then the connection matrix $\bar{\mathbf{K}}^{(\alpha)\to(\beta)}$ between the sequences of Laguerre polynomials $\{\bar{L}_n^{(\alpha)}\}_{n\in\mathbb{N}_0}$ and $\{\bar{L}_n^{(\beta)}\}_{n\in\mathbb{N}_0}$ is upper $(\alpha - \beta)$-generator representable semiseparable.*

PROOF. The matrix $\bar{\mathbf{K}}^{(\alpha)\to(\beta)}$ is a product of $\alpha - \beta$ many upper (1)-generator representable semiseparable matrices,

$$\bar{\mathbf{K}}^{(\alpha)\to(\beta)} = \bar{\mathbf{K}}^{(\beta+1)\to(\beta)} \cdot \bar{\mathbf{K}}^{(\beta+2)\to(\beta+1)} \cdot \ldots \cdot \bar{\mathbf{K}}^{(\alpha-1)\to(\alpha-2)} \cdot \bar{\mathbf{K}}^{(\alpha)\to(\alpha-1)},$$

and is as such upper $(\alpha - \beta)$-generator representable semiseparable; see Theorem 2.7 on page 63. □

We have now identified important special cases, more precisely, those that represent integer steps from α to β which give rise to rank-structured connection matrices $\mathbf{K}^{(\alpha)\to(\beta)}$. The structuredness of these matrices can be exploited to apply the connection matrix to a vector with $\mathcal{O}(n)$ operations instead of $\mathcal{O}(n^2)$.

Remark 1.47 The situation for the connection coefficients between different Laguerre polynomials is typical and the theme repeats in the case of Jacobi polynomials, albeit in a slightly more complicated way; see Section 1.4.2.

As shown in Figure 1.2, any connection between different sequences of Laguerre polynomials, $\{\bar{L}_n^{(\alpha)}\}_{n\in\mathbb{N}_0}$ and $\{\bar{L}_n^{(\beta)}\}_{n\in\mathbb{N}_0}$, can be decomposed into two parts: a transformation from the parameter α to a certain α' and then from α' to the actual target β, where α' has been determined such that $\alpha' - \alpha$ is an integer and $|\beta - \alpha'| < 1$.

The purpose of such a decomposition becomes clear when one is interested in applying the connection matrix $\bar{\mathbf{K}}^{(\alpha)\to(\beta)}$ efficiently. This matrix can be factored into the product of the heavily structured matrix $\bar{\mathbf{K}}^{(\alpha)\to(\alpha')}$ (which is either banded or semiseparable) and the matrix $\bar{\mathbf{K}}^{(\alpha')\to(\beta)}$, i.e.,

$$\bar{\mathbf{K}}^{(\alpha)\to(\beta)} = \bar{\mathbf{K}}^{(\alpha')\to(\beta)}\,\bar{\mathbf{K}}^{(\alpha)\to(\alpha')}.$$

The matrix $\bar{\mathbf{K}}^{(\alpha')\to(\beta)}$, as we shall see in Chapter 3, can also be applied efficiently.

This completes the important results about the connection coefficients in the Laguerre case. Let us turn to the second and arguably more important class of Jacobi polynomials. The situation here is more complex since Jacobi polynomials $P_n^{(\alpha,\beta)}$ carry to parameters $\alpha, \beta > -1$. However, if one parameter is kept fixed, then the connection coefficients have a similar behavior, as observed for the Laguerre polynomials, with respect to structuredness of the connection matrices.

Jacobi polynomials with one parameter kept fixed

Definition 1.48 *Let $\{\bar{P}_n^{(\alpha,\beta)}\}_{n\in\mathbb{N}_0}$ with $\alpha, \beta > -1$ and $\{\bar{P}_n^{(\gamma,\delta)}\}_{n\in\mathbb{N}_0}$ with $\gamma, \delta > -1$ be two families of monic Jacobi polynomials. Then the connection coefficients $\bar{\kappa}_{i,j}$ in the formula*

$$\bar{P}_j^{(\alpha,\beta)} = \sum_{i=0}^{j} \bar{\kappa}_{i,j} \bar{P}_i^{(\gamma,\delta)}, \qquad j = 0, 1, \ldots, \tag{1.34}$$

are denoted $\bar{\kappa}_{i,j} = \kappa_{i,j}^{\mathrm{J},(\alpha,\beta)\to(\gamma,\delta)}$ or just $\bar{\kappa}_{i,j}^{(\alpha,\beta)\to(\gamma,\delta)}$ if it is clear that the corresponding polynomials are the Jacobi polynomials. The corresponding connection matrix is denoted $\bar{\mathbf{K}}^{\mathrm{J},(\alpha,\beta)\to(\gamma,\delta)}$ or $\bar{\mathbf{K}}^{(\alpha,\beta)\to(\gamma,\delta)}$ for short.

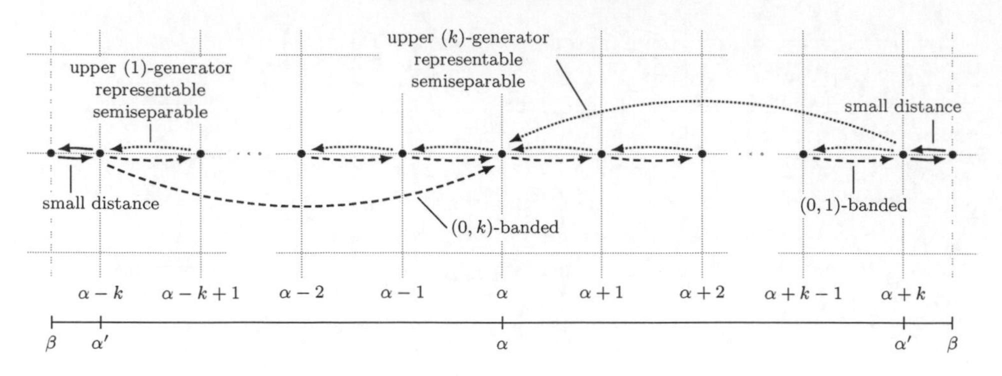

Figure 1.2: The connection between Laguerre polynomials: If the parameter α in the polynomials $L_n(\alpha)$ is increased or decreased in unit steps then the connection matrices are $(0,1)$-banded or upper (1)-generator representable semiseparable, respectively. Integer steps of length k lead to $(0,k)$-banded or upper (k)-generator representable semiseparable matrices. General steps can be decomposed into an integer length step and a second step that spans only a small distance.

As said before, the connection between sequences of Jacobi polynomials is somewhat symmetric to the Laguerre case. But unlike Laguerre polynomials, Jacobi polynomials carry two parameters, α and β. It is therefore natural to start with the cases where one of these parameters remains fixed. A useful observation is that once an expression with respect to one parameter has been obtained, the identity $P_n^{(\alpha,\beta)}(x) = (-1)^n P_n^{(\beta,\alpha)}(-x)$ readily allows to obtain the corresponding expressions with the roles of α and β exchanged. This is made precise in the following result.

Lemma 1.49 *The connection coefficients for monic Jacobi polynomials satisfy the identity*

$$\bar{\kappa}_{i,j}^{(\alpha,\beta)\to(\gamma,\delta)} = (-1)^{i+j}\bar{\kappa}_{i,j}^{(\beta,\alpha)\to(\delta,\gamma)}.$$

PROOF. Take the connection coefficients $\bar{\kappa}_{i,j}^{(\alpha,\beta)\to(\gamma,\delta)}$ which satisfy

$$\bar{P}_j^{(\alpha,\beta)}(x) = \sum_{i=0}^{j} \bar{\kappa}_{i,j}^{(\alpha,\beta)\to(\gamma,\delta)}\,\bar{P}_i^{(\gamma,\delta)}(x), \qquad j = 0,1,\ldots.$$

Then use (1.21) to obtain

$$\bar{P}_j^{(\beta,\alpha)}(-x) = \sum_{i=0}^{j} (-1)^{i+j}\bar{\kappa}_{i,j}^{(\alpha,\beta)\to(\gamma,\delta)}\,\bar{P}_i^{(\delta,\gamma)}(-x), \qquad j = 0,1,\ldots.$$

□

The following result is found in [4, p. 63] for Jacobi polynomials in the standard normalization, and in [59, p. 305] for monic Jacobi polynomials, in the latter with the roles of α and β exchanged due to a different definition.

Theorem 1.50 *Let $\alpha, \gamma, \beta > -1$. Then the connection coefficients between the sequences of Jacobi polynomials $\{\bar{P}_n^{(\alpha,\beta)}\}_{n\in\mathbb{N}_0}$ and $\{\bar{P}_n^{(\gamma,\beta)}\}_{n\in\mathbb{N}_0}$ are given by*

$$\begin{aligned}\bar{\kappa}_{i,j}^{(\alpha,\beta)\to(\gamma,\beta)} &= \frac{1}{\Gamma(\alpha-\gamma)} \frac{2^j}{2^i} \frac{\Gamma(j+1)}{\Gamma(i+1)} \frac{\Gamma(j+\beta+1)}{\Gamma(i+\beta+1)} \frac{\Gamma(2i+\gamma+\beta+2)}{\Gamma(2j+\alpha+\beta+1)} \\ &\quad \times \frac{\Gamma(j+i+\alpha+\beta+1)}{\Gamma(j+i+\gamma+\beta+2)} \frac{\Gamma(j-i+\alpha-\gamma)}{\Gamma(j-i+1)}\end{aligned}$$

with $0 \le i \le j$.

Proof. Verify that the recurrence formula (1.27) and initial conditions (1.28) are satisfied. □

Now, we are in a position to use Lemma 1.49 to derive the corresponding expression for the connection coefficients when the parameter α is kept fixed.

Corollary 1.51 *Let $\alpha, \beta, \delta > -1$. Then the connection coefficients between the sequences of Jacobi polynomials $\{\bar{P}_n^{(\alpha,\beta)}\}_{n\in\mathbb{N}_0}$ and $\{\bar{P}_n^{(\alpha,\delta)}\}_{n\in\mathbb{N}_0}$ are given by*

$$\begin{aligned}\bar{\kappa}_{i,j}^{(\alpha,\beta)\to(\alpha,\delta)} &= (-1)^{i+j} \bar{\kappa}_{i,j}^{(\beta,\alpha)\to(\delta,\alpha)} \\ &= \frac{(-1)^{i+j}}{\Gamma(\beta-\delta)} \frac{2^j}{2^i} \frac{\Gamma(j+1)}{\Gamma(i+1)} \frac{\Gamma(j+\alpha+1)}{\Gamma(i+\alpha+1)} \frac{\Gamma(2i+\alpha+\delta+2)}{\Gamma(2j+\alpha+\beta+1)} \\ &\quad \times \frac{\Gamma(j+i+\alpha+\beta+1)}{\Gamma(j+i+\alpha+\delta+2)} \frac{\Gamma(j-i+\beta-\delta)}{\Gamma(j-i+1)},\end{aligned}$$

with $0 \le i \le j$.

As for the Laguerre case, important special cases are obtained for integer changes in one of the parameters α and β.

Corollary 1.52 *Let $-1 < \alpha < \gamma$ and $-1 < \beta < \delta$. If $\gamma - \alpha$ is a positive integer, then the connection coefficients between the sequences of Jacobi polynomials $\{\bar{P}_n^{(\alpha,\beta)}\}_{n\in\mathbb{N}_0}$ and $\{\bar{P}_n^{(\gamma,\beta)}\}_{n\in\mathbb{N}_0}$ are given by*

$$\bar{\kappa}_{i,j}^{(\alpha,\beta)\to(\gamma,\beta)} = \begin{cases} (-1)^{i+j} \dfrac{2^j}{2^i} \dfrac{\Gamma(j+1)}{\Gamma(i+1)} \dfrac{\Gamma(j+\beta+1)}{\Gamma(i+\beta+1)} & \\ \quad \times \dbinom{\gamma-\alpha}{j-i} \dfrac{\Gamma(2i+\gamma+\beta+2)}{\Gamma(2j+\alpha+\beta+1)} & \\ \quad \times \dfrac{\Gamma(j+i+\alpha+\beta+1)}{\Gamma(j+i+\gamma+\beta+2)}, & \text{if } 0 \le i \le j \le j+\gamma-\alpha, \\ 0, & \text{else,} \end{cases}$$

and if $\delta-\beta$ is a positive integer, then the connection coefficients between the sequences of Jacobi polynomials $\{\bar{P}_n^{(\alpha,\beta)}\}_{n\in\mathbb{N}_0}$ and $\{\bar{P}_n^{(\alpha,\delta)}\}_{n\in\mathbb{N}_0}$ are given by

$$\bar{\kappa}_{i,j}^{(\alpha,\beta)\to(\alpha,\delta)} = \begin{cases} \dfrac{2^j}{2^i}\dfrac{\Gamma(j+1)}{\Gamma(i+1)}\dfrac{\Gamma(j+\alpha+1)}{\Gamma(i+\alpha+1)} \\ \quad\times\dbinom{\delta-\beta}{j-i}\dfrac{\Gamma(2i+\alpha+\delta+2)}{\Gamma(2j+\alpha+\beta+1)} \\ \quad\times\dfrac{\Gamma(j+i+\alpha+\beta+1)}{\Gamma(j+i+\alpha+\delta+2)}, & \text{if } 0\le i\le j\le i+\delta-\beta, \\ 0, & \text{else.} \end{cases}$$

This implies that the connection matrix

$$\bar{\mathbf{K}}^{(\alpha,\beta)\to(\gamma,\beta)} = \begin{pmatrix} \overbrace{1 \quad * \quad \dots \quad *}^{\gamma-\alpha+1} & & & \\ & 1 \quad * & \ddots & \\ & \quad 1 \quad * & & * \\ & & 1 \quad \ddots & \vdots \\ & & \ddots & * \\ & & & 1 \end{pmatrix}$$

is strictly $(0,\gamma-\alpha)$-banded. Similarly, the matrix $\bar{\mathbf{K}}^{(\alpha,\beta)\to(\alpha,\delta)}$ is strictly $(0,\delta-\beta)$-banded; *see Definition 2.2 on page 60.*

PROOF. Use Theorem 1.50 and Corollary 1.51 in conjunction with Lemma 1.40. □

As seen before, unit changes to the parameters result in even more simpler structured connection matrices.

Corollary 1.53 *Let $\alpha,\beta>-1$. Then the connection coefficients between the sequences of Jacobi polynomials $\{\bar{P}_n^{(\alpha,\beta)}\}_{n\in\mathbb{N}_0}$ and $\{\bar{P}_n^{(\alpha+1,\beta)}\}_{n\in\mathbb{N}_0}$, or $\{\bar{P}_n^{(\alpha,\beta)}\}_{n\in\mathbb{N}_0}$ and $\{\bar{P}_n^{(\alpha,\beta+1)}\}_{n\in\mathbb{N}_0}$, respectively, are given by*

$$\bar{\kappa}_{i,j}^{(\alpha,\beta)\to(\alpha+1,\beta)} = \begin{cases} 1, & \text{if } j=i, \\ \dfrac{-2j(j+\beta)}{(2j+\alpha+\beta)(2j+\alpha+\beta+1)}, & \text{if } j=i+1, \\ 0, & \text{else,} \end{cases}$$

$$\bar{\kappa}_{i,j}^{(\alpha,\beta)\to(\alpha,\beta+1)} = \begin{cases} 1, & \text{if } j=i, \\ \dfrac{2j(j+\alpha)}{(2j+\alpha+\beta)(2j+\alpha+\beta+1)}, & \text{if } j=i+1, \\ 0, & \text{else.} \end{cases}$$

Thus, the connection matrix

$$\bar{\mathbf{K}}^{(\alpha,\beta)\to(\alpha+1,\beta)} = \begin{pmatrix} 1 & * & & & \\ & 1 & * & & \\ & & 1 & \ddots & \\ & & & \ddots & * \\ & & & & 1 \end{pmatrix}$$

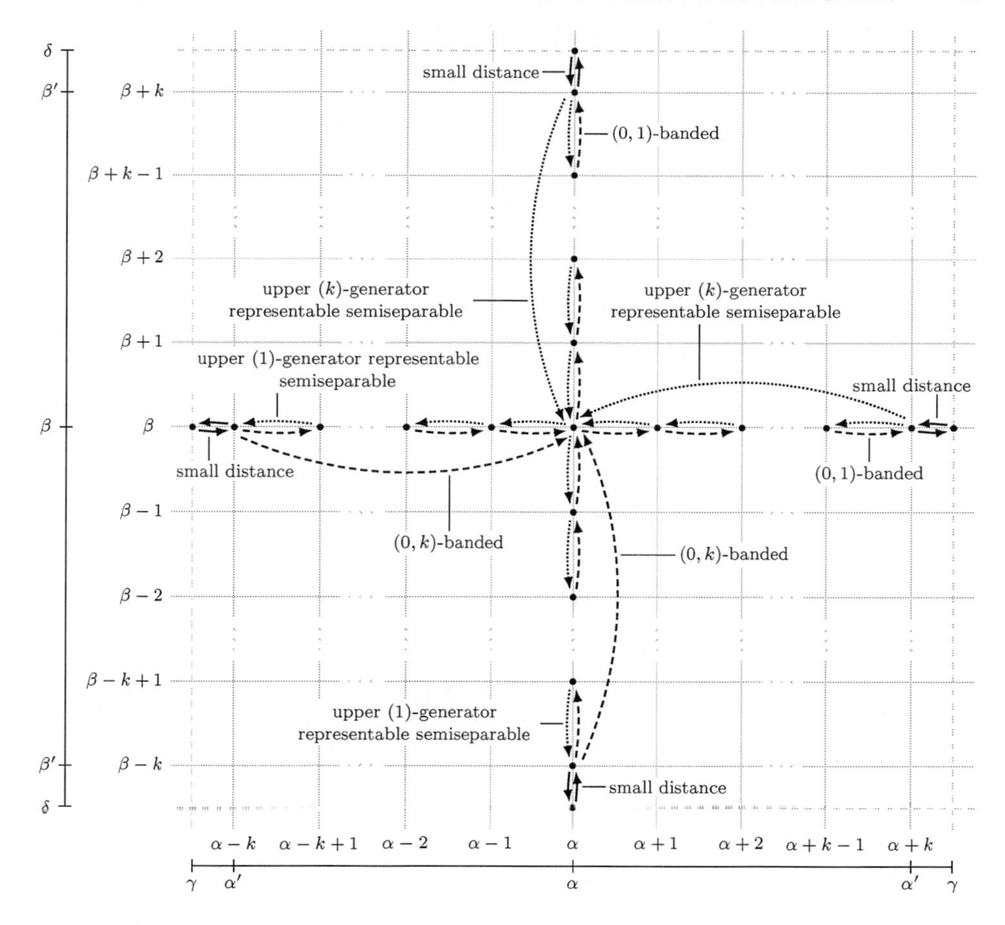

Figure 1.3: The connection between Jacobi polynomials where one parameter is kept fixed. If either α or β are increased or decreased in unit steps, then the connection matrices are $(0,1)$-banded or upper (1)-generator representable semiseparable, respectively. Integer steps of length k lead to $(0,k)$-banded or upper (k)-generator representable semiseparable matrices. General steps can be decomposed into a step of integer length and a second step that spans only a small distance.

is strictly $(0,1)$*-banded. Similarly, the matrix* $\bar{\mathbf{K}}^{(\alpha,\beta)\to(\alpha,\beta+1)}$ *is* $(0,1)$*-banded; see Definition* 2.2 *on page* 60*.*

PROOF. The proof is a direct consequence of the preceding Corollary. □
Also similar to the Laguerre case, let us start in the other direction with the cases $\gamma = \alpha - 1$ and $\delta = \beta - 1$, respectively.

Corollary 1.54 *Let* $\alpha > 0$ *and* $\beta > -1$*. Then the connection coefficients between the sequences of Jacobi polynomials* $\{\bar{P}_n^{(\alpha,\beta)}\}_{n\in\mathbb{N}_0}$ *and* $\{\bar{P}_n^{(\alpha-1,\beta)}\}_{n\in\mathbb{N}_0}$ *are given by*

$$\bar{\kappa}_{i,j}^{(\alpha,\beta)\to(\alpha-1,\beta)} = \frac{2^j}{2^i}\frac{\Gamma(j+1)}{\Gamma(i+1)}\frac{\Gamma(j+\beta+1)}{\Gamma(i+\beta+1)}\frac{\Gamma(2i+\alpha+\beta+1)}{\Gamma(2j+\alpha+\beta+1)},$$

with $0 \le i \le j$. *Let* $\alpha > -1$ *and* $\beta > 0$. *Then the connection coefficients between the sequences of Jacobi polynomials* $\{\bar{P}_n^{(\alpha,\beta)}\}_{n\in\mathbb{N}_0}$ *and* $\{\bar{P}_n^{(\alpha,\beta-1)}\}_{n\in\mathbb{N}_0}$ *are given by*

$$\bar{\kappa}_{i,j}^{(\alpha,\beta)\to(\alpha,\beta-1)} = (-1)^{i+j}\frac{2^j}{2^i}\frac{\Gamma(j+1)}{\Gamma(i+1)}\frac{\Gamma(j+\alpha+1)}{\Gamma(i+\alpha+1)}\frac{\Gamma(2i+\alpha+\beta+1)}{\Gamma(2j+\alpha+\beta+1)},$$

with $0 \le i \le j$. *As a consequence, the matrices*

$$\bar{\mathbf{K}}^{(\alpha,\beta)\to(\alpha-1,\beta)} = \mathrm{triu}\left(\mathbf{u}\,\mathbf{v}^{\mathrm{T}}\right),$$

with

$$\mathbf{u} = \left(\frac{\Gamma(2i+\alpha+\beta+1)}{2^i\Gamma(i+1)\Gamma(i+\beta+1)}\right)_{i=0}^n, \qquad \mathbf{v} = \left(\frac{2^j\Gamma(j+1)\Gamma(j+\beta+1)}{\Gamma(2j+\alpha+\beta+1)}\right)_{j=0}^n,$$

and

$$\bar{\mathbf{K}}^{(\alpha,\beta)\to(\alpha,\beta-1)} = \mathrm{triu}\left(\mathbf{u}\,\mathbf{v}^{\mathrm{T}}\right),$$

with

$$\mathbf{u} = \left(\frac{(-1)^i\Gamma(2i+\alpha+\beta+1)}{2^i\Gamma(i+1)\Gamma(i+\alpha+1)}\right)_{i=0}^n, \qquad \mathbf{v} = \left(\frac{2^j\Gamma(j+1)\Gamma(j+\alpha+1)}{(-1)^j\Gamma(2j+\alpha+\beta+1)}\right)_{j=0}^n,$$

are upper (1)*-generator representable semiseparable; see Definition* 2.6 *on page* 62.

PROOF. The result is a direct consequence of Theorem 1.50 and Corollary 1.51. □

Now, following the same principle as before, larger integer steps result in higher order semiseparable matrices.

Corollary 1.55 *Let* $\alpha > \gamma > -1$ *and* $\beta > -1$ *such that* $\alpha - \gamma$ *is a positive integer. Then the connection matrix* $\bar{\mathbf{K}}^{(\alpha,\beta)\to(\gamma,\beta)}$ *between the sequences of Jacobi polynomials* $\{\bar{P}_n^{(\alpha,\beta)}\}_{n\in\mathbb{N}_0}$ *and* $\{\bar{P}_n^{(\gamma,\beta)}\}_{n\in\mathbb{N}_0}$ *is upper* $(\alpha-\gamma)$*-generator representable semiseparable. Analogously, the connection matrix* $\bar{\mathbf{K}}^{(\alpha,\beta)\to(\alpha,\delta)}$ *is upper* $(\beta-\delta)$*-generator representable semiseparable, if for* $\beta > \delta > -1$ *and* $\alpha > -1$*, the value* $\beta - \delta$ *is a positive integer.*

PROOF. We can use the same argument as in Corollary 1.46 on page 21. □

Jacobi polynomials – general case

The connection between Jacobi polynomials where both parameters are varied is more involved. Askey [4, p. 62] uses a relatively direct approach to calculate the expressions for the connection coefficients in the general case, using the Rodrigues formula and a derivative identity; see also Section 1.5. He arrives at a formula that is equivalent to the one shown below, but for Jacobi polynomials in the standard normalization. From that, the special cases with one parameter fixed are derived.

Another way is to combine the results for changes in one parameter, which have been obtained in the last section, to obtain expressions for changes in both parameters. This can be done by using the identities

$$\bar{\mathbf{K}}^{(\alpha,\beta)\to(\gamma,\delta)} = \bar{\mathbf{K}}^{(\gamma,\beta)\to(\gamma,\delta)}\,\bar{\mathbf{K}}^{(\alpha,\beta)\to(\gamma,\beta)},$$
$$\bar{\mathbf{K}}^{(\alpha,\beta)\to(\gamma,\delta)} = \bar{\mathbf{K}}^{(\alpha,\delta)\to(\gamma,\delta)}\,\bar{\mathbf{K}}^{(\alpha,\beta)\to(\alpha,\delta)}.$$

The following theorem combines these to express the connection coefficients in the general case through the coefficients for changes in a single parameter. A similar result is given in [59, p. 306].

Lemma 1.56 *Let $\alpha, \gamma, \beta, \delta > -1$. Then the connection coefficients between the sequences of Jacobi polynomials $\{\bar{P}_n^{(\alpha,\beta)}\}_{n\in\mathbb{N}_0}$ and $\{\bar{P}_n^{(\gamma,\delta)}\}_{n\in\mathbb{N}_0}$ satisfy the identity*

$$\bar{\kappa}_{i,j}^{(\alpha,\beta)\to(\gamma,\delta)} = \sum_{k=i}^{j} \bar{\kappa}_{i,k}^{(\gamma,\beta)\to(\gamma,\delta)} \bar{\kappa}_{k,j}^{(\alpha,\beta)\to(\gamma,\beta)}. \tag{1.35}$$

PROOF. Consider connection the coefficients between the three sequences $\{\bar{P}_n^{(\alpha,\beta)}\}_{n\in\mathbb{N}_0}$, $\{\bar{P}_n^{(\gamma,\beta)}\}_{n\in\mathbb{N}_0}$, and $\{\bar{P}_n^{(\gamma,\delta)}\}_{n\in\mathbb{N}_0}$ as given in Theorems 1.50 and 1.51. Then

$$\begin{aligned}
\bar{P}_j^{(\alpha,\beta)} &= \sum_{k=0}^{j} \bar{\kappa}_{k,j}^{(\alpha,\beta)\to(\gamma,\beta)} \bar{P}_k^{(\gamma,\beta)} \\
&= \sum_{k=0}^{j} \bar{\kappa}_{k,j}^{(\alpha,\beta)\to(\gamma,\beta)} \sum_{i=0}^{k} \bar{\kappa}_{i,k}^{(\gamma,\beta)\to(\gamma,\delta)} \bar{P}_i^{(\gamma,\delta)} \\
&= \sum_{i=0}^{j} \sum_{k=i}^{j} \bar{\kappa}_{i,k}^{(\gamma,\beta)\to(\gamma,\delta)} \bar{\kappa}_{k,j}^{(\alpha,\beta)\to(\gamma,\beta)} \bar{P}_i^{(\gamma,\delta)}.
\end{aligned}$$

Thus, we have

$$\bar{\kappa}_{i,j}^{(\alpha,\beta)\to(\gamma,\delta)} = \sum_{k=i}^{j} \bar{\kappa}_{i,k}^{(\gamma,\beta)\to(\gamma,\delta)} \bar{\kappa}_{k,j}^{(\alpha,\beta)\to(\gamma,\beta)}.$$

Analogous results hold for differently normalized variants. □

Theorem 1.57 *Let $\alpha, \gamma, \beta, \delta > -1$. Then the connection coefficients between the sequences of Jacobi polynomials $\{\bar{P}_n^{(\alpha,\beta)}\}_{n\in\mathbb{N}_0}$ and $\{\bar{P}_n^{(\gamma,\delta)}\}_{n\in\mathbb{N}_0}$ are given by*

$$\begin{aligned}
\bar{\kappa}_{i,j}^{(\alpha,\beta)\to(\gamma,\delta)} = & \frac{(-1)^i}{\Gamma(\alpha-\gamma)\Gamma(\beta-\delta)} \frac{2^j}{2^i} \frac{\Gamma(j+1)}{\Gamma(i+1)} \frac{\Gamma(j+\beta+1)}{\Gamma(i+\gamma+1)} \frac{\Gamma(2i+\gamma+\delta+2)}{\Gamma(2j+\alpha+\beta+1)} \\
& \times \sum_{k=i}^{j} (-1)^k (2k+\gamma+\beta+1) \frac{\Gamma(k+\gamma+1)}{\Gamma(k+\beta+1)} \frac{\Gamma(j+k+\alpha+\beta+1)}{\Gamma(j+k+\gamma+\beta+2)} \\
& \times \frac{\Gamma(k+i+\gamma+\beta+1)}{\Gamma(k+i+\gamma+\delta+2)} \frac{\Gamma(j-k+\alpha-\gamma)}{\Gamma(j-k+1)} \frac{\Gamma(k-i+\beta-\delta)}{\Gamma(k-i+1)}.
\end{aligned} \tag{1.36}$$

PROOF. We insert the explicit expressions for the connection coefficients in (1.35). This leads to

$$\begin{aligned}
&\bar{\kappa}_{i,j}^{(\alpha,\beta)\to(\gamma,\delta)} \\
&= \sum_{k=i}^{j} \bar{\kappa}_{i,k}^{(\gamma,\beta)\to(\gamma,\delta)} \bar{\kappa}_{k,j}^{(\alpha,\beta)\to(\gamma,\beta)} \\
&= \sum_{k=i}^{j} \frac{(-1)^{i+k}}{\Gamma(\beta-\delta)} \frac{2^k}{2^i} \frac{\Gamma(k+1)}{\Gamma(i+1)} \frac{\Gamma(k+\gamma+1)}{\Gamma(i+\gamma+1)} \frac{\Gamma(2i+\gamma+\delta+2)}{\Gamma(2k+\gamma+\beta+1)} \frac{\Gamma(k+i+\gamma+\beta+1)}{\Gamma(k+i+\gamma+\delta+2)} \\
&\quad \times \frac{\Gamma(k-i+\beta-\delta)}{\Gamma(k-i+1)} \frac{1}{\Gamma(\alpha-\gamma)} \frac{2^j}{2^k} \frac{\Gamma(j+1)}{\Gamma(k+1)} \frac{\Gamma(j+\beta+1)}{\Gamma(k+\beta+1)} \frac{\Gamma(2k+\gamma+\beta+2)}{\Gamma(2j+\alpha+\beta+1)} \\
&\quad \times \frac{\Gamma(j+k+\alpha+\beta+1)}{\Gamma(j+k+\gamma+\beta+2)} \frac{\Gamma(j-k+\alpha-\gamma)}{\Gamma(j-k+1)}.
\end{aligned}$$

Then, we rearrange and simplify this expression to the desired result. □

Remark 1.58 Askey [4, p. 63] gives a formula for Jacobi polynomials in the standard normalization which involves the hypergeometric function. Using Lemma 1.37, one obtains

$$\begin{aligned}\bar{\kappa}_{i,j}^{(\alpha,\beta)\to(\gamma,\delta)} &= \frac{2^j}{2^i}\frac{\Gamma(j+1)}{\Gamma(i+1)}\frac{\Gamma(j+\alpha+1)}{\Gamma(i+\alpha+1)}\frac{\Gamma(j+i+\alpha+\beta+1)}{\Gamma(2j+\alpha+\beta+1)\Gamma(j-i+1)} \\ &\times {}_3F_2\left(\begin{matrix} i-j, j+i+\alpha+\beta+1, i+\gamma+1 \\ i+\alpha+1, 2i+\gamma+\delta+2\end{matrix}; 1\right)\end{aligned}$$

which is equivalent to (1.36) and where ${}_3F_2$ denotes the hypergeometric function.

These formulae are usually too complicated to be used in numerical practice. However, there are other special cases, more precisely, when both parameters α and β are changed the same way, that again lead to particularly structured connection matrices. To consider these cases is only natural, given that we have already considered integer changes in one parameter at a time. Let us look at the case $\gamma = \alpha + 1$ and $\delta = \beta + 1$ first.

Lemma 1.59 *Let $\alpha, \beta > -1$. Then the connection coefficients between the sequences of Jacobi polynomials $\{\bar{P}_n^{(\alpha,\beta)}\}_{n\in\mathbb{N}_0}$ and $\{\bar{P}_n^{(\alpha+1,\beta+1)}\}_{n\in\mathbb{N}_0}$ satisfy the identity*

$$\bar{\kappa}_{i,j}^{(\alpha,\beta)\to(\alpha+1,\beta+1)} = \begin{cases} 1, & \text{if } j = i, \\ \dfrac{2j(j+\alpha+1)}{(2j+\alpha+\beta+1)(2j+\alpha+\beta+2)} \\ \qquad - \dfrac{2j(j+\beta)}{(2j+\alpha+\beta)(2j+\alpha+\beta+1)}, & \text{if } j = i+1, \\ \dfrac{-4(j-1)j(j+\alpha)(j+\beta)}{(2j+\alpha+\beta-1)(2j+\alpha+\beta)^2(2j+\alpha+\beta+1)}, & \text{if } j = i+2, \\ 0, & \text{else.} \end{cases}$$

Thus, the connection matrix

$$\bar{\mathbf{K}}^{(\alpha,\beta)\to(\alpha+1,\beta+1)} = \begin{pmatrix} 1 & * & * & & \\ & 1 & * & \ddots & \\ & & 1 & \ddots & * \\ & & & \ddots & * \\ & & & & 1 \end{pmatrix}$$

is strictly $(0,2)$*-banded; see Definition* 2.2 *on page* 60.

PROOF. It is clear that the connection matrix $\bar{\mathbf{K}}^{(\alpha,\beta)\to(\alpha+1,\beta+1)}$ must be $(0,2)$-banded, since it is the product of the two $(0,1)$-banded matrices,

$$\bar{\mathbf{K}}^{(\alpha,\beta)\to(\alpha+1,\beta+1)} = \bar{\mathbf{K}}^{(\alpha+1,\beta)\to(\alpha+1,\beta+1)}\bar{\mathbf{K}}^{(\alpha,\beta)\to(\alpha+1,\beta)}.$$

To prove the explicit expressions for the entries, we use Lemma 1.56 which gives

$$\bar{\kappa}_{i,j}^{(\alpha,\beta)\to(\alpha+1,\beta+1)} = \sum_{k=i}^{j} \bar{\kappa}_{i,k}^{(\alpha+1,\beta)\to(\alpha+1,\beta+1)} \bar{\kappa}_{k,j}^{(\alpha,\beta)\to(\alpha+1,\beta)}$$

$$= \begin{cases} 1, & \text{if } j = i, \\ \bar{\kappa}_{j-1,j-1}^{(\alpha+1,\beta)\to(\alpha+1,\beta+1)} \bar{\kappa}_{j-1,j}^{(\alpha,\beta)\to(\alpha+1,\beta)} \\ \qquad + \bar{\kappa}_{j-1,j}^{(\alpha+1,\beta)\to(\alpha+1,\beta+1)} \bar{\kappa}_{j,j}^{(\alpha,\beta)\to(\alpha+1,\beta)}, & \text{if } j = i+1, \\ \kappa_{j-2,j-1}^{(\alpha+1,\beta)\to(\alpha+1,\beta+1)} \kappa_{j-1,j}^{(\alpha,\beta)\to(\alpha+1,\beta)}, & \text{if } j = i+2, \\ 0, & \text{else.} \end{cases}$$

We insert the explicit expressions obtained in Corollary 1.53 to complete the proof. □

The result shows that we are able to carry over results for changes to a single parameter, α or β, somewhat naturally to situations for symmetric changes to both parameters by using Lemma 1.56. Of course, one is also interested in the case when $\gamma = \alpha - 1$ and $\delta = \beta - 1$. But before stating the result, we need the following lemma about the calculation of a particular sum.

Lemma 1.60 *Let $\alpha, \beta > 0$, and $i, j \in \mathbb{N}_0$ with $i \leq j$. Then*

$$\sum_{k=i}^{j} (-1)^k (2k + \alpha + \beta) \frac{\Gamma(k+\alpha)}{\Gamma(k+\beta+1)} = (-1)^j \frac{\Gamma(j+\alpha+1)}{\Gamma(j+\beta+1)} + (-1)^i \frac{\Gamma(i+\alpha)}{\Gamma(i+\beta)}.$$

PROOF. Let us start with the case $i = j$. Then the sum consists of a single term,

$$\sum_{k=i}^{i} (-1)^k (2k + \alpha + \beta) \frac{\Gamma(k+\alpha)}{\Gamma(k+\beta+1)} = (-1)^i (2i + \alpha + \beta) \frac{\Gamma(i+\alpha)}{\Gamma(i+\beta+1)}$$

$$= (-1)^i ((i+\alpha) + (i+\beta)) \frac{\Gamma(i+\alpha)}{\Gamma(i+\beta+1)} = (-1)^i \frac{\Gamma(i+\alpha+1)}{\Gamma(i+\beta+1)} + (-1)^i \frac{\Gamma(i+\alpha)}{\Gamma(i+\beta)}.$$

Thus, the formula is true for $i = j$. Now, fix an arbitrary $i \geq 0$ and prove the general case by induction over j. The initial case $i = j$ has already been verified. So we fix $j \geq i$ arbitrary and suppose that the formula is true for this particular j; this is the induction hypothesis. Then, using the induction hypothesis, the sum for $j + 1$ satisfies

$$\sum_{k=i}^{j+1} (-1)^k (2k + \alpha + \beta) \frac{\Gamma(k+\alpha)}{\Gamma(k+\beta+1)}$$

$$= \sum_{k=i}^{j} (-1)^k (2k + \alpha + \beta) \frac{\Gamma(k+\alpha)}{\Gamma(k+\beta+1)} + (-1)^{j+1} (2j + \alpha + \beta + 2) \frac{\Gamma(j+\alpha+1)}{\Gamma(j+\beta+2)}$$

$$= (-1)^j \frac{\Gamma(j+\alpha+1)}{\Gamma(j+\beta+1)} + (-1)^i \frac{\Gamma(i+\alpha)}{\Gamma(i+\beta)} + (-1)^{j+1} (2j + \alpha + \beta + 2) \frac{\Gamma(j+\alpha+1)}{\Gamma(j+\beta+2)}$$

$$= (-1)^{j+1} \frac{\Gamma(j+\alpha+2)}{\Gamma(j+\beta+2)} \left(\frac{2j + \alpha + \beta + 2}{j + \alpha + 1} - \frac{j + \beta + 1}{j + \alpha + 1} \right) + (-1)^i \frac{\Gamma(i+\alpha)}{\Gamma(i+\beta)}$$

$$= (-1)^{j+1} \frac{\Gamma(j+\alpha+2)}{\Gamma(j+\beta+2)} + (-1)^i \frac{\Gamma(i+\alpha)}{\Gamma(i+\beta)},$$

which proves the result. □

Now we are ready to state an explicit expression for the connection coefficients between Jacobi polynomials when $\gamma = \alpha - 1$ and $\delta = \beta - 1$.

Theorem 1.61 *Let $\alpha, \beta > 0$. Then the connection coefficients between the sequences of Jacobi polynomials $\{\bar{P}_n^{(\alpha,\beta)}\}_{n\in\mathbb{N}_0}$ and $\{\bar{P}_n^{(\alpha-1,\beta-1)}\}_{n\in\mathbb{N}_0}$ are given by*

$$\begin{aligned}&\bar{\kappa}_{i,j}^{(\alpha,\beta)\to(\alpha-1,\beta-1)}\\ &= \frac{2^j}{2^i}\frac{\Gamma(j+1)}{\Gamma(i+1)}\frac{\Gamma(2i+\alpha+\beta)}{\Gamma(2j+\alpha+\beta+1)}\left((-1)^{i+j}\frac{\Gamma(j+\alpha+1)}{\Gamma(i+\alpha)}+\frac{\Gamma(j+\beta+1)}{\Gamma(i+\beta)}\right).\end{aligned}$$

PROOF. We take the result from Lemma 1.56 for $\gamma = \alpha - 1$ and $\delta = \beta - 1$,

$$\kappa_{i,j}^{(\alpha,\beta)\to(\alpha-1,\beta-1)} = \sum_{k=i}^{j} \kappa_{i,k}^{(\alpha-1,\beta)\to(\alpha-1,\beta-1)}\kappa_{k,j}^{(\alpha,\beta)\to(\alpha-1,\beta)},$$

and replace $\bar{\kappa}_{i,k}^{(\alpha-1,\beta)\to(\alpha-1,\beta-1)}$ and $\bar{\kappa}_{k,j}^{(\alpha,\beta)\to(\alpha-1,\beta)}$ by the expressions obtained in Corollary 1.54. Then

$$\begin{aligned}&\bar{\kappa}_{i,j}^{(\alpha,\beta)\to(\alpha-1,\beta-1)}\\ &=\sum_{k=i}^{j}(-1)^{i+k}\frac{2^k\Gamma(k+1)\Gamma(k+\alpha)\Gamma(2i+\alpha+\beta)}{2^i\Gamma(i+1)\Gamma(i+\alpha)\Gamma(2k+\alpha+\beta)}\frac{2^j\Gamma(j+1)\Gamma(j+\beta+1)\Gamma(2k+\alpha+\beta+1)}{2^k\Gamma(k+1)\Gamma(k+\beta+1)\Gamma(2j+\alpha+\beta+1)}\\ &=(-1)^i\frac{2^j}{2^i}\frac{\Gamma(j+1)}{\Gamma(i+1)}\frac{\Gamma(j+\beta+1)}{\Gamma(i+\alpha)}\frac{\Gamma(2i+\alpha+\beta)}{\Gamma(2j+\alpha+\beta+1)}\sum_{k=i}^{j}(-1)^k(2k+\alpha+\beta)\frac{\Gamma(k+\alpha)}{\Gamma(k+\beta+1)}.\end{aligned}$$

Now, we replace the sum by the expression obtained in Lemma 1.60,

$$\begin{aligned}\bar{\kappa}_{i,j}^{(\alpha,\beta)\to(\alpha-1,\beta-1)} =&(-1)^i\frac{2^j}{2^i}\frac{\Gamma(j+1)}{\Gamma(i+1)}\frac{\Gamma(j+\beta+1)}{\Gamma(i+\alpha)}\frac{\Gamma(2i+\alpha+\beta)}{\Gamma(2j+\alpha+\beta+1)}\\ &\times\left((-1)^j\frac{\Gamma(j+\alpha+1)}{\Gamma(j+\beta+1)}+(-1)^i\frac{\Gamma(i+\alpha)}{\Gamma(i+\beta)}\right).\end{aligned}$$

This is easily simplified to the desired expression. □

Corollary 1.62 *Let $\alpha, \beta > 0$. Then the connection matrix $\bar{\mathbf{K}}^{(\alpha,\beta)\to(\alpha-1,\beta-1)}$ can be written as*

$$\bar{\mathbf{K}}^{(\alpha,\beta)\to(\alpha-1,\beta-1)} = \operatorname{triu}(\mathbf{u}\,\mathbf{v}^{\mathrm{T}}) + \operatorname{triu}(\mathbf{w}\,\mathbf{z}^{\mathrm{T}}),$$

with

$$\begin{aligned}\mathbf{u} &= \left((-1)^i\frac{\Gamma(2i+\alpha+\beta)}{2^i\Gamma(i+1)\Gamma(i+\alpha)}\right)_{i=0}^{n}, & \mathbf{v} &= \left((-1)^j\frac{2^j\Gamma(j+1)\Gamma(j+\alpha+1)}{\Gamma(2j+\alpha+\beta+1)}\right)_{j=0}^{n},\\ \mathbf{w} &= \left(\frac{\Gamma(2i+\alpha+\beta)}{2^i\Gamma(i+1)\Gamma(i+\beta)}\right)_{i=0}^{n}, & \mathbf{z} &= \left(\frac{2^j\Gamma(j+1)\Gamma(j+\beta+1)}{\Gamma(2j+\alpha+\beta+1)}\right)_{j=0}^{n}.\end{aligned}$$

The matrix $\bar{\mathbf{K}}^{(\alpha,\beta)\to(\alpha+1,\beta+1)}$ is upper (2)-generator representable semiseparable; see Definition 2.6 on page 62.

Corollary 1.63 *Let $\alpha, \beta, \gamma, \delta > 0$ such that $\alpha - \gamma = \beta - \delta$ is a positive integer. Then the connection matrix $\bar{\mathbf{K}}^{(\alpha,\beta)\to(\gamma,\delta)}$ is upper $\big(2(\gamma-\alpha)\big)$-generator representable semiseparable.*

The results in this section allow us to make Figure 1.3 complete. Although the connection between arbitrary Jacobi polynomials is not a simple expression, changing both parameters symmetrically leads to easier special cases. This is shown in Figure 1.4.

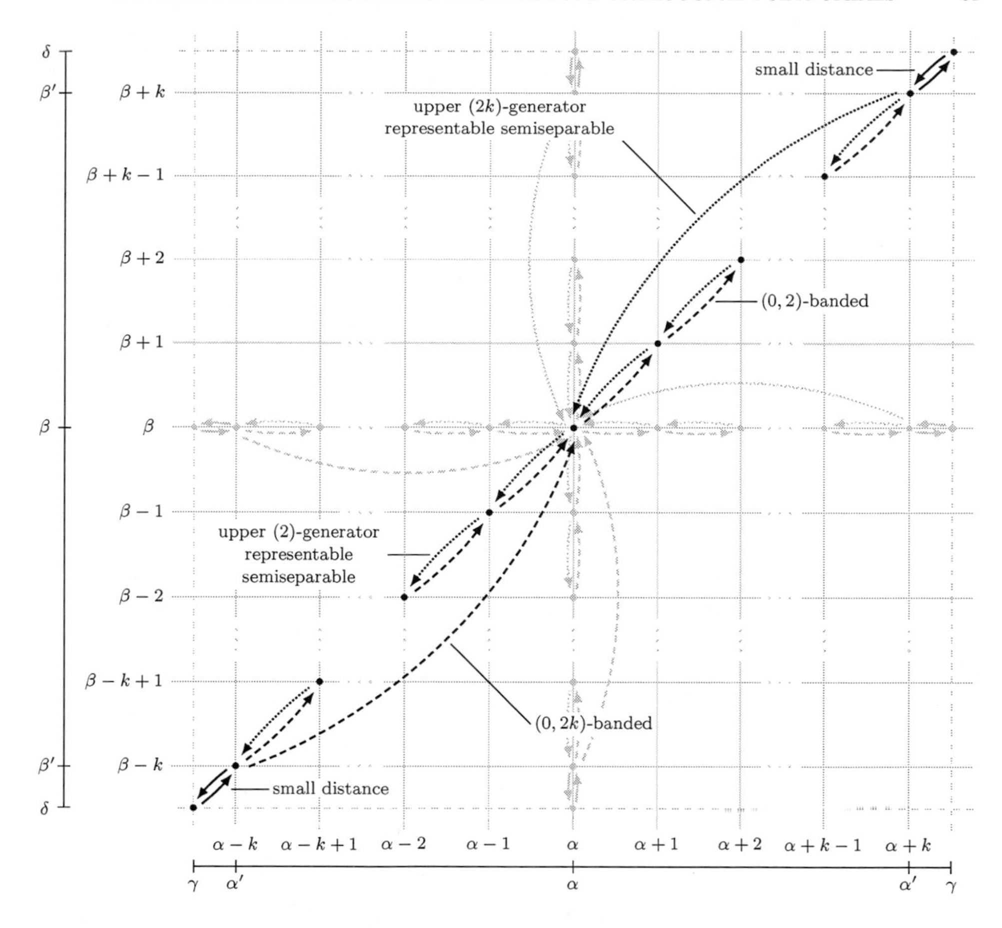

Figure 1.4: The connection between Jacobi polynomials if both parameters, α and β, in the polynomials $P_n^{(\alpha,\beta)}$ are changed symmetrically: If α and β are simultaneously increased or decreased in unit steps, then the connection matrices are $(0,2)$-banded or upper (2)-generator representable semiseparable, respectively. Integer steps of length k have $(0,2k)$-banded or upper $(2k)$-generator representable semiseparable matrices. General steps can be decomposed into an integer length step and a second step that spans only a small distance.

Gegenbauer polynomials

Gegenbauer polynomials are a special case of the Jacobi polynomials $P_n^{(\alpha,\beta)}$ with $\alpha = \beta$ and a different normalization; see Section 1.3.2. Since they carry only a single parameter, the results for the connection coefficients will be somewhat similar to those for Laguerre polynomials. The following is a first general result. We recall that since Gegenbauer polynomials are orthogonal with respect to a symmetric measure, the connection coefficients $\kappa_{i,j}$ are zero if $i + j$ is odd. This is understood in the following.

Definition 1.64 *Let* $\{\bar{C}_n^{(\alpha)}\}_{n\in\mathbb{N}_0}$ *with* $\alpha > -1/2$ *and* $\{\bar{C}_n^{(\beta)}\}_{n\in\mathbb{N}_0}$ *with* $\beta > -1/2$ *be two families of* monic *Gegenbauer polynomials. Then the connection coefficients* $\bar{\kappa}_{i,j}$ *in the formula*

$$\bar{C}_j^{(\alpha)} = \sum_{i=0}^{j} \bar{\kappa}_{i,j}\bar{C}_i^{(\beta)}, \qquad j = 0, 1, \ldots,$$

are denoted $\bar{\kappa}_{i,j} = \bar{\kappa}_{i,j}^{\mathrm{G},(\alpha)\to(\beta)}$ *or* $\bar{\kappa}_{i,j}^{(\alpha)\to(\beta)}$ *if clear that we mean the Gegenbauer polynomials. The corresponding connection matrix is denoted* $\bar{\mathbf{K}}^{\mathrm{G},(\alpha)\to(\beta)}$ *or* $\bar{\mathbf{K}}^{(\alpha)\to(\beta)}$ *for short.*

The following formula is found in [4, p. 359], [36], [73, p. 99], and also [59, p. 298].

Theorem 1.65 *Let* $\alpha, \beta > -1/2$. *Then the connection coefficients between the sequences of Gegenbauer polynomials* $\{\bar{C}_n^{(\alpha)}\}_{n\in\mathbb{N}_0}$ *and* $\{\bar{C}_n^{(\beta)}\}_{n\in\mathbb{N}_0}$ *are given by*

$$\bar{\kappa}_{i,j}^{(\alpha)\to(\beta)} = \frac{1}{\Gamma(\alpha-\beta)}\frac{2^i}{2^j}\frac{\Gamma(i+\beta+1)}{\Gamma(i+1)}\frac{\Gamma(j+1)}{\Gamma(j+\alpha)}\frac{\Gamma\left(\frac{j-i}{2}+\alpha-\beta\right)}{\Gamma\left(\frac{j-i}{2}+1\right)}\frac{\Gamma\left(\frac{j+i}{2}+\alpha\right)}{\Gamma\left(\frac{j+i}{2}+\beta+1\right)},$$

with $i + j$ *even and* $i \le j$.

Proof. It can be verified that the recurrence formula (1.27) and the initial conditions (1.28) are satisfied. □

Remark 1.66 The proof found in [59] is rather complicated. A more elegant proof is given in [2] by noting that when $\alpha = \beta$ and $\gamma = \delta$,

$$\begin{aligned}\bar{\kappa}_{i,j}^{(\alpha)\to(\gamma)} &= \frac{2^j}{2^i}\frac{\Gamma(j+1)}{\Gamma(i+1)}\frac{\Gamma(j+\alpha+1)}{\Gamma(i+\alpha+1)}\frac{\Gamma(j+i+2\alpha+1)}{\Gamma(2j+2\alpha+1)\Gamma(j-i+1)}\\ &\quad\times {}_3F_2\left(\begin{matrix} i-j, j+i+2\alpha+1, i+\gamma+1 \\ i+\alpha+1, 2i+2\gamma+2\end{matrix}; 1\right).\end{aligned}$$

In this case, the hypergeometric function is explicitly summable [2, p. 148, Theorem 3.5.5],

$${}_3F_2\left(\begin{matrix} a, b, c \\ (a+b+1)/2, 2c\end{matrix}; 1\right) = \frac{\Gamma(1/2)\Gamma\left(c+\frac{1}{2}\right)\Gamma\left(\frac{a+b+1}{2}\right)\Gamma\left(c-\frac{a+b-1}{2}\right)}{\Gamma\left(\frac{a+1}{2}\right)\Gamma\left(\frac{b+1}{2}\right)\Gamma\left(c-\frac{a-1}{2}\right)\Gamma\left(c-\frac{b-1}{2}\right)}.$$

Combining the two results, one obtains the desired identity.

The next natural step is to look at the case when the difference between β and α is a positive integer.

Corollary 1.67 *Let* $-1/2 < \alpha < \beta$ *such that* $\beta - \alpha$ *is a positive integer. Then the connection coefficients between the sequences of Gegenbauer polynomials* $\{\bar{C}_n^{(\alpha)}\}_{n\in\mathbb{N}_0}$ *and* $\{\bar{C}_n^{(\beta)}\}_{n\in\mathbb{N}_0}$ *are given by*

$$\bar{\kappa}_{i,j}^{(\alpha)\to(\beta)} = \begin{cases} (-1)^{(j-i)/2}\binom{\beta-\alpha}{(j-i)/2}\dfrac{2^i}{2^j} & \\ \quad\times\dfrac{\Gamma(i+\beta+1)}{\Gamma(i+1)}\dfrac{\Gamma(j+1)}{\Gamma(j+\alpha)}\dfrac{\Gamma\left(\frac{j+i}{2}+\alpha\right)}{\Gamma\left(\frac{j+i}{2}+\beta+1\right)}, & \text{if } i+j \text{ even and } i \le j \le i+2(\beta-\alpha), \\ 0, & \text{else.}\end{cases}$$

This implies that the connection matrix

$$\bar{\mathbf{K}}^{(\alpha)\to(\beta)} = \begin{pmatrix} \overbrace{1 \; 0 \; * \; 0 \; * \; \dots \; *}^{2(\beta-\alpha)+1} & & & & & & & \\ & 1 & 0 & * & 0 & * & \ddots & \\ & & 1 & 0 & * & 0 & * & * \\ & & & 1 & 0 & * & 0 & \ddots & \vdots \\ & & & & 1 & 0 & * & \ddots & * \\ & & & & & 1 & 0 & \ddots & 0 \\ & & & & & & 1 & \ddots & * \\ & & & & & & & \ddots & 0 \\ & & & & & & & & 1 \end{pmatrix}$$

is checkerboard-like $(0, 2(\beta-\alpha))$*-banded; see Definition* 2.2 *on page* 60 *and Definition* 2.9 *on page* 63.

Corollary 1.68 *Let* $\alpha > -1/2$. *Then the connection coefficients between the sequences of Gegenbauer polynomials* $\{\bar{C}_n^{(\alpha)}\}_{n\in\mathbb{N}_0}$ *and* $\{\bar{C}_n^{(\alpha+1)}\}_{n\in\mathbb{N}_0}$ *are given by*

$$\bar{\kappa}_{i,j}^{(\alpha)\to(\alpha+1)} = \begin{cases} 1, & \text{if } j = i, \\ \dfrac{-(j-1)j}{4(j+\alpha-1)(j+\alpha)}, & \text{if } j = i+2, \\ 0, & \text{else.} \end{cases}$$

Thus, the connection matrix

$$\bar{\mathbf{K}}^{(\alpha)\to(\alpha+1)} = \begin{pmatrix} 1 & 0 & * & & \\ & 1 & 0 & \ddots & \\ & & 1 & \ddots & * \\ & & & \ddots & 0 \\ & & & & 1 \end{pmatrix}$$

is checkerboard-like $(0, 2)$*-banded; see Definition* 2.2 *on page* 60 *and Definition* 2.9 *on page* 63.

PROOF. The proof is a direct consequence of the previous result. □

Let us also observe the case when the parameter α is decreased in an integer step.

Corollary 1.69 *Let* $\alpha > 1/2$. *Then the connection coefficients between the sequences of Gegenbauer polynomials* $\{\bar{C}_n^{(\alpha)}\}_{n\in\mathbb{N}_0}$ *and* $\{\bar{C}_n^{(\alpha-1)}\}_{n\in\mathbb{N}_0}$ *are given by*

$$\bar{\kappa}_{i,j}^{(\alpha)\to(\alpha-1)} = \frac{2^i}{2^j} \frac{\Gamma(j+1)}{\Gamma(j+\alpha)} \frac{\Gamma(i+\alpha)}{\Gamma(i+1)}, \qquad \text{with } i+j \text{ even and } i \le j.$$

Thus, the connection matrix

$$\bar{\mathbf{K}}^{(\alpha)\to(\alpha-1)} = \operatorname{triuc}(\mathbf{u}\,\mathbf{v}^{\mathrm{T}})$$

with

$$\mathbf{u} = \left(\frac{2^i\Gamma(i+\alpha)}{\Gamma(i+1)} \right)_{i=0}^{n}, \quad \mathbf{v} = \left(\frac{\Gamma(j+1)}{2^j\Gamma(j+\alpha)} \right)_{j=0}^{n}$$

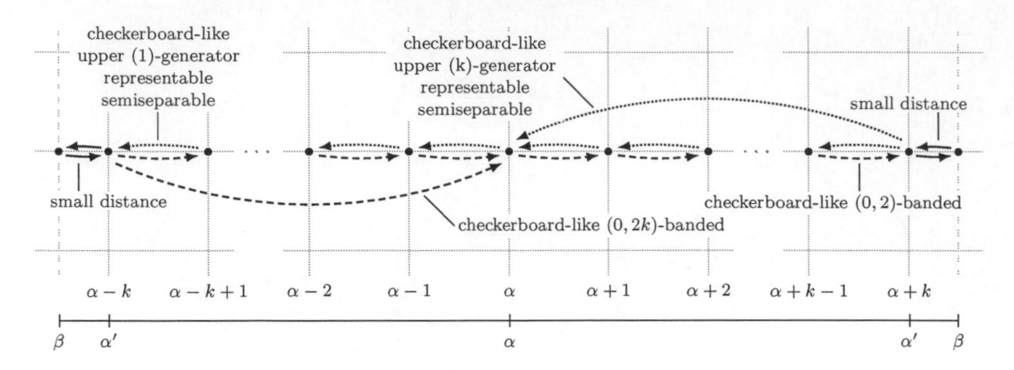

Figure 1.5: The connection between Gegenbauer polynomials: If the parameter α in $C_n^{(\alpha)}$ is increased or decreased in unit steps, then the connection matrices are checkerboard-like $(0, 2)$-banded or upper (1)-generator representable semiseparable, respectively. Integer steps of length k lead to checkerboard-like $(0, 2k)$-banded or upper (k)-generator representable semiseparable matrices. General steps can be decomposed into an integer length step and a second step that spans only a small distance.

is checkerboard-like upper (1)*-generator representable semiseparable.*

PROOF. The proof is a direct consequence of Theorem 1.65. □

Corollary 1.70 *Let* $-1/2 < \beta < \alpha$ *such that* $\alpha - \beta$ *is a positive integer. Then the connection matrix* $\bar{\mathbf{K}}^{(\alpha)\to(\beta)}$ *between the sequences of Gegenbauer polynomials* $\{\bar{C}_n^{(\alpha)}\}_{n\in\mathbb{N}_0}$ *and* $\{\bar{C}_n^{(\beta)}\}_{n\in\mathbb{N}_0}$ *is checkerboard-like upper* $(\alpha - \beta)$*-generator representable semiseparable.*

PROOF. The matrix $\bar{\mathbf{K}}^{(\alpha)\to(\beta)}$ is a product of $(\alpha-\beta)$ checkerboard-like upper (1)-generator representable semiseparable matrices,

$$\bar{\mathbf{K}}^{(\alpha)\to(\beta)} = \bar{\mathbf{K}}^{(\beta+1)\to(\beta)} \cdot \bar{\mathbf{K}}^{(\beta+2)\to(\beta+1)} \cdot \ldots \cdot \bar{\mathbf{K}}^{(\alpha-1)\to(\alpha-2)} \cdot \bar{\mathbf{K}}^{(\alpha)\to(\alpha-1)}$$

and is therefore checkerboard-like upper $(\alpha - \beta)$-generator representable semiseparable. See Corollary 9.57 in [78, p. 432]. □

The connection between Gegenbauer polynomials is illustrated in Figure 1.5.

Chebyshev and Legendre polynomials

The results of the previous section enable us to obtain the connection coefficients between Chebyshev polynomials of first and second kind and Legendre polynomials, since these are, up to normalization, identical to the Gegenbauer polynomials. Note that the monic and normalized Gegenbauer polynomials $\{\bar{C}_n^{(\alpha)}\}_{n\in\mathbb{N}_0}$ are also well defined for $\alpha = 0$, in contrast to the standard normalization.

Theorem 1.71 *The connection coefficients between the sequences of Chebyshev polynomials*

$$\{\bar{T}_n\}_{n\in\mathbb{N}_0} = \{\bar{P}_n^{(-1/2,-1/2)}\}_{n\in\mathbb{N}_0} = \{\bar{C}_n^{(0)}\}_{n\in\mathbb{N}_0}$$

and

$$\{\bar{U}_n\}_{n\in\mathbb{N}_0} = \{\bar{P}_n^{(1/2,1/2)}\}_{n\in\mathbb{N}_0} = \{\bar{C}_n^{(1)}\}_{n\in\mathbb{N}_0}$$

are given by

$$\bar{\kappa}_{i,j}^{(-1/2,-1/2)\to(1/2,1/2)} = \begin{cases} 1, & \text{if } j = i, \\ -1/4, & \text{if } j = i+2, \\ 0, & \text{else}, \end{cases}$$

$$\bar{\kappa}_{i,j}^{(1/2,1/2)\to(-1/2,-1/2)} = 2^{i-j}, \qquad \text{with } i+j \text{ even}.$$

PROOF. The proof is obtained from Corollary 1.68 for $\alpha = 0$ and Corollary 1.69 for $\alpha = 1$. Alternatively, we can use Lemma 1.59 for $\alpha = \beta = -1/2$ and Theorem 1.61 for $\alpha = \beta = 1/2$. □

Theorem 1.72 *The connection coefficients between the Chebyshev polynomials of first kind*

$$\{\bar{T}_n\}_{n\in\mathbb{N}_0} = \{\bar{P}_n^{(-1/2,-1/2)}\}_{n\in\mathbb{N}_0} = \{\bar{C}_n^{(0)}\}_{n\in\mathbb{N}_0}$$

and the Legendre polynomials

$$\{\bar{P}_n\}_{n\in\mathbb{N}_0} = \{\bar{P}_n^{(0,0)}\}_{n\in\mathbb{N}_0} = \{\bar{C}_n^{(1/2)}\}_{n\in\mathbb{N}_0}$$

are given by

$$\bar{\kappa}_{i,j}^{(-1/2,-1/2)\to(0,0)} = \frac{-j}{2\sqrt{\pi}} \frac{2^i}{2^j} \frac{\Gamma\left(i+\frac{3}{2}\right)}{\Gamma(i+1)} \frac{\Gamma\left(\frac{j-i}{2}-\frac{1}{2}\right)}{\Gamma\left(\frac{j-i}{2}+1\right)} \frac{\Gamma\left(\frac{i+j}{2}\right)}{\Gamma\left(\frac{j+i}{2}+\frac{3}{2}\right)},$$

$$\bar{\kappa}_{i,j}^{(0,0)\to(-1/2,-1/2)} = \frac{1}{\sqrt{\pi}} \frac{2^i}{2^j} \frac{\Gamma(j+1)}{\Gamma\left(j+\frac{1}{2}\right)} \frac{\Gamma\left(\frac{j-i}{2}+\frac{1}{2}\right)}{\Gamma\left(\frac{j-i}{2}+1\right)} \frac{\Gamma\left(\frac{j+i}{2}+\frac{1}{2}\right)}{\Gamma\left(\frac{j+i}{2}+1\right)}.$$

Theorem 1.73 *The connection coefficients between the Chebyshev polynomials of first kind*

$$\{\bar{U}_n\}_{n\in\mathbb{N}_0} = \{\bar{P}_n^{(1/2,1/2)}\}_{n\in\mathbb{N}_0} = \{\bar{C}_n^{(1)}\}_{n\in\mathbb{N}_0}$$

and the Legendre polynomials

$$\{\bar{P}_n\}_{n\in\mathbb{N}_0} = \{\bar{P}_n^{(0,0)}\}_{n\in\mathbb{N}_0} = \{\bar{C}_n^{(1/2)}\}_{n\in\mathbb{N}_0}$$

are given by

$$\bar{\kappa}_{i,j}^{(0,0)\to(1/2,1/2)} = \frac{-(i+1)}{2\sqrt{\pi}} \frac{2^i}{2^j} \frac{\Gamma(j+1)}{\Gamma\left(j+\frac{1}{2}\right)} \frac{\Gamma\left(\frac{j-i}{2}-\frac{1}{2}\right)}{\Gamma\left(\frac{j-i}{2}+1\right)} \frac{\Gamma\left(\frac{j+i}{2}+\frac{1}{2}\right)}{\Gamma\left(\frac{j+i}{2}+2\right)},$$

$$\bar{\kappa}_{i,j}^{(1/2,1/2)\to(0,0)} = \frac{1}{\sqrt{\pi}} \frac{2^i}{2^j} \frac{\Gamma\left(i+\frac{3}{2}\right)}{\Gamma(i+1)} \frac{\Gamma\left(\frac{j-i}{2}+\frac{1}{2}\right)}{\Gamma\left(\frac{j-i}{2}+1\right)} \frac{\Gamma\left(\frac{j+i}{2}+1\right)}{\Gamma\left(\frac{j+i}{2}+\frac{3}{2}\right)}.$$

1.5 Derivative identities

An important property of classical orthogonal polynomials is that they are linked by a single differential equation. As we have seen in Lemma 1.29 on page 7, derivatives of classical polynomials are classical themselves. The Rodrigues formula makes it easy to identify these derivatives.

Lemma 1.74 *Let $n \geq 0$. Then the following identities hold:*

$$\frac{\mathrm{d}}{\mathrm{d}x}\bar{H}_n = n\bar{H}_{n-1}, \tag{1.37}$$
$$\frac{\mathrm{d}}{\mathrm{d}x}\bar{L}_n^{(\alpha)} = n\bar{L}_{n-1}^{(\alpha+1)},$$
$$\frac{\mathrm{d}}{\mathrm{d}x}\bar{P}_n^{(\alpha,\beta)} = n\bar{P}_{n-1}^{(\alpha+1,\beta+1)}. \tag{1.38}$$

PROOF. The proof follows from the Rodrigues formula (1.16) on page 8. □

These are well-known results. For our purposes, we need to express the derivatives of a classical polynomial in the same family it belongs to. For example, the first derivative of a Laguerre polynomial $L_n^{(\alpha)}$ should be expressed as a linear combination of the polynomials $L_i^{(\alpha)}$ instead of the polynomials $L_i^{(\alpha+1)}$, $i = 0, 1, \dots, n-1$. Of course, this can be done by using the connection coefficients obtained in the last section. At this point, the chief observation is that the resulting expressions have a very simple structure.

Theorem 1.75 *For the monic Laguerre polynomials $\{\bar{L}_n^{(\alpha)}\}_{n\in\mathbb{N}_0}$, we have*

$$\frac{\mathrm{d}}{\mathrm{d}x}\bar{L}_n^{(\alpha)} = \bar{A}_n \sum_{i=0}^{n-1} \bar{B}_i \bar{L}_i^{(\alpha)},$$

with

$$\bar{A}_n := (-1)^{n+1}\Gamma(n+1), \qquad \bar{B}_i := \frac{(-1)^i}{\Gamma(i+1)}.$$

Similarly, for the monic Jacobi polynomials $\bar{P}_n^{(\alpha,\beta)}$, we have

$$\frac{\mathrm{d}}{\mathrm{d}x}\bar{P}_n^{(\alpha,\beta)} = \bar{A}_n \sum_{i=0}^{n-1} \bar{B}_i \bar{P}_i^{(\alpha,\beta)} + \bar{C}_n \sum_{i=0}^{n-1} \bar{D}_i \bar{P}_i^{(\alpha,\beta)},$$

with

$$\begin{aligned}
\bar{A}_n &= (-1)^{n+1}\Gamma(n+\alpha+1)\bar{A}'_n, & \bar{B}_i &= \frac{(-1)^i}{\Gamma(i+\alpha+1)}\bar{B}'_i, \\
\bar{C}_n &= \Gamma(n+\beta+1)\bar{A}'_n, & \bar{D}_i &= \frac{1}{\Gamma(i+\beta+1)}\bar{B}'_i, \\
\bar{A}'_n &= \frac{2^{n-1}\Gamma(n+1)}{\Gamma(2n+\alpha+\beta+1)}, & \bar{B}'_i &= \frac{\Gamma(2i+\alpha+\beta+2)}{2^i\Gamma(i+1)}.
\end{aligned} \tag{1.39}$$

PROOF. For the Laguerre polynomials, we use Lemma 1.74 in conjunction with Corollary 1.45. The result for Jacobi polynomials follows similarly from Lemma 1.74 together with Theorem 1.61. □

The derivative identity obtained for the monic Jacobi polynomials easily translates into the special cases for the Chebyshev polynomials of first and second kind, the Legendre polynomials, and the Gegenbauer polynomials.

Corollary 1.76 *Assume $n \in \mathbb{N}_0$ and define*

$$\chi = \chi(n) := \begin{cases} 1, & \text{if } n \text{ even}, \\ 0, & \text{if } n \text{ odd}. \end{cases}$$

Then the following identities hold:

$$\begin{aligned}
\frac{\mathrm{d}}{\mathrm{d}x}\bar{T}_n &= \frac{n}{2^n}\sum_{i=0}^{n-1} 2^i\left((-1)^{n+i+1}+1\right)T_i \\
&= \frac{n}{2^n}\sum_{i=0}^{\lfloor (n-1)/2\rfloor} 2^{2i+\chi+1}T_{2i+\chi}, \\
\frac{\mathrm{d}}{\mathrm{d}x}\bar{U}_n &= \frac{1}{2^n}\sum_{i=0}^{n-1}(i+1)2^i\left((-1)^{n+i+1}+1\right)U_i \\
&= \frac{1}{2^n}\sum_{i=0}^{\lfloor (n-1)/2\rfloor} 2^{2i+\chi+1}(2i+\chi+1)U_{2i+\chi}, \\
\frac{\mathrm{d}}{\mathrm{d}x}\bar{C}_n^{(\alpha)} &= \frac{\Gamma(n+1)}{2^n\Gamma(n+\alpha)}\sum_{i=0}^{n-1}\frac{2^i\Gamma(i+\alpha+1)}{\Gamma(i+1)}\left((-1)^{n+i-1}+1\right)\bar{C}_i^{(\alpha)} \\
&= \frac{\Gamma(n+1)}{2^n\Gamma(n+\alpha)}\sum_{i=0}^{\lfloor (n-1)/2\rfloor}\frac{2^{2i+\chi+1}\Gamma(2i+\chi+\alpha+1)}{\Gamma(2i+\chi+1)}\bar{C}_{2i+\chi}^{(\alpha)}.
\end{aligned}$$

1.6 Classical associated functions

There does not seem to be a widely accepted definition of *associated functions* in the context of classical orthogonal polynomials. The one used in this text is based on a modification to the Rodrigues formula for classical orthogonal polynomials. It is motivated by the attempt to generalize some results obtained for the well-known associated Legendre functions, see, e.g., [2, p. 456], to functions of similar type. It seems appropriate to give a definition of associated functions via a Rodrigues-type formula derived from the one for classical orthogonal polynomials.
It can be shown that the obtained functions are identical, at least up to multiplicative constants, to those used in many fields of mathematical physics. For example, the *associated Gegenbauer functions* are important on higher dimensional spheres, and the definition given below coincides with the one used in [62]. It is also possible to define the more general *associated Jacobi functions* which play a role in harmonic analysis on the cross product of two spheres; see [48, 49, 50, 51, 75]. Further types of associated functions, like *associated Laguerre functions* and *generalized associated Jacobi functions*, are also given below. Generally, the theory developed in this section involves classes of functions that are well-known in certain circles, but that are usually not found in the standard references about classical orthogonal polynomials.

1.6.1 General theory. The following definition introduces associated functions that are derived from the Rodrigues formula for classical orthogonal polynomials.

Definition 1.77 *Let $\{p_n\}_{n\in\mathbb{N}_0}$ be a sequence of classical polynomials, orthogonal with respect to an absolutely continuous measure $\mathrm{d}\lambda(x) = w(x)\mathrm{d}x$, given by the Rodrigues formula*

$$p_n^{(m)}(x) = \frac{A_{n,m}B_n}{\sigma^m(x)w(x)}\frac{\mathrm{d}^{n-m}}{\mathrm{d}x^{n-m}}\left(\sigma^n(x)w(x)\right),$$

which satisfy a differential equation of hypergeometric type

$$\sigma(x)p_n''(x) + \tau(x)p_n'(x) + \lambda_n p_n(x) = 0.$$

Then the corresponding associated functions of order m, *denoted* $\{p_n^m\}_{n\in\mathbb{N}_0,|m|\leq n}$ *are defined by the Rodrigues-type formula*

$$p_n^m(x) := C_{n,\mu}\sigma^{\mu/2}(x)p_n^{(\mu)}(x) = \frac{A_{n,\mu}B_nC_{n,\mu}}{\sigma^{\mu/2}(x)w(x)}\frac{\mathrm{d}^{n-\mu}}{\mathrm{d}x^{n-\mu}}(\sigma^n(x)w(x)) \tag{1.40}$$

for $n = \mu, \mu+1, \ldots$, *with* $\mu = \mu(m) := |m|$ *and the normalizing factor*

$$C_{n,\mu} = \left(\frac{\int_{\mathbb{R}}\left(p_n(x)\right)^2 \mathrm{d}\lambda(x)}{\int_{\mathbb{R}}\left(p_n^{(\mu)}(x)\right)^2 \mathrm{d}\lambda^{(\mu)}(x)}\right)^{1/2}, \tag{1.41}$$

and where $\mathrm{d}\lambda^{(\mu)}(x) := \sigma^{\mu}(x)w(x)\mathrm{d}x$ *is the measure of orthogonality for the polynomials* $\{p_n^{(\mu)}\}_{n\in\mathbb{N}_0,\mu\leq n}$.

Remark 1.78 According to Definition 1.77, we have $p_n^m = p_n^{-m}$. This redundancy is intended, as associated functions are often used in combination with other functions that also depend on the order m and where the sign of m matters. Moreover, in the literature, the associated functions are often defined to incorporate a factor of $(-1)^m$ or similar. The normalizing factor $C_{n,\mu}$ is introduced for convenience; see Lemma 1.80 below.
If the polynomials p_n are the Hermite polynomials H_n, then $\sigma(x) = 1$, and it follows from (1.40) and (1.37) that the corresponding associated functions H_n^m are identical to the Hermite polynomials H_n. For the rest of this text we will therefore only treat the Laguerre and the Jacobi case.

The classical associated functions are functions that are either polynomials or almost polynomials. To see this, recall that we have $\sigma(x) = x$ for Laguerre polynomials and $\sigma(x) = 1 - x^2$ for the Jacobi polynomials, respectively. Then from (1.40) it follows that the parity of μ, that is, the parity of the order m, decides whether the polynomial part $p_n^{(\mu)}(x)$ on the right-hand side is multiplied by an integer or a half-integer power of either x or $1-x^2$. This means that associated functions are not more than a factor of $\sqrt{x}$ or $\sqrt{1-x^2}$ away from the space of polynomials $\mathbb{P}$. This is summarized in the following result. The proof follows immediately from (1.40).

Lemma 1.79 *Let* $\{p_n^m\}_{n\in\mathbb{N}_0,\mu\leq n}$ *be a sequence of associated functions. If the order* m *is even, then* μ *is even and* p_n^m *is a polynomial of degree*

$$\frac{\mu}{2}\deg(\sigma) + n - \mu = n + \mu\left(\frac{\deg(\sigma)}{2} - 1\right).$$

If the order m *is odd, then* μ *is odd and* p_n^m *is a polynomial of degree*

$$\frac{\mu-1}{2}\deg(\sigma) + n - \mu = n + \mu\left(\frac{\deg(\sigma)}{2} - 1\right) - \deg(\sigma)/2$$

multiplied by $\sqrt{\sigma(x)}$. *Here,* $\deg(\sigma)$ *denotes the exponent of the highest power of* x *that appears in* σ.

Associated functions have a lot in common with the sequence of orthogonal polynomials they are derived from. An easy to verify consequence of Definition 1.77 is that associated functions p_n^m are orthogonal with respect to the same inner product.

Lemma 1.80 *Let* $\{p_n\}_{n\in\mathbb{N}_0}$ *be a sequence of classical polynomials, orthogonal with respect to an absolutely continuous measure* $\mathrm{d}\lambda(x) = w(x)\mathrm{d}x$. *Then the associated functions of order* m, $\{p_n^m\}_{n\in\mathbb{N}_0,\mu\leq n}$, *are also orthogonal with respect to the measure* $\mathrm{d}\lambda(x)$ *and satisfy*

$$\int_{\mathbb{R}}\left(p_n^m(x)\right)^2 \mathrm{d}\lambda(x) = \int_{\mathbb{R}}\left(p_n(x)\right)^2 \mathrm{d}\lambda(x).$$

PROOF. Using (1.40) and (1.41), a direct calculation shows that

$$\begin{aligned}\int_{\mathbb{R}} p_n^m(x)p_k^m(x)\,\mathrm{d}\lambda(x) &= C_{n,\mu}C_{k,\mu}\int_{\mathbb{R}} p_n^{(\mu)}(x)p_k^{(\mu)}(x)\,\mathrm{d}\lambda^{(\mu)}(x)\\ &= \delta_{n,k}\int_{\mathbb{R}}\left(p_n(x)\right)^2\,\mathrm{d}\lambda(x).\end{aligned}$$

□

The notion of leading coefficients k_n and squared norms h_n can be adopted to the associated functions. The following definition introduces these quantities.

Definition 1.81 *Let $\{p_n^m\}_{n\in\mathbb{N}_0,\mu\le n}$ be a sequence of classical associated functions. Then k_n^m denotes the factor in front of the highest power of x that appears in p_n^m, that is,*

$$p_n^m(x) = k_n^m x^r + \cdots, \qquad \text{with } r = \mu\left(\frac{\deg(\sigma)}{2}-1\right)+n,$$

and h_n^m denotes the squared norm of p_n^m with respect to the measure $\mathrm{d}\lambda(x)$,

$$h_n^m := \int_{\mathbb{R}}\left(p_n^m(x)\right)^2\,\mathrm{d}\lambda(x).$$

Also, it is convenient to define, respectively, monic and orthonormal variants of associated functions.

Definition 1.82 *Let $\{p_n^m\}_{n\in\mathbb{N}_0,\mu\le n}$ be a sequence of associated functions. Then we denote by $\{\bar{p}_n^m\}_{n\in\mathbb{N}_0,\mu\le n}$ the corresponding monic associated functions*

$$\bar{p}_n^m(x) := \frac{p_n^m(x)}{k_n^m} = x^r + \cdots,$$

with leading coefficient $\bar{k}_n^m = 1$. The normalized associated functions $\{\tilde{p}_n^m\}_{n\in\mathbb{N}_0,\mu\le n}$ are defined by

$$\tilde{p}_n^m(x) := \frac{p_n^m(x)}{\sqrt{h_n^m}},$$

and we have $\tilde{h}_n^m = 1$.

Lemma 1.4, which allows to calculate squared norms of the monic variants and leading coefficients of the orthonormal variants for orthogonal polynomials, easily translates to the situation involving associated functions. The proof is entirely analogous.

Lemma 1.83 *Let $\{p_n^m\}_{n\in\mathbb{N}_0,\mu\le n}$ be a sequence of associated functions. Then*

$$\bar{h}_n^m = \frac{h_n^m}{\left(k_n^m\right)^2}, \qquad \tilde{k}_n^m = \frac{k_n^m}{\sqrt{h_n^m}}.$$

Associated functions also satisfy a three-term recurrence, because the polynomials $p_n^{(\mu)}$ used in (1.40) also do.

Lemma 1.84 *Let the assumptions of Definition 1.77 hold. Furthermore, assume that the polynomials $\{p_n^{(\mu)}\}_{n\in\mathbb{N}_0,\mu\le n}$ satisfy the three-term recurrence formula and initial conditions*

$$\begin{aligned} p_{n+1}^{(\mu)}(x) &= (a_n x - b_n)p_n^{(\mu)}(x) - c_n p_{n-1}^{(\mu)}(x), \qquad n = \mu,\mu+1,\ldots,\\ p_{\mu-1}^{(\mu)}(x) &= 0, \qquad p_\mu^{(\mu)}(x) = k_0. \end{aligned} \tag{1.42}$$

Then the associated functions $\{p_n^m\}_{n\in\mathbb{N}_0,\mu\le n}$ *satisfy the three-term recurrence and initial conditions*

$$p_{n+1}^m(x) = (a_n^m x - b_n^m)p_n^m(x) - c_n^m p_{n-1}^m(x), \qquad n = \mu, \mu+1, \ldots,$$
$$p_{\mu-1}^m(x) = 0, \qquad p_\mu^m(x) = C_{\mu,\mu}\sigma^{\mu/2}(x)k_0,$$

with

$$a_n^m = \frac{C_{n+1,\mu}}{C_{n,\mu}}a_n, \quad b_n^m = \frac{C_{n+1,\mu}}{C_{n,\mu}}b_n, \qquad c_n^m = \frac{C_{n+1,\mu}}{C_{n-1,\mu}}c_n.$$

PROOF. The initial conditions can be verified directly using (1.40) and (1.42). Now, assume $\mu \le n$. Then

$$\begin{aligned} p_{n+1}^m(x) &= C_{n+1,\mu}\sigma^{\mu/2}(x)p_{n+1}^{(\mu)}(x) \\ &= C_{n+1,\mu}\sigma^{\mu/2}(x)\left((a_n x - b_n)p_n^{(\mu)}(x) - c_n p_{n-1}^{(\mu)}(x)\right) \\ &= \left(\frac{C_{n+1,\mu}}{C_{n,\mu}}a_n x - \frac{C_{n+1,\mu}}{C_{n,\mu}}b_n\right)p_n^m(x) - \frac{C_{n+1,\mu}}{C_{n-1,\mu}}c_n p_{n-1}^m(x). \end{aligned}$$

□

Since the classical associated functions are closely related to the classical orthogonal polynomials, it is not surprising that they are solutions to similar differential equations.

Theorem 1.85 *Let* $\{p_n\}_{n\in\mathbb{N}_0}$ *be a sequence of classical orthogonal polynomials that are solutions to the hypergeometric differential equation*

$$\sigma(x)y''(x) + \tau(x)y'(x) + \lambda_n y(x) = 0,$$

with certain $\sigma \in \mathbb{P}_2$, $\sigma \in \mathbb{P}_1$, *and* $\lambda_n \in \mathbb{R}$. *Then the corresponding associated functions* $\{p_n^m\}_{n\in\mathbb{N}_0,\mu\le n}$ *satisfy the* hypergeometric-like differential equation

$$\sigma(x)y''(x) + \tau(x)y'(x) + \big(\lambda_n + f_\mu(x)\big)y(x) = 0, \quad y = p_n^m, \tag{1.43}$$

with

$$f_\mu(x) := \mu\tau' + \frac{\mu(\mu-2)}{2}\sigma'' - \frac{\mu}{2}\frac{\sigma'(x)}{\sigma(x)}\left(\left(\frac{\mu}{2}-1\right)\sigma'(x) + \tau(x)\right).$$

PROOF. Let $y = p_n^m$ and $\tilde{y} = p_n^{(\mu)}$. Now, we use (1.40) and verify

$$\begin{aligned} \tau(x)\frac{\mathrm{d}}{\mathrm{d}x}y(x) =& C_{n,\mu}\sigma^{\mu/2}(x)\left(\frac{\mu\sigma'(x)}{2\sigma(x)}\tau(x)\tilde{y}(x) + \tau(x)\tilde{y}'(x)\right), \\ \sigma(x)\frac{\mathrm{d}^2}{\mathrm{d}x^2}y(x) =& C_{n,\mu}\sigma^{\mu/2}(x)\Bigg(\left(\frac{\mu\sigma'(x)}{2\sigma(x)}\left(\frac{\mu}{2}-1\right)\sigma'(x) + \frac{\mu}{2}\sigma''\right)\tilde{y}(x) \\ &+ \mu\sigma'(x)\tilde{y}'(x) + \sigma(x)\tilde{y}''(x)\Bigg). \end{aligned}$$

Plugging this into the left-hand side of (1.43), we obtain

$$\begin{aligned} &\sigma(x)y''(x) + \tau(x)y'(x) + \big(\lambda_n + f_\mu(x)\big)y(x) \\ &= C_{n,\mu}\sigma^{\mu/2}(x)\left(\sigma(x)\tilde{y}''(x) + \big(\tau(x) + \mu\sigma'(x)\big)\tilde{y}'(x) + \left(\lambda_n + \mu\tau' + \tfrac{\mu(\mu-1)}{2}\sigma''\right)\tilde{y}(x)\right). \end{aligned}$$

This part vanishes since it is equivalent to the differential equation satisfied by the polynomials $\{p_n^{(\mu)}\}_{n\in\mathbb{N}_0,\mu\le n}$; cf. Lemma 1.29. Thus, the associated functions $\{p_n^m\}_{n\in\mathbb{N}_0,\mu\le n}$ satisfy the differential equation (1.43). □

1.6.2 Generalized associated Jacobi functions. While the previous definition applies to all types of classical orthogonal polynomials, the following class of generalized associated functions is defined exclusively for the Jacobi polynomials and their descendants.

Definition 1.86 *Let* $\{P_n^{(\alpha,\beta)}\}_{n\in\mathbb{N}_0}$ *be the Jacobi polynomials defined by the Rodrigues formula*

$$P_n^{(\alpha,\beta)}(x) = \frac{(-1)^n}{2^n\Gamma(n+1)}(1-x)^{-\alpha}(1+x)^{-\beta}\frac{\mathrm{d}^n}{\mathrm{d}x^n}\left((1-x)^{n+\alpha}(1+x)^{n+\beta}\right).$$

Furthermore, assume $m, m' \in \mathbb{Z}$ *and define the symbols*

$$n^* := \max\{|m|, |m'|\}, \qquad \mu := |m'-m|, \qquad \nu := |m'+m|.$$

Notice that we have $n^* = \frac{\mu+\nu}{2}$. *Then for* $n = n^*, n^*+1, \ldots,$ *the* generalized associated Jacobi functions $P_n^{(\alpha,\beta),m,m'}$ of orders m and m' *are defined by the Rodrigues-type formula*

$$\begin{aligned} P_n^{(\alpha,\beta),m,m'}(x) &:= C_{n,\mu,\nu}(1-x)^{\mu/2}(1+x)^{\nu/2}P_{n-n^*}^{(\alpha+\mu,\beta+\nu)}(x) \\ &= \frac{(-1)^{n-n^*}C_{n,\mu,\nu}}{2^{n-n^*}\Gamma(n-n^*+1)(1-x)^{\alpha+\mu/2}(1+x)^{\beta+\nu/2}} \\ &\quad\times \frac{\mathrm{d}^{n-n^*}}{\mathrm{d}x^{n-n^*}}\left((1-x)^{n-n^*+\alpha+\mu}(1+x)^{n-n^*+\beta+\nu}\right), \end{aligned} \tag{1.44}$$

with

$$\begin{aligned} C_{n,\mu,\nu} :=& 2^{-n^*}\left(\frac{\Gamma(n+\alpha+1)\Gamma(n+\beta+1)}{\Gamma(n+1)\Gamma(n+\alpha+\beta+1)}\right. \\ &\left.\times\frac{\Gamma(n-n^*+1)\Gamma(n+n^*+\alpha+\beta+1)}{\Gamma(n-n^*+\alpha+\mu+1)\Gamma(n-n^*+\beta+\nu+1)}\right)^{1/2}. \end{aligned}$$

Again, the normalizing factor $C_{n,\mu,\nu}$ is chosen such that the norm of $P_n^{(\alpha,\beta),m,m'}(x)$ is equal to that of the polynomial $P_n^{(\alpha,\beta)}(x)$ with respect to the measure of orthogonality.

Lemma 1.87 *The generalized associated Jacobi functions* $\{P_n^{(\alpha,\beta),m,m'}\}_{n\in\mathbb{N}_0,n^*\leq n}$ *of orders* m *and* m' *are orthogonal on the interval* $[-1,1]$ *with respect to the measure* $\mathrm{d}\lambda(x) = (1-x)^\alpha(1+x)^\beta\mathrm{d}x$ *and satisfy*

$$\int_{-1}^{1}\left(P_n^{(\alpha,\beta),m,m'}(x)\right)^2\mathrm{d}\lambda(x) = \int_{-1}^{1}\left(P_n^{(\alpha,\beta)}(x)\right)^2\mathrm{d}\lambda(x).$$

PROOF. The constant $C_{n,\mu,\nu}$ must satisfy

$$C_{n,\mu,\nu} = \left(\frac{\int_{-1}^{1}\left(P_n^{(\alpha,\beta)}(x)\right)^2(1-x)^\alpha(1+x)^\beta\,\mathrm{d}x}{\int_{-1}^{1}\left(P_{n-n^*}^{(\alpha+\mu,\beta+\nu)}(x)\right)^2(1-x)^{\alpha+\mu}(1+x)^{\beta+\nu}\,\mathrm{d}x}\right)^{1/2}.$$

This is indeed true, since

$$\int_{-1}^{1}\left(P_n^{(\alpha,\beta)}(x)\right)^2(1-x)^\alpha(1+x)^\beta\,\mathrm{d}x = \frac{2^{\alpha+\beta+1}}{(2n+\alpha+\beta+1)}\frac{\Gamma(n+\alpha+1)\Gamma(n+\beta+1)}{\Gamma(n+1)\Gamma(n+\alpha+\beta+1)}$$

$$\begin{aligned}
&\int_{-1}^{1} \left(P_{n-n^*}^{(\alpha+\mu,\beta+\nu)}(x)\right)^2 (1-x)^{\alpha+\mu}(1+x)^{\beta+\nu}\,\mathrm{d}x \\
&= \frac{2^{\alpha+\mu+\beta+\nu+1}}{(2(n-n^*)+\alpha+\mu+\beta+\nu+1)} \frac{\Gamma(n-n^*+\alpha+\mu+1)\Gamma(n-n^*+\beta+\nu+1)}{\Gamma(n-n^*+1)\Gamma(n-n^*+\alpha+\mu+\beta+\nu+1)} \\
&= \frac{2^{2n^*+\alpha+\beta+1}}{(2n+\alpha+\beta+1)} \frac{\Gamma(n-n^*+\alpha+\mu+1)\Gamma(n-n^*+\beta+\nu+1)}{\Gamma(n-n^*+1)\Gamma(n+n^*+\alpha+\beta+1)}.
\end{aligned}$$

The desired result is then verified by

$$\begin{aligned}
&\int_{-1}^{1} P_n^{(\alpha,\beta),m,m'}(x)P_k^{(\alpha,\beta),m,m'}(x)\,(1-x)^{\alpha}(1+x)^{\beta}\mathrm{d}x \\
&= C_{n,\mu,\nu}C_{k,\mu,\nu}\int_{-1}^{1} P_{n-n^*}^{(\alpha+\mu,\beta+\nu)}(x)P_{k-n^*}^{(\alpha+\mu,\beta+\nu)}(x)\,(1-x)^{\alpha+\mu}(1+x)^{\beta+\nu}\,\mathrm{d}x \\
&= \delta_{n,k}\int_{-1}^{1} \left(P_n^{(\alpha,\beta)}(x)\right)^2 (1-x)^{\alpha}(1+x)^{\beta}\,\mathrm{d}x.
\end{aligned}$$

□

The notion of leading coefficients, squared norms, and monic and orthonormal functions can be adapted to the generalized associated Jacobi functions. Similar to Lemma 1.84, one verifies that a three-term recurrence is satisfied by the generalized associated Jacobi functions.

Lemma 1.88 *The generalized associated Jacobi functions $P_n^{(\alpha,\beta),m,m'}(x)$ satisfy the three-term recurrence*

$$P_{n+1}^{(\alpha,\beta),m,m'}(x) = \left(a_n^{m,m'}x - b_n^{m,m'}\right)P_n^{(\alpha,\beta),m,m'}(x) - c_n^{m,m'}P_{n-1}^{(\alpha,\beta),m,m'}(x),$$

and initial conditions

$$\begin{aligned}
P_{n^*-1}^{(\alpha,\beta),m,m'}(x) &= 0, \\
P_{n^*}^{(\alpha,\beta),m,m'}(x) &= 2^{-n^*}\left(\frac{\Gamma(n^*+\alpha+1)\Gamma(n^*+\beta+1)\Gamma(2n^*+\alpha+\beta+1)}{\Gamma(n^*+1)\Gamma(n^*+\alpha+\beta+1)\Gamma(\alpha+\mu+1)\Gamma(\beta+\nu+1)}\right)^{1/2} \\
&\quad\times (1-x)^{\mu/2}(1+x)^{\nu/2},
\end{aligned}$$

with

$$\begin{aligned}
a_n^{m,m'} &= \left(\frac{(n+\alpha+1)(n+\beta+1)}{(n-n^*+\alpha+\mu+1)(n-n^*+\beta+\nu+1)(n-n^*+1)(n+1)}\right)^{1/2} \\
&\quad\times \frac{(2n+\alpha+\beta+1)(2n+\alpha+\beta+2)}{2\left((n+\alpha+\beta+1)(n+n^*+\alpha+\beta+1)\right)^{1/2}}, \\
b_n^{m,m'} &= \left(\frac{(n+\alpha+1)(n+\beta+1)}{(n-n^*+\alpha+\mu+1)(n-n^*+\beta+\nu+1)(n-n^*+1)(n+1)}\right)^{1/2} \\
&\quad\times \frac{(2n+\alpha+\beta+1)}{(2n+\alpha+\beta)} \frac{(\beta+\nu-\alpha-\mu)(2n^*+\alpha+\beta)}{2\left((n+\alpha+\beta+1)(n+n^*+\alpha+\beta+1)\right)^{1/2}},
\end{aligned}$$

and

$$c_n^{m,m'} = \frac{2n+\alpha+\beta+2}{2n+\alpha+\beta}\left(\frac{(n-n^*)}{(n-n^*+1)}\frac{(n+\alpha)(n+\alpha+1)(n+\beta)(n+\beta+1)}{n(n+1)(n+\alpha+\beta)(n+\alpha+\beta+1)}\right.$$
$$\left.\times\frac{(n+n^*+\alpha+\beta)}{(n+n^*+\alpha+\beta+1)}\frac{(n-n^*+\alpha+\mu)}{(n-n^*+\alpha+\mu+1)}\frac{(n-n^*+\beta+\nu)}{(n-n^*+\beta+\nu+1)}\right)^{1/2}.$$

Not surprisingly, there is also a hypergeometric-like differential equation satisfied by the generalized associated Jacobi functions.

Theorem 1.89 *The Jacobi polynomials $\{P_n^{(\alpha,\beta)}\}_{n\in\mathbb{N}_0}$ are solutions to the Jacobi differential equation of hypergeometric type*

$$\sigma(x)y''(x)+\tau(x)y'(x)+\lambda_n y(x)=0,\quad y=P_n^{(\alpha,\beta)},$$

with

$$\sigma(x)=1-x^2,\qquad \tau(x)=-(\alpha+\beta+2)x+\beta-\alpha,\qquad \lambda_n=n(n+\alpha+\beta+1).$$

The generalized associated Jacobi functions $P_n^{(\alpha,\beta),m,m'}$ satisfy the hypergeometric-like differential equation

$$\sigma(x)y''(x)+\tau(x)y'(x)+\big(\lambda_n+f_{\mu,\nu}(x)\big)y(x)=0,\quad y=P_n^{(\alpha,\beta),m,m'},\tag{1.45}$$

with

$$f_{\mu,\nu}(x):=-\left(\frac{\mu(2\alpha+\mu)}{2(1-x)}+\frac{\nu(2\beta+\nu)}{2(1+x)}\right).$$

PROOF. Let $y=P_n^{(\alpha,\beta),m,m'}$ and $\tilde{y}=P_{n-n^*}^{(\alpha+\mu,\beta+\nu)}$. Respecting (1.44), we write down the terms appearing in (1.45):

$$\tau(x)\frac{\mathrm{d}}{\mathrm{d}x}y(x)=C_{n,\mu,\nu}(1-x)^{\mu/2}(1+x)^{\nu/2}\tau(x)\left(\frac{1}{2}\left(\frac{\nu}{1+x}-\frac{\mu}{1-x}\right)\tilde{y}(x)+\tilde{y}'(x)\right),$$

$$\sigma(x)\frac{\mathrm{d}^2}{\mathrm{d}x^2}y(x)=C_{n,\mu,\nu}(1-x)^{\mu/2}(1+x)^{\nu/2}$$
$$\times\left(\frac{1}{4}\left(\mu(\mu-2)\frac{1+x}{1-x}-2\mu\nu+\nu(\nu-2)\frac{1-x}{1+x}\right)\tilde{y}(x)\right.$$
$$\left.+\,(\nu(1-x)-\mu(1+x))\tilde{y}'(x)+\sigma(x)\tilde{y}''(x)\right).$$

Plugging this into (1.45) and grouping terms after $\tilde{y}$, $\tilde{y}'$, and $\tilde{y}''$, we get

$$\begin{aligned}&\sigma(x)y''(x)+\tau(x)y'(x)+\big(\lambda_n+f_{\mu,\nu}(x)\big)y(x)\\=&(1-x)^{\mu/2}(1+x)^{\nu/2}\Big(\sigma(x)\tilde{y}''(x)+\big(\tau(x)+\nu(1-x)-\mu(1+x)\big)\tilde{y}'(x)+\gamma\tilde{y}(x)\Big),\end{aligned}\tag{1.46}$$

where

$$\gamma=\lambda_n+\frac{\tau(x)}{2}\left(\frac{\nu}{1+x}-\frac{\mu}{1-x}\right)+\frac{1}{4}\left(\mu(\mu-2)\frac{1+x}{1-x}-2\mu\nu+\nu(\nu-2)\frac{1-x}{1+x}\right)+f_{\mu,\nu}(x).$$

The plan for the rest of the proof is to manipulate the second line of (1.46) to identify terms appearing in the hypergeometric-like differential equation satisfied by the polynomials $\tilde{y}=P_{n-n^*}^{(\alpha+\mu,\beta+\nu)}$ which reads

$$\sigma(x)\tilde{y}''(x)+\tilde{\tau}(x)\tilde{y}'(x)+\tilde{\lambda}_{n-n^*}\tilde{y}(x)=0,\tag{1.47}$$

with

$$\tilde{\tau}(x) = -(\alpha+\mu+\beta+\nu+2)x+\beta+\nu-\alpha-\mu, \quad \tilde{\lambda}_{n-n^*} = (n-n^*)(n-n^*+\alpha+\mu+\beta+\nu+1).$$

We start with the term $\tau(x) + \nu(1-x) - \mu(1+x)$ which can be expanded to

$$\tau(x) + \nu(1-x) - \mu(1+x) = -(\alpha+\mu+\beta+\nu+2)x + \beta + \nu - \alpha - \mu = \tilde{\tau}(x).$$

Next, we recall that $n^* = \frac{\mu+\nu}{2}$ and rewrite λ_n as

$$\begin{aligned} \lambda_n &= n(n+\alpha+\beta+1) \\ &= (n - n^* + n^*)(n - n^* + n^* + \alpha + \beta + 1) \\ &= \tilde{\lambda}_{n-n^*} + \frac{\mu+\nu}{2}\left(\frac{\mu+\nu}{2} + \alpha + \beta + 1\right). \end{aligned}$$

Then it can be verified by elementary manipulations that

$$\begin{aligned} &\frac{\mu+\nu}{2}\left(\frac{\mu+\nu}{2} + \alpha + \beta + 1\right) + \frac{\tau(x)}{2}\left(\frac{\nu}{1+x} - \frac{\mu}{1-x}\right) \\ &\qquad + \frac{1}{4}\left(\mu(\mu-2)\frac{1+x}{1-x} - 2\mu\nu + \nu(\nu-2)\frac{1-x}{1+x}\right) + f_{\mu,\nu}(x) = 0. \end{aligned}$$

This implies that (1.46) is equivalent to

$$\begin{aligned} &\sigma(x)y''(x) + \tau(x)y'(x) + \left(\lambda_n + f_{\mu,\nu}(x)\right)y(x) \\ &= (1-x)^{\mu/2}(1+x)^{\nu/2}\left(\sigma(x)\tilde{y}''(x) + \tilde{\tau}(x)\tilde{y}'(x) + \tilde{\lambda}_{n-n^*}\tilde{y}(x)\right). \end{aligned}$$

The right-hand side vanishes by (1.47) which completes the proof. □

In the following, examples are given for the different families of associated functions, including the corresponding Rodrigues-type formula, differential equation, and three-term-recurrence. The reader may want continue with Section 1.6.4 and return to this section as needed.

1.6.3 Examples.

Associated Laguerre functions

Laguerre polynomials $\{L_n^{(\alpha)}\}_{n\in\mathbb{N}_0}$ are defined by the Rodrigues formula

$$L_n^{(\alpha)}(x) = \frac{\mathrm{e}^x x^{-\alpha}}{\Gamma(n+1)} \frac{\mathrm{d}^n}{\mathrm{d}x^n}\left(x^{n+\alpha}\mathrm{e}^{-x}\right),$$

and satisfy the differential equation

$$xy''(x) + (1+\alpha-x)y'(x) + ny(x) = 0, \qquad \text{with } y = L_n^{(\alpha)}.$$

The associated Laguerre functions $\{L_n^{(\alpha),m}\}_{n\in\mathbb{N}_0,\mu\le n}$ are defined by the Rodrigues-type formula

$$\begin{aligned} L_n^{(\alpha),m}(x) &= \left(\frac{\Gamma(n-\mu+1)}{\Gamma(n+1)}\right)^{1/2} x^{\mu/2}\frac{\mathrm{d}^\mu}{\mathrm{d}x^\mu}L_n^{(\alpha)}(x) \\ &= (-1)^\mu\left(\frac{\Gamma(n-\mu+1)}{\Gamma(n+1)}\right)^{1/2} x^{\mu/2}L_{n-\mu}^{(\alpha+\mu)}(x) \\ &= \frac{(-1)^\mu \mathrm{e}^x x^{-\alpha-\mu/2}}{\sqrt{\Gamma(n+1)\Gamma(n-\mu+1)}}\frac{\mathrm{d}^{n-\mu}}{\mathrm{d}x^{n-\mu}}\left(x^{n+\alpha}\mathrm{e}^{-x}\right). \end{aligned}$$

Associated Laguerre functions are solutions to the associated Laguerre differential equation

$$xy''(x) + (1+\alpha-x)y'(x) + \left(n - \frac{\mu(2(x+\alpha)+\mu)}{4x}\right) y(x) = 0,$$

with $y = L_n^{(\alpha),m}$. They satisfy the three-term recurrence and initial conditions

$$L_{n+1}^{(\alpha),m}(x) = (a_n^m x - b_n^m) L_n^{(\alpha),m}(x) - c_n^m L_{n-1}^{(\alpha),m}(x), \qquad n = \mu, \mu+1, \ldots,$$

$$L_{\mu-1}^{(\alpha),m}(x) = 0, \qquad L_\mu^{(\alpha),m}(x) = \frac{(-1)^\mu x^{\mu/2}}{\sqrt{\Gamma(\mu+1)}},$$

where

$$a_n^m = -\frac{1}{\sqrt{(n+1)(n-\mu+1)}},$$

$$b_n^m = -\frac{2n+\alpha-\mu+1}{\sqrt{(n+1)(n-\mu+1)}},$$

$$c_n^m = (n+\alpha)\sqrt{\frac{n-\mu}{n(n+1)(n-\mu+1)}}.$$

Associated Jacobi functions

The Jacobi polynomials $\{P_n^{(\alpha,\beta)}\}_{n\in\mathbb{N}_0}$ are defined by the Rodrigues formula

$$P_n^{(\alpha,\beta)}(x) = \frac{(-1)^n}{2^n\Gamma(n+1)}(1-x)^{-\alpha}(1+x)^{-\beta}\frac{\mathrm{d}^n}{\mathrm{d}x^n}\left((1-x)^{n+\alpha}(1+x)^{n+\beta}\right),$$

and satisfy the differential equation

$$\sigma(x)y''(x) + \tau(x)y'(x) + \lambda_n y(x) = 0,$$

with $y = P_n^{(\alpha,\beta)}$, $\sigma(x) = 1-x^2$, $\tau(x) = -(\alpha+\beta+2)x + \beta - \alpha$, and $\lambda_n = n(n+\alpha+\beta+1)$. The associated Jacobi functions $\{P_n^{(\alpha,\beta),m}\}_{n\in\mathbb{N}_{0,\mu\leq n}}$ are thus defined by the formula

$$P_n^{(\alpha,\beta),m}(x) = C_{n,\mu}(1-x^2)^{\mu/2}\frac{\mathrm{d}^\mu}{\mathrm{d}x^\mu}P_n^{(\alpha,\beta)}(x),$$

with a certain normalising constant $C_{n,\mu}$ given by (1.41). The identity

$$\frac{\mathrm{d}^\mu}{\mathrm{d}x^\mu}P_n^{(\alpha,\beta)}(x) = \frac{\Gamma(n+\mu+\alpha+\beta+1)}{2^\mu\Gamma(n+\alpha+\beta+1)}P_{n-\mu}^{(\alpha+\mu,\beta+\mu)}(x)$$

can be obtained from (1.38) on page 36 using that the leading coefficient k_n of the polynomial $P_n^{(\alpha,\beta)}$ is given by

$$k_n = \frac{\Gamma(2n+\alpha+\beta+1)}{2^n\Gamma(n+1)\Gamma(n+\alpha+\beta+1)}.$$

This reveals that the associated Jacobi functions $P_n^{(\alpha,\beta),m}$ are a special case of generalized associated Jacobi functions $P_n^{(\alpha,\beta),m,m'}$, with $m=0$ and m' renamed to m; cf. (1.44). Thus, the constant $C_{n,\mu}$ is

$$C_{n,\mu} = \left(\frac{\Gamma(n-\mu+1)\Gamma(n+\alpha+\beta+1)}{\Gamma(n+1)\Gamma(n+\mu+\alpha+\beta+1)}\right)^{1/2},$$

and the Rodrigues-type formula for the associated Jacobi functions $P_n^{(\alpha,\beta),m}$ reads

$$P_n^{(\alpha,\beta),m}(x) = 2^{-\mu}\left(\frac{\Gamma(n-\mu+1)\Gamma(n+\mu+\alpha+\beta+1)}{\Gamma(n+1)\Gamma(n+\alpha+\beta+1)}\right)^{1/2}(1-x^2)^{\mu/2}P_{n-\mu}^{(\alpha+\mu,\beta+\mu)}(x)$$
$$= \frac{(-1)^{n-\mu}}{2^n}\left(\frac{\Gamma(n+\mu+\alpha+\beta+1)}{\Gamma(n-\mu+1)\Gamma(n+1)\Gamma(n+\alpha+\beta+1)}\right)^{1/2}$$
$$\times (1-x)^{-\alpha-\mu/2}(1+x)^{-\beta-\mu/2}\frac{\mathrm{d}^{n-\mu}}{\mathrm{d}x^{n-\mu}}\left((1-x)^{n+\alpha}(1+x)^{n+\beta}\right).$$

They satisfy the associated Jacobi differential equation

$$\sigma(x)y''(x)+\tau(x)y'(x)+\left(\lambda_n-\frac{\mu(2\alpha+\mu)}{2(1-x)}-\frac{\mu(2\beta+\mu)}{2(1+x)}\right)y(x)=0,$$

with $y = P_n^{(\alpha,\beta),m}$, and the three-term recurrence and initial conditions

$$P_{n+1}^{(\alpha,\beta),m}(x) = (a_n^m x - b_n^m)P_n^{(\alpha,\beta),m}(x) - c_n^m P_{n-1}^{(\alpha,\beta),m}(x), \qquad n=\mu,\mu+1,\dots,$$
$$P_{\mu-1}^{(\alpha,\beta),m}(x)=0,$$
$$P_\mu^{(\alpha,\beta),m}(x) = 2^{-\mu}\left(\frac{\Gamma(2\mu+\alpha+\beta+1)}{\Gamma(\mu+1)\Gamma(\mu+\alpha+\beta+1)}\right)^{1/2}(1-x^2)^{\mu/2},$$

where

$$a_n^m = \frac{(2n+\alpha+\beta+1)(2n+\alpha+\beta+2)}{2\sqrt{(n+1)(n-\mu+1)(n+\alpha+\beta+1)(n+\mu+\alpha+\beta+1)}},$$
$$b_n^m = \frac{2n+\alpha+\beta+1}{2n+\alpha+\beta}\,\frac{(\beta-\alpha)(2\mu+\alpha+\beta)}{2\sqrt{(n+1)(n-\mu+1)(n+\alpha+\beta+1)(n+\mu+\alpha+\beta+1)}},$$
$$c_n^m = \frac{2n+\alpha+\beta+2}{2n+\alpha+\beta}\,\frac{(n+\alpha)(n+\beta)}{\sqrt{n(n+1)(n+\alpha+\beta)(n+\alpha+\beta+1)}}$$
$$\times\left(\frac{(n-\mu)(n+\mu+\alpha+\beta)}{(n-\mu+1)(n+\mu+\alpha+\beta+1)}\right)^{1/2}.$$

Associated Gegenbauer functions

The associated Gegenbauer functions $\{C_n^{(\alpha),m}\}_{n\in\mathbb{N}_0,\mu\le n}$ with $\alpha > -1/2$ and $\alpha \neq 0$ are a special case of associated Jacobi functions with a different normalization,

$$C_n^{(\alpha),m}(x) := \frac{\Gamma(\alpha+1/2)\Gamma(n+2\alpha)}{\Gamma(n+\alpha+1/2)\Gamma(2\alpha)}P_n^{(\alpha-1/2,\alpha-1/2),m}(x).$$

The Rodrigues-type formula therefore is

$$C_n^{(\alpha),m}(x) = \left(\frac{\Gamma(n-\mu+1)\Gamma(n+2\alpha)}{\Gamma(n+1)\Gamma(n+\mu+2\alpha)}\right)^{1/2}(1-x^2)^{\mu/2}\frac{\mathrm{d}^\mu}{\mathrm{d}x^\mu}C_n^{(\alpha)}(x)$$
$$= \frac{(-1)^{n-\mu}\Gamma(\alpha+1/2)}{2^n\Gamma(2\alpha)\Gamma(n+\alpha+1/2)}\left(\frac{\Gamma(n+2\alpha)\Gamma(n+\mu+2\alpha)}{\Gamma(n-\mu+1)\Gamma(n+1)}\right)^{1/2}$$
$$\times\left(1-x^2\right)^{1/2-\alpha-\mu/2}\frac{\mathrm{d}^{n-\mu}}{\mathrm{d}x^{n-\mu}}\left(\left(1-x^2\right)^{n+\alpha-1/2}\right).$$

They satisfy the associated Gegenbauer differential equation

$$(1-x^2)y''(x)-(2\alpha+1)xy'(x)+\left(n(n+2\alpha)-\frac{\mu(2\alpha+\mu-1)}{1-x^2}\right)y(x)=0,$$

with $y = C_n^{(\alpha),m}$. The three-term recurrence and initial conditions are

$$C_{n+1}^{(\alpha),m}(x) = (a_n^m x - b_n^m)C_n^{(\alpha),m}(x) - c_n^m C_{n-1}^{(\alpha),m}(x), \qquad n = \mu, \mu+1, \dots,$$
$$C_{\mu-1}^{(\alpha),m}(x) = 0,$$
$$C_\mu^{(\alpha),m}(x) = \frac{2^{-\alpha}}{\Gamma(\alpha)}\left(\frac{2\sqrt{\pi}\,\Gamma(\mu+\alpha)\Gamma(\mu+2\alpha)}{\Gamma(\mu+\alpha+1/2)}\right)^{1/2}(1-x^2)^{\mu/2},$$

where

$$a_n^m = \frac{2(n+\alpha)(n+2\alpha)^{1/2}}{\sqrt{(n+1)(n-\mu+1)(n+\mu+2\alpha)}},$$
$$b_n^m = 0,$$
$$c_n^m = \sqrt{\frac{(n+2\alpha-1)(n+2\alpha)(n-\mu)(n+\mu+2\alpha-1)}{n(n+1)(n-\mu+1)(n+\mu+2\alpha)}}.$$

Generalized associated Gegenbauer functions

The generalized associated Gegenbauer functions $\{C_n^{(\alpha),m,m'}\}_{n\in\mathbb{N}_0, n^*\leq n}$ with $\alpha > -1/2$ and[2] $\alpha \neq 0$ are a special case of generalized associated Jacobi functions with a different normalization,

$$\begin{aligned}C_n^{(\alpha),m,m'}(x) &:= \frac{\Gamma(\alpha+1/2)\Gamma(n+2\alpha)}{\Gamma(n+\alpha+1/2)\Gamma(2\alpha)}P_n^{(\alpha-1/2,\alpha-1/2),m,m'}(x)\\ &= C_{n,\mu,\nu}(1-x)^{\mu/2}(1+x)^{\nu/2}P_{n-n^*}^{(\alpha+\mu-1/2,\alpha+\nu-1/2)}(x),\end{aligned}$$

with

$$C_{n,\mu,\nu} := \frac{\Gamma(\alpha+1/2)}{2^{n^*}\Gamma(2\alpha)}\left(\frac{\Gamma(n-n^*+1)\Gamma(n+2\alpha)\Gamma(n+n^*+2\alpha)}{\Gamma(n+1)\Gamma(n-n^*+\alpha+\mu+1/2)\Gamma(n-n^*+\alpha+\nu+1/2)}\right)^{1/2}.$$

The Rodrigues-type formula is

$$\begin{aligned}C_n^{(\alpha),m,m'}(x) = &\left(\frac{\Gamma(n-n^*+1)\Gamma(n+2\alpha)\Gamma(n+n^*+2\alpha)}{\Gamma(n+1)\Gamma(n-n^*+\alpha+\mu+1/2)\Gamma(n-n^*+\alpha+\nu+1/2)}\right)^{1/2}\\ &\times \frac{(-1)^{n-n^*}\Gamma(\alpha+1/2)}{2^n\Gamma(2\alpha)\Gamma(n-n^*+1)}(1-x)^{-\alpha-\mu/2+1/2}(x+1)^{-\alpha-\nu/2+1/2}\\ &\times \frac{\mathrm{d}^{n-n^*}}{\mathrm{d}x^{n-n^*}}\left((1-x)^{n-n^*+\alpha+\mu-1/2}(1+x)^{n-n^*+\alpha+\nu-1/2}\right).\end{aligned}$$

They satisfy the generalized associated Gegenbauer differential equation

$$(1-x^2)y''(x) - (2\alpha+1)xy'(x) + \left(n(n+2\alpha) - \frac{\mu(2\alpha+\mu-1)}{2(1-x)} - \frac{\nu(2\alpha+\nu-1)}{2(1+x)}\right)y(x) = 0,$$

with $y = C_n^{(\alpha),m,m'}$. The three-term recurrence and initial conditions are

$$C_{n+1}^{(\alpha),m,m'}(x) = (a_n^{m,m'}x - b_n^{m,m'})C_n^{(\alpha),m,m'}(x) - c_n^{m,m'}C_{n-1}^{(\alpha),m,m'}(x),$$

[2] The monic and normalized variants are well-defined even for $\alpha = 0$.

with $n = n^*, n^*+1, \ldots$, and

$$
\begin{aligned}
C_{n^*-1}^{(\alpha),m,m'}(x) &= 0,\\
C_{n^*}^{(\alpha,),m,m'}(x) &= \frac{\Gamma(\alpha+1/2)}{2^{n^*}\Gamma(2\alpha)}\left(\frac{\Gamma(n^*+2\alpha)\Gamma(2n^*+2\alpha)}{\Gamma(n^*+1)\Gamma(\alpha+\mu+1/2)\Gamma(\alpha+\nu+1/2)}\right)^{1/2}\\
&\quad\times(1-x)^{\mu/2}(1+x)^{\nu/2},
\end{aligned}
$$

where

$$
\begin{aligned}
a_n^{m,m'} &= \frac{(n+\alpha)(2n+2\alpha+1)(n+2\alpha)^{1/2}}{\sqrt{(n+1)(n-n^*+1)(n+n^*+2\alpha)}}\\
&\quad\times\frac{1}{\sqrt{(n-n^*+\alpha+\mu+1/2)(n-n^*+\alpha+\nu+1/2)}},\\
b_n^{m,m'} &= \frac{(\nu-\mu)(n+\alpha)(2n^*+2\alpha-1)(n+2\alpha)^{1/2}}{(2n+2\alpha-1)\sqrt{(n+1)(n-n^*+1)(n+n^*+2\alpha)}}\\
&\quad\times\frac{1}{\sqrt{(n-n^*+\alpha+\mu+1/2)(n-n^*+\alpha+\nu+1/2)}},\\
c_n^{m,m'} &= \left(\frac{(n+2\alpha-1)(n+2\alpha)(n-n^*)(n+n^*+2\alpha-1)}{(n-n^*+1)n(n+1)(n+n^*+2\alpha)}\right)^{1/2}\\
&\quad\times\left(\frac{(n-n^*+\alpha+\mu-1/2)(n-n^*+\alpha+\nu-1/2)}{(n-n^*+\alpha+\mu+1/2)(n-n^*+\alpha+\nu+1/2)}\right)^{1/2}\frac{2n+2\alpha+1}{2n+2\alpha-1}.
\end{aligned}
$$

Associated Legendre functions

The associated Legendre functions $\{P_n^m\}_{n\in\mathbb{N}_0,\mu\le n}$ are a special case of associated Jacobi functions $P_n^{(\alpha,\beta),m}$ for $\alpha=\beta=0$,

$$P_n^m(x) := P_n^{(0,0),m}(x) = C_{n,\mu}(1-x^2)^{\mu/2}\frac{\mathrm{d}^\mu}{\mathrm{d}x^\mu}P_n(x),$$

with

$$C_{n,\mu} = \left(\frac{\Gamma(n-\mu+1)}{\Gamma(n+\mu+1)}\right)^{1/2}.$$

The corresponding Rodrigues-type formula is

$$P_n^m(x) = \frac{(-1)^{n-\mu}}{2^n\sqrt{\Gamma(n-\mu+1)\Gamma(n+\mu+1)}}\left(1-x^2\right)^{-\mu/2}\frac{\mathrm{d}^{n-\mu}}{\mathrm{d}x^{n-\mu}}\left((1-x^2)^n\right).$$

Associated Legendre functions satisfy the associated Legendre differential equation

$$(1-x^2)y''(x) - 2xy'(x) + \left(n(n+1) - \frac{\mu^2}{1-x^2}\right)y(x) = 0,$$

with $y = P_n^m$. The three-term recurrence and initial conditions are

$$
\begin{aligned}
P_{n+1}^m(x) &= (a_n^m x - b_n^m)P_n^m(x) - c_n^m P_{n-1}^m(x), \qquad n = \mu, \mu+1, \ldots,\\
P_{\mu-1}^m(x) &= 0, \qquad P_\mu^m(x) = \frac{\sqrt{\Gamma(2\mu+1)}}{2^\mu\Gamma(\mu+1)}(1-x^2)^{\mu/2},
\end{aligned}
$$

where

$$a_n^m = \frac{2n+1}{\sqrt{(n-\mu+1)(n+\mu+1)}}, \quad b_n^m = 0, \quad c_n^m = \left(\frac{(n-\mu)(n+\mu)}{(n-\mu+1)(n+\mu+1)}\right)^{1/2}.$$

Generalized associated Legendre functions

The generalized associated Legendre functions $\{P_n^{m,m'}\}_{n\in\mathbb{N}_0,n^*\leq n}$ are generalized associated Jacobi functions $P_n^{(\alpha,\beta),m,m'}$ with $\alpha=\beta=0$, that is,

$$P_n^{m,m'}(x) := P_n^{(0,0),m,m'}(x) = C_{n,\mu,\nu}(1-x)^{\mu/2}(1+x)^{\nu/2}P_{n-n^*}^{(\mu,\nu)}(x), \tag{1.48}$$

with

$$C_{n,\mu,\nu} := 2^{-n^*}\left(\frac{\Gamma(n-n^*+1)\Gamma(n+n^*+1)}{\Gamma(n-n^*+\mu+1)\Gamma(n-n^*+\nu+1)}\right)^{1/2}.$$

The corresponding Rodrigues-type formula is

$$\begin{aligned}P_n^{m,m'}(x) = {} & \frac{(-1)^{n-n^*}}{2^n}\left(\frac{\Gamma(n+n^*+1)}{\Gamma(n-n^*+1)\Gamma(n-n^*+\mu+1)\Gamma(n-n^*+\nu+1)}\right)^{1/2}\\ & \times(1-x)^{-\mu/2}(1+x)^{-\nu/2}\frac{\mathrm{d}^{n-n^*}}{\mathrm{d}x^{n-n^*}}\left((1-x)^{n-n^*+\mu}(1+x)^{n-n^*+\nu}\right).\end{aligned}$$

Generalized associated Legendre functions satisfy the generalized associated Legendre differential equation

$$(1-x^2)y''(x)-2xy'(x)+\left(n(n+1)-\frac{\mu^2}{2(1-x)}-\frac{\nu^2}{2(1+x)}\right)y(x)=0,$$

with $y=P_n^{m,m'}$. The three-term recurrence and initial conditions are

$$\begin{aligned}&P_{n+1}^{m,m'}(x) = (a_n^{m,m'}x-b_n^{m,m'})P_n^{m,m'}(x)-c_n^{m,m'}P_{n-1}^{m,m'}(x), \qquad n=n^*,n^*+1,\ldots,\\ &P_{n^*}^{m,m'}(x)=0, \qquad P_{n^*}^{m,m'}(x)=2^{-n^*}\left(\frac{\Gamma(2n^*+1)}{\Gamma(\mu+1)\Gamma(\nu+1)}\right)^{1/2}(1-x)^{\mu/2}(1+x)^{\nu/2},\end{aligned}$$

where

$$\begin{aligned}a_n^{m,m'} &= \frac{(n+1)(2n+1)}{\sqrt{(n-n^*+1)(n+n^*+1)(n-n^*+\mu+1)(n-n^*+\nu+1)}},\\ b_n^{m,m'} &= \frac{2n+1}{2n}\frac{n^*(\nu-\mu)}{\sqrt{(n-n^*+1)(n+n^*+1)(n-n^*+\mu+1)(n-n^*+\nu+1)}},\\ c_n^{m,m'} &= \frac{n+1}{n}\left(\frac{(n-n^*)(n+n^*)(n-n^*+\mu)(n-n^*+\nu)}{(n-n^*+1)(n+n^*+1)(n-n^*+\mu+1)(n-n^*+\nu+1)}\right)^{1/2}.\end{aligned}$$

Associated Chebyshev functions of first kind

Associated Chebyshev functions of first kind $\{T_n^m\}_{n\in\mathbb{N}_0,\mu\leq n}$ are a special case of associated Jacobi functions $P_n^{(\alpha,\beta),m}$ with a particular normalization,

$$T_n^m(x) := \frac{\Gamma(1/2)\Gamma(n+1)}{\Gamma(n+1/2)}P_n^{(-1/2,-1/2),m}(x).$$

The Rodrigues-type formula is

$$\begin{aligned}T_n^m(x) &= C_{n,\mu}(1-x^2)^{\mu/2}\frac{\mathrm{d}^\mu}{\mathrm{d}x^\mu}T_n(x)\\ &= \frac{(-1)^{n-\mu}\Gamma(1/2)}{2^n\Gamma(n+1/2)}\left(\frac{n\Gamma(n+\mu)}{\Gamma(n-\mu+1)}\right)^{1/2}(1-x^2)^{1/2-\mu/2}\frac{\mathrm{d}^{n-\mu}}{\mathrm{d}x^{n-\mu}}\left((1-x^2)^{n-1/2}\right),\end{aligned}$$

with

$$C_{n,\mu} = \left(\frac{\Gamma(n-\mu+1)}{n\Gamma(n+\mu)}\right)^{1/2}.$$

They satisfy the associated Chebyshev differential equation of first kind

$$(1-x^2)y''(x) - xy'(x) + \left(n^2 - \frac{\mu(\mu-1)}{1-x^2}\right) y(x) = 0,$$

with $y = T_n^m$. The three-term recurrence and initial conditions are

$$T_{n+1}^m(x) = (a_n^m x - b_n^m) T_n^m(x) - c_n^m T_{n-1}^m(x), \qquad n = \mu, \mu+1, \dots,$$

$$T_{\mu-1}^m(x) = 0, \qquad T_\mu^m(x) = \left(\frac{(1+\delta_{n,0})\Gamma(1/2)\Gamma(\mu+1)}{2\Gamma(\mu+1/2)}\right)^{1/2} (1-x^2)^{\mu/2},$$

where

$$a_n^m = \frac{2}{1+\delta_{n,0}} \left(\frac{n(n+1)}{(n+\mu)(n-\mu+1)}\right)^{1/2},$$

$$b_n^m = 0,$$

$$c_n^m = \left(\frac{(n+1)(n+\mu-1)(n-\mu)}{(n-1)(n+\mu)(n-\mu+1)}\right)^{1/2}.$$

Generalized associated Chebyshev functions of first kind

The generalized associated Chebyshev functions of first kind $\{T_n^{m,m'}\}_{n\in\mathbb{N}_0, n^*\le n}$ are a special case of generalized associated Jacobi functions $P_n^{(\alpha,\beta),m,m'}$ with a different normalization,

$$\begin{aligned} T_n^{m,m'}(x) &:= \frac{\Gamma(1/2)\Gamma(n+1)}{\Gamma(n+1/2)} P_n^{(-1/2,-1/2),m,m'}(x) \\ &= C_{n,\mu,\nu} (1-x)^{\mu/2} (1+x)^{\nu/2} P_{n-n^*}^{(\mu-1/2,\nu-1/2)}(x), \qquad \text{with } n \ge 1, \end{aligned}$$

and

$$T_0^{0,0}(x) := 1,$$

where

$$C_{n,\mu,\nu} := \frac{\Gamma(1/2)}{2^{n^*}} \left(\frac{n\Gamma(n-n^*+1)\Gamma(n+n^*)}{\Gamma(n-n^*+\mu+1/2)\Gamma(n-n^*+\nu+1/2)}\right)^{1/2}.$$

The Rodrigues-type formula becomes

$$\begin{aligned} T_n^{m,m'}(x) = &\left(\frac{n\Gamma(n+n^*)}{\Gamma(n-n^*+1)\Gamma(n-n^*+\mu+1/2)\Gamma(n-n^*+\nu+1/2)}\right)^{1/2} \\ &\times \frac{(-1)^{n-n^*}\Gamma(1/2)}{2^n} (1-x)^{1/2-\mu/2} (1+x)^{1/2-\nu/2} \\ &\times \frac{\mathrm{d}^{n-n^*}}{\mathrm{d}x^{n-n^*}} \left((1-x)^{n-n^*+\mu-1/2} (1+x)^{n-n^*+\nu-1/2}\right). \end{aligned}$$

They satisfy the generalized associated Chebyshev differential equation of first kind

$$(1-x^2)y''(x) - xy'(x) + \left(n^2 - \frac{\mu(\mu-1)}{2(1-x)} - \frac{\nu(\nu-1)}{2(1+x)}\right) y(x) = 0,$$

with $y = T_n^{m,m'}$. The three-term recurrence and initial conditions are

$$T_{n+1}^{m,m'}(x) = (a_n^{m,m'}x - b_n^{m,m'})T_n^{m,m'}(x) - c_n^{m,m'}T_{n-1}^{m,m'}(x), \quad n = n^*, n^*+1, \dots,$$

$$T_{n^*-1}^{m,m'}(x) = 0,$$

$$T_{n^*}^{m,m'}(x) = \left(\frac{(1+\delta_{n,0})\Gamma(1/2)\Gamma(n^*+1)\Gamma(n^*+1/2)}{2\Gamma(\mu+1/2)\Gamma(\nu+1/2)}\right)^{1/2}(1-x)^{\mu/2}(1+x)^{\nu/2},$$

where

$$a_n^{m,m'} = \frac{2n+1}{1+\delta_{n,0}}\left(\frac{n(n+1)}{(n+n^*)(n-n^*+1)(n-n^*+\mu+1/2)(n-n^*+\nu+1/2)}\right)^{1/2},$$

$$b_n^{m,m'} = \frac{(\nu-\mu)(2n^*-1)}{(2n-1)} \times \left(\frac{n(n+1)}{(n+n^*)(n-n^*+1)(n-n^*+\mu+1/2)(n-n^*+\nu+1/2)}\right)^{1/2},$$

$$c_n^{m,m'} = \frac{2n+1}{2n-1}\left(\frac{(n+1)(n+n^*-1)(n-n^*)(n-n^*+\mu-1/2)(n-n^*+\nu-1/2)}{(n-1)(n+n^*)(n-n^*+1)(n-n^*+\mu+1/2)(n-n^*+\nu+1/2)}\right)^{1/2}.$$

Associated Chebyshev functions of second kind

The associated Chebyshev functions of second kind are a special case of associated Jacobi functions $P_n^{(\alpha,\beta),m}$ with a particular normalization,

$$U_n^m(x) := \frac{\Gamma(3/2)\Gamma(n+2)}{\Gamma(n+3/2)}P_n^{(1/2,1/2),m}(x).$$

Thus, the Rodrigues-type formula is

$$\begin{aligned} U_n^m(x) &= C_{n,\mu}(1-x^2)^{\mu/2}\frac{\mathrm{d}^\mu}{\mathrm{d}x^\mu}U_n(x) \\ &= \frac{(-1)^{n-\mu}\Gamma(3/2)}{2^n\Gamma(n+3/2)}\left(\frac{(n+1)\Gamma(n+\mu+2)}{\Gamma(n-\mu+1)}\right)^{1/2}(1-x^2)^{-1/2-\mu/2} \\ &\quad\times \frac{\mathrm{d}^{n-\mu}}{\mathrm{d}x^{n-\mu}}\left((1-x^2)^{n+1/2}\right), \end{aligned}$$

with

$$C_{n,\mu} = \left(\frac{(n+1)\Gamma(n-\mu+1)}{\Gamma(n+\mu+2)}\right)^{1/2}.$$

They satisfy the associated Chebyshev differential equation of second kind

$$(1-x^2)y''(x) - 3xy'(x) + \left(n(n+2) - \frac{\mu(\mu+1)}{1-x^2}\right)y(x) = 0.$$

The three-term recurrence and initial conditions are

$$U_{n+1}^m(x) = (a_n^m x - b_n^m)U_n^m(x) - c_n^m U_{n-1}^m(x), \quad n = n^*, n^*+1, \dots,$$

$$U_{\mu-1}^m(x) = 0, \qquad U_\mu^m(x) = \left(\frac{\Gamma(3/2)\Gamma(\mu+2)}{\Gamma(\mu+3/2)}\right)^{1/2}(1-x^2)^{\mu/2},$$

where

$$a_n^m = 2\left(\frac{(n+1)(n+2)}{(n-\mu+1)(n+\mu+2)}\right)^{1/2}, \quad b_n^m = 0, \quad c_n^m = \left(\frac{(n+2)(n-\mu)(n+\mu+1)}{n(n-\mu+1)(n+\mu+2)}\right)^{1/2}.$$

Generalized associated Chebyshev functions of second kind

The generalized associated Chebyshev functions of second kind $\{U_n^{m,m'}\}_{n\in\mathbb{N}_0, n^*\leq n}$ are a special case of generalized associated Jacobi functions $P_n^{(\alpha,\beta),m,m'}$ with a different normalization

$$U_n^{m,m'}(x) := \frac{\Gamma(3/2)\Gamma(n+2)}{\Gamma(n+3/2)} P_n^{(1/2,1/2),m,m'}(x)$$
$$= C_{n,\mu,\nu}(1-x)^{\mu/2}(1+x)^{\nu/2} P_{n-n^*}^{(\mu+1/2,\nu+1/2)}(x),$$

with

$$C_{n,\mu,\nu} := \frac{\Gamma(3/2)}{2^{n^*}} \left(\frac{(n+1)\Gamma(n-n^*+1)\Gamma(n+n^*+2)}{\Gamma(n-n^*+\mu+3/2)\Gamma(n-n^*+\nu+3/2)} \right)^{1/2}.$$

The Rodrigues-type formula becomes

$$U_n^{m,m'}(x) = \left(\frac{(n+1)\Gamma(n+n^*+2)}{\Gamma(n-n^*+1)\Gamma(n-n^*+\mu+3/2)\Gamma(n-n^*+\nu+3/2)} \right)^{1/2}$$
$$\times \frac{(-1)^{n-n^*}\Gamma(3/2)}{2^n}(1-x)^{-1/2-\mu/2}(1+x)^{-1/2-\nu/2}$$
$$\times \frac{\mathrm{d}^{n-n^*}}{\mathrm{d}x^{n-n^*}} \left((1-x)^{n-n^*+\mu+1/2}(1+x)^{n-n^*+\nu+1/2} \right).$$

They satisfy the generalized associated Chebyshev differential equation of second kind

$$(1-x^2)y''(x) - 3xy'(x) + \left(n(n+2) - \frac{\mu(\mu+1)}{2(1-x)} - \frac{\nu(\nu+1)}{2(1+x)} \right) y(x) = 0,$$

with $y = U_n^{m,m'}$. The three-term recurrence and initial conditions are

$$U_{n+1}^{m,m'}(x) = (a_n^{m,m'}x - b_n^{m,m'})U_n^{m,m'}(x) - c_n^{m,m'}U_{n-1}^{m,m'}(x), \qquad n = n^*, n^*+1, \ldots,$$
$$U_{n^*-1}^{m,m'}(x) = 0,$$
$$U_{n^*}^{m,m'}(x) = \left(\frac{\Gamma(3/2)\Gamma(n^*+2)\Gamma(n^*+3/2)}{\Gamma(\mu+3/2)\Gamma(\nu+3/2)} \right)^{1/2} (1-x)^{\mu/2}(1+x)^{\nu/2},$$

where

$$a_n^{m,m'} = (2n+3) \left(\frac{(n+1)(n+2)}{(n+n^*+2)(n-n^*+1)(n-n^*+\mu+3/2)(n-n^*+\nu+3/2)} \right)^{1/2},$$

$$b_n^{m,m'} = \frac{(\nu-\mu)(2n^*+1)}{2n+1}$$
$$\times \left(\frac{(n+1)(n+2)}{(n+n^*+2)(n-n^*+1)(n-n^*+\mu+3/2)(n-n^*+\nu+3/2)} \right)^{1/2},$$

$$c_n^{m,m'} = \frac{2n+3}{2n+1} \left(\frac{(n+2)(n+n^*+1)(n-n^*)(n-n^*+\mu+1/2)(n-n^*+\nu+1/2)}{n(n+n^*+2)(n-n^*+1)(n-n^*+\mu+3/2)(n-n^*+\nu+3/2)} \right)^{1/2}.$$

1.6.4 Connection coefficients for associated functions. Somewhat similar to the situation for orthogonal polynomials, associated functions can be represented through other associated functions. But only conversions between associated functions of different orders, say, m and $\hat{m}$, that are derived from the *same* sequence of orthogonal polynomials will be investigated. Such mapping does not exist for every admissible pair of orders m and $\hat{m}$, rather these parameters have to satisfy a simple condition. Generally, the transformation works only from larger $\mu = |m|$ to smaller $\hat{\mu} = |\hat{m}|$. The second condition is that $m + \hat{m}$ has to be an even number, that is, the orders m and $\hat{m}$ have to share the same parity. The following Lemma introduces this situation.

Lemma 1.90 *Let $\{p_n^m\}_{n\in\mathbb{N}_0,\mu\leq n}$ and $\{p_n^{\hat{m}}\}_{n\in\mathbb{N}_0,\hat{\mu}\leq n}$ be two sequences of associated functions, that are derived from the same sequence of orthogonal polynomials with $m, \hat{m} \in \mathbb{N}_0$ such that $m + \hat{m}$ is even and $|\hat{m}| < |m|$. Then every associated function p_j^m with $j \geq \mu$ can be represented as a linear combination of the form*

$$p_j^m = \sum_{i=\hat{\mu}}^{N} \kappa_{i,j} p_i^{\hat{m}}, \qquad \text{where } N = j - (\mu - \hat{\mu})\big(1 - \deg(\sigma)/2\big). \tag{1.49}$$

Note that the upper limit of the sum is always an integer and thus well defined.

PROOF. By definition, the space spanned by the functions $\{p_n^m\}_{n\in\mathbb{N}_0,\mu\leq n}$ contains all functions that can be represented by $\sigma^{\mu/2}(x)$ times a polynomial in x of some degree. In particular, the associated function p_j^m, with $\mu \leq j$, is identical to $\sigma^{\mu/2}(x)$ times a polynomial of degree $j - \mu$:,

$$p_j^m(x) = C_{j,\mu}\sigma^{\mu/2}(x)p_j^{(\mu)}(x),$$

with a certain normalization constant $C_{n,\mu}$; cf. Definition 1.77. This can be written as

$$p_j^m(x) = C_{j,\mu}\sigma^{\hat{\mu}/2}(x)\Big(\sigma^{\frac{\mu-\hat{\mu}}{2}}(x)p_j^{(\mu)}(x)\Big).$$

We observe that the expression in parentheses is a proper polynomial of degree $(\mu - \hat{\mu})\deg(\sigma)/2 + j - \mu$. This is now ready to be expanded into the associated functions $\{p_n^{\hat{m}}\}_{n\in\mathbb{N}_0,\hat{\mu}\leq n}$. Using Lemma 1.79, we verify that only those functions up to $n = \hat{\mu} + j + (\mu - \hat{\mu})\deg(\sigma)/2 - \mu$ are needed. □

We call the coefficients $\kappa_{i,j}$ in (1.49) the connection coefficients between the sequences of associated functions $\{p_n^m\}_{n\in\mathbb{N}_0,\mu\leq n}$ and $\{p_n^{\hat{m}}\}_{n\in\mathbb{N}_0,\hat{\mu}\leq n}$. The following definition is to formalize this denotation.

Definition 1.91 *The associated functions $\{p_n^m\}_{n\in\mathbb{N}_0,\mu\leq n}$ are called the* source functions *and the associated functions $\{p_n^{\hat{m}}\}_{n\in\mathbb{N}_0,\hat{\mu}\leq n}$ are called the* target functions. *The coefficients $\kappa_{i,j}$ are called the* connection coefficients *between $\{p_n^m\}_{n\in\mathbb{N}_0,\mu\leq n}$ and $\{p_n^{\hat{m}}\}_{n\in\mathbb{N}_0,\hat{\mu}\leq n}$. For $i < \hat{\mu}$, $j < \mu$, or $j < i + (\mu - \hat{\mu})(1 - \deg(\sigma)/2)$, we define $\kappa_{i,j} = 0$.*

Here are two versions of Lemma 1.6.4 that are specialized to the associated Laguerre and the associated Jacobi functions, respectively.

Lemma 1.92 *Let $\{L_n^{(\alpha),m}\}_{n\in\mathbb{N}_0,\mu\leq n}$ and $\{L_n^{(\alpha),\hat{m}}\}_{n\in\mathbb{N}_0,\hat{\mu}\leq n}$ be two sequences of associated Laguerre functions with $\alpha > -1$ and $m, \hat{m} \in \mathbb{N}_0$, such that $m + \hat{m}$ is even and $|\hat{\mu}| < |\mu|$. Then every associated Laguerre function $L_j^{(\alpha),m}$ with $j \geq \mu$, can be represented as a linear combination of the form*

$$L_j^{(\alpha),m} = \sum_{i=\hat{\mu}}^{j-(\mu-\hat{\mu})/2} \kappa_{i,j} L_i^{(\alpha),\hat{m}}.$$

Lemma 1.93 *Let* $\{P_n^{(\alpha,\beta),m}\}_{n\in\mathbb{N}_0,\mu\leq n}$ *and* $\{P_n^{(\alpha,\beta),\hat{m}}\}_{n\in\mathbb{N}_0,\hat{\mu}\leq n}$ *be two sequences of associated Jacobi functions with* $\alpha,\beta > -1$ *and* $m,\hat{m}\in\mathbb{N}_0$, *such that* $m+\hat{m}$ *is even and* $|\hat{\mu}|<|\mu|$. *Then every associated Jacobi function* $P_j^{(\alpha,\beta),m}$ *with* $j\geq\mu$, *can be represented as a linear combination of the form*

$$P_j^{(\alpha,\beta),m} = \sum_{i=\hat{\mu}}^{j} \kappa_{i,j} P_i^{(\alpha,\beta),\hat{m}}.$$

For generalized associated Jacobi functions, the situation is a bit more complicated due to the increased number of parameters. But, the general principles remain intact.

Lemma 1.94 *Let* $\{P_n^{(\alpha,\beta),m,m'}\}_{n\in\mathbb{N}_0,n^*\leq n}$ *and* $\{P_n^{(\alpha,\beta),\hat{m},\hat{m}'}\}_{n\in\mathbb{N}_0,\hat{n}^*\leq n}$ *with* $\alpha,\beta>-1$ *and* $m,m',\hat{m},\hat{m}'\in\mathbb{N}_0$, *be two sequences of generalized associated Jacobi functions, such that* $m+\hat{m}+m'+\hat{m}'$ *is even and* $\hat{\mu}<\mu$ *as well as* $\hat{\nu}<\nu$. *Then every generalized associated Jacobi function* $P_j^{(\alpha,\beta),m,m'}$ *with* $j\geq n^*$ *can be represented as a linear combination of the form*

$$P_j^{(\alpha,\beta),m,m'} = \sum_{i=\hat{n}^*}^{j} \kappa_{i,j} P_i^{(\alpha,\beta),\hat{m},\hat{m}'}.$$

Proof. The space spanned by the functions $\{P_n^{(\alpha,\beta),m,m'}\}_{n\in\mathbb{N}_0,n^*\leq n}$ contains all functions that can be represented by $(1-x)^{\mu/2}(1+x)^{\nu/2}$ times a polynomial of some degree. In particular, the function $P_j^{(\alpha,\beta),m,m'}$ is given by

$$P_j^{(\alpha,\beta),m,m'}(x) = C_{j,\mu,\nu}(1-x)^{\mu/2}(1+x)^{\nu/2}P_{n-n^*}^{(\alpha+\mu,\beta+\nu)}(x),$$

with the normalization constant $C_{n,\mu,\nu}$ from Definition 1.86. We can write this as

$$P_j^{(\alpha,\beta),m,m'}(x) = C_{j,\mu,\nu}(1-x)^{\hat{\mu}/2}(1+x)^{\hat{\nu}/2}\left((1-x)^{(\mu-\hat{\mu})/2}(1+x)^{(\nu-\hat{\nu})/2}P_{j-n^*}^{(\alpha+\mu,\beta+\nu)}(x)\right),$$

and observe that $(1-x)^{(\mu-\hat{\mu})/2}(1+x)^{(\nu-\hat{\nu})/2}P_{j-n^*}^{(\alpha+\mu,\beta+\nu)}(x)$ is a proper polynomial of degree $(\mu-\hat{\mu}+\nu-\hat{\nu})/2+j-n^*=j-\hat{n}^*$, a fact that follows from $\mu+\nu=2n^*$ and $\hat{\mu}+\hat{\nu}=2\hat{n}^*$. This makes the function $P_j^{(\alpha,\beta),m,m'}(x)$ equivalent to a factor of $(1-x)^{\hat{\mu}/2}(1+x)^{\hat{\nu}/2}$ times a polynomial of degree $j-\hat{n}^*$. This can be represented using the functions $\{P_n^{(\alpha,\beta),\hat{m},\hat{m}'}\}_{n\in\mathbb{N}_0,\hat{n}^*\leq n}$ where only those up to degree j are needed. □

The connection coefficients between associated functions of different orders can be identified with inner products of the form $\langle\cdot,\cdot\rangle_{\mathrm{d}\lambda}$, where $\mathrm{d}\lambda$ is the corresponding measure of orthogonality. Note, that in contrast to plain orthogonal polynomials, this measure of orthogonality is the same for all associated functions which are derived from the same sequence of orthogonal polynomials. The connection coefficients are clearly given by

$$\kappa_{i,j} = \frac{\langle p_i^{\hat{m}}, p_j^{m}\rangle_{\mathrm{d}\lambda}}{\langle p_i^{\hat{m}}, p_i^{\hat{m}}\rangle_{\mathrm{d}\lambda}}, \tag{1.50}$$

where p_j^m and $p_i^{\hat{m}}$ are the respective associated functions. A similar expression holds for the generalized associated Jacobi functions. In this text, explicit expressions for the connection coefficients between associated functions of different orders will not be derived. The reason is that the resulting expressions are not simple enough to be exploitable for numerical computations. This is different to the situation for classical orthogonal polynomials. Here, the respective expressions can be efficiently evaluated. To show that the situation for associated functions still bears similarity to the polynomial case, the following theorem shows that the connection coefficients are be generated by a recurrence. This is similar

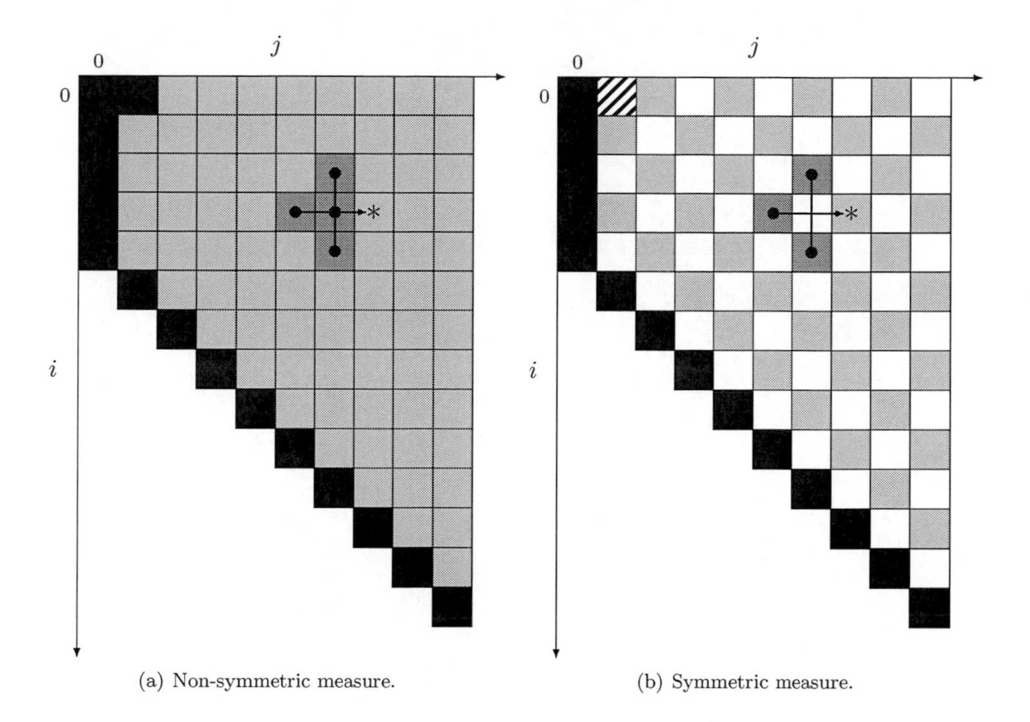

(a) Non-symmetric measure.

(b) Symmetric measure.

Figure 1.6: Schematic representation of the recurrence for the connection coefficients $\kappa_{i,j}$ for classical associated functions. Non-shaded areas represent coefficients that always vanish. Black squares represent entries given by the initial conditions; these are $\kappa_{i,i}$, $i = 0, 1, \ldots$, and $\kappa_{0,1}$. Gray squares stand for the rest of the coefficients that are determined by the three-term recurrence. For the computation of one of these coefficients $\kappa_{i,j}$ (represented by $*$), the entries $\kappa_{i-1,j-1}$, $\kappa_{i,j-1}$, $\kappa_{i,j-2}$, and $\kappa_{i+1,j-1}$ (represented by $\bullet$) have to be known. In the case of a symmetric measure, the recurrence is simpler since the dependence on $\kappa_{i,j-1}$ is removed. Also, the coefficient $\kappa_{0,1}$ is known to be zero. The rest of the coefficients forms a checkerboard pattern.

to Theorem 1.35 for orthogonal polynomials. The situation for the connection matrices between (generalized) associated functions is illustrated in Figure 1.6.

Theorem 1.95 *Let $\{p_n^m\}_{n\in\mathbb{N}_0,\mu\leq n}$ be a sequence of associated functions that satisfy the three-term recurrence and initial conditions*

$$\begin{aligned} p_{n+1}^m(x) &= (a_n^m x - b_n^m)p_n^m(x) - c_n^m p_{n-1}^m(x), \quad n = \mu, \mu+1, \ldots, \\ p_{\mu-1}^m(x) &= 0, \quad p_\mu^m(x) = C_{\mu,\mu}\sigma^{\mu/2}(x)k_0, \end{aligned} \tag{1.51}$$

with the leading coefficients k_n^m. Let $\{p_n^{\hat{m}}\}_{n\in\mathbb{N}_0,\hat{\mu}\leq n}$ be another sequence of associated functions derived from the same sequence of orthogonal polynomials such that $m + \hat{m}$ is even and $\hat{\mu} < \mu$. Both sequences of associated functions are orthogonal with respect to a certain measure $\mathrm{d}\lambda$ which induces an inner product $\langle\cdot,\cdot\rangle$. Denote by $\hat{h}_n^{\hat{m}} = \langle \hat{p}_n^{\hat{m}}, \hat{p}_n^{\hat{m}}\rangle$ the squared norm of the function $\hat{p}_n^{\hat{m}}$ and by $\hat{k}_n^{\hat{m}}$ its leading coefficient. Furthermore, let the *corresponding three-term recurrence and initial conditions be given by*

$$\begin{aligned} p_{n+1}^{\hat{m}}(x) &= (a_n^{\hat{m}} x - b_n^{\hat{m}}) p_n^{\hat{m}}(x) - c_n^{\hat{m}} p_{n-1}^{\hat{m}}(x), \quad n = \hat{\mu}, \hat{\mu}+1, \dots, \\ p_{\hat{\mu}-1}^{\hat{m}}(x) &= 0, \quad p_{\hat{\mu}}^{\hat{m}}(x) = C_{\hat{\mu},\hat{\mu}} \sigma^{\hat{\mu}/2}(x) \hat{k}_0. \end{aligned} \tag{1.52}$$

Then the connection coefficients $\kappa_{i,j}$ in the formula

$$p_j^m = \sum_{i=\hat{\mu}}^{N} \kappa_{i,j} p_i^{\hat{m}}, \qquad \text{with } N = j - (\mu - \hat{\mu})(1 - \deg(\sigma)/2),$$

satisfy the recurrence formula

$$\begin{aligned} \kappa_{i,j} = a_{j-1}^m & \left(\frac{1}{a_i^{\hat{m}}} \frac{h_{i+1}^{\hat{m}}}{h_i^{\hat{m}}} \kappa_{i+1,j-1} + \frac{b_i^{\hat{m}}}{a_i^{\hat{m}}} \kappa_{i,j-1} + \frac{c_i^{\hat{m}}}{a_i^{\hat{m}}} \frac{h_{i+1}^{\hat{m}}}{h_{i-1}^{\hat{m}}} \kappa_{i-1,j-1} \right) \\ & - b_{j-1}^m \kappa_{i,j-1} - c_{j-1}^m \kappa_{i,j-2}, \end{aligned} \tag{1.53}$$

for $\hat{\mu} \le i$, $\mu \le j$, and $i \le j - (\mu - \hat{\mu})(1 - \deg(\sigma)/2)$ with the initial conditions

$$\kappa_{i,j} = \begin{cases} \dfrac{1}{h_{\hat{\mu}}^{\hat{m}}} \displaystyle\int_{\mathbb{R}} p_{\hat{\mu}}^{\hat{m}}(x) p_{\mu+1}^m(x) \, \mathrm{d}\lambda(x), & \text{if } i = \hat{\mu} \text{ and } j = \mu + 1, \\ \dfrac{1}{h_i^{\hat{m}}} \displaystyle\int_{\mathbb{R}} p_i^{\hat{m}}(x) p_\mu^m(x) \, \mathrm{d}\lambda(x), & \text{if } j = \mu \text{ and } \hat{\mu} \le i \le \hat{\mu} + (\mu - \hat{\mu}) \deg(\sigma)/2, \\ \dfrac{k_{i+(\mu-\hat{\mu})(1-\deg(\sigma)/2)}^m}{\hat{k}_i^{\hat{m}}}, & \text{if } i = j - (\mu - \hat{\mu})\left(1 - \deg(\sigma)/2\right). \end{cases} \tag{1.54}$$

PROOF. The proof is similar to that of Theorem 1.35. To prove the recurrence formula (1.53) we use (1.50) and the three-term recurrence (1.51),

$$\kappa_{i,j} = \frac{\langle p_i^{\hat{m}}, p_j^m \rangle_{\mathrm{d}\lambda}}{h_i^{\hat{m}}} = \frac{1}{h_i^{\hat{m}}} \left(a_{j-1}^m \langle p_i^{\hat{m}}, x p_{j-1}^m \rangle - b_{j-1}^m \langle p_i^{\hat{m}}, p_{j-1}^m \rangle - c_{j-1}^m \langle p_i^{\hat{m}}, p_{j-2}^m \rangle \right). \tag{1.55}$$

With the shift property of the inner product, that is, $\langle \cdot, x \cdot \rangle = \langle x \cdot, \cdot \rangle$ and the three-term recurrence (1.52), we obtain

$$\langle x p_i^{\hat{m}}, p_{j-1}^m \rangle = \frac{1}{a_i^{\hat{m}}} \langle p_{i+1}^{\hat{m}}, p_{j-1}^m \rangle + \frac{b_i^{\hat{m}}}{a_i^{\hat{m}}} \langle p_i^{\hat{m}}, p_{j-1}^m \rangle + \frac{c_i^{\hat{m}}}{a_i^{\hat{m}}} \langle p_{i-1}^{\hat{m}}, p_{j-1}^m \rangle.$$

We combine this result with (1.55) and identify the remaining inner products with the connection coefficients, e.g., $\langle p_i^{\hat{m}}, p_{j-1}^m \rangle = h_i^{\hat{m}} \kappa_{i,j-1}$.
For the initial conditions (1.54), we assume $j = i + \gamma$ with $\gamma = (\mu - \hat{\mu})\left(1 - \deg(\sigma)/2\right)$. Then the coefficient $\kappa_{i,i+\gamma}$ has the form

$$\begin{aligned} \kappa_{i,i+\gamma} &= \frac{1}{h_i^{\hat{m}}} \int_{\mathbb{R}} p_i^{\hat{m}}(x) p_{i+\gamma}^m(x) \, \mathrm{d}\lambda(x), \\ &= \frac{1}{h_i^{\hat{m}}} \int_{\mathbb{R}} C_{i,\hat{\mu}} \sigma^{\hat{\mu}/2}(x) p_i^{(\hat{\mu})}(x) \, C_{i+\gamma,\mu} \sigma^{\mu/2}(x) p_{i+\gamma}^{(\mu)}(x) \, \mathrm{d}\lambda(x) \\ &= \frac{1}{h_i^{\hat{m}}} \int_{\mathbb{R}} C_{i,\hat{\mu}} p_i^{(\hat{\mu})}(x) \, C_{i+\gamma,\mu} p_{i+\gamma}^{(\mu)}(x) \sigma^{(\mu-\hat{\mu})/2}(x) \, \mathrm{d}\lambda^{(\hat{\mu})}(x), \end{aligned}$$

with the measure $\mathrm{d}\lambda^{(\hat{\mu})}(x) = \sigma^{\hat{\mu}}(x) \mathrm{d}\lambda(x)$. Note that the polynomial $p_i^{(\hat{\mu})}(x)$ of degree $i - \hat{\mu}$ is orthogonal to every polynomial of strictly smaller degree with respect to the measure

$\mathrm{d}\lambda^{(\mu)}$. This allows to replace the expression

$$C_{i+\gamma,\mu}p_{i+\gamma}^{(\mu)}(x)\sigma^{(\hat{\mu}-\mu)/2}(x)$$

which is a polynomial of degree $i+\gamma-\mu+(\hat{\mu}-\mu)\deg(\sigma)/2 = i-\hat{\mu}$ with the polynomial

$$\frac{k_{i+\gamma}^{m}}{k_{i}^{\hat{m}}}C_{i,\hat{\mu}}p_{i}^{(\hat{\mu})}(x)$$

of same degree without changing the integral. Thus, we obtain

$$\kappa_{i,i+\gamma} = \frac{k_{i+\gamma}^{m}}{\hat{k}_{i}^{\hat{m}}}\frac{\langle p_{i}^{\hat{m}}, p_{i}^{\hat{m}}\rangle}{h_{i}^{\hat{m}}} = \frac{k_{i+\gamma}^{m}}{\hat{k}_{i}^{\hat{m}}}.$$

The rest of the initial conditions follows directly from the definition of the connection coefficients. □

In case that the measure $\mathrm{d}\lambda$ is symmetric, the recurrence formula is simplified.

Corollary 1.96 *Let $\{p_n^m\}_{n\in\mathbb{N}_0,\mu\le n}$ and $\{p_n^{\hat{m}}\}_{n\in\mathbb{N}_0,\hat{\mu}\le n}$ be two sequences of associated functions that satisfy the assumptions of Theorem 1.95 and that are orthogonal with respect to a symmetric measure. Then the connection coefficients $\kappa_{i,j}$ satisfy the recurrence formula*

$$\kappa_{i,j} = a_{j-1}^{m}\left(\frac{1}{a_{i}^{\hat{m}}}\frac{h_{i+1}^{\hat{m}}}{h_{i}^{\hat{m}}}\kappa_{i+1,j-1} + \frac{c_{i}^{\hat{m}}}{a_{i}^{\hat{m}}}\frac{\hat{h}_{i-1}^{\hat{m}}}{h_{i}^{\hat{m}}}\kappa_{i-1,j-1}\right) - c_{j-1}^{m}\kappa_{i,j-2}, \tag{1.56}$$

for $\hat{\mu}\le i$, $\mu\le j$, and $i\le j-(\mu-\hat{\mu})(1-\deg(\sigma)/2)$ with the initial conditions

$$\kappa_{i,j} = \begin{cases} \dfrac{1}{h_i^{\hat{m}}}\displaystyle\int_{\mathbb{R}} p_i^{\hat{m}}(x)p_\mu^m(x)\,\mathrm{d}\lambda(x), & \text{if } j=\mu \text{ and } \hat{\mu}\le i\le\hat{\mu}+(\mu-\hat{\mu})\deg(\sigma)/2,\\ 0, & \text{if } i=\mu \text{ and } j=\mu+1,\\ \dfrac{k_{i+(\mu-\hat{\mu})(1-\deg(\sigma)/2)}^{m}}{\hat{k}_i^{\hat{m}}}, & \text{if } i=j-(\mu-\hat{\mu})\,(1-\deg(\sigma)/2). \end{cases} \tag{1.57}$$

This implies $\kappa_{i,j}=0$ if $i+j$ is odd.

PROOF. The proof is a direct consequence of Lemma 1.17 which implies $b_n^m = b_n^{\hat{m}} = 0$ in (1.51), (1.52), and the proof of Theorem 1.95. □

1.7 Fast polynomial transforms

Fast polynomial transform are the main theme of this thesis and the techniques developed in the following two chapters deliver algorithms that enable these type of transforms. Still, fast polynomial transforms are more a consequence of these algorithms rather than the primary object under investigation.

A *discrete polynomial transform* is a generalization of the usual discrete Fourier transform to a more general set of basis functions. In our case, these are the classical orthogonal polynomials or the classical (generalized) associated functions. A *fast polynomial transform* is a fast algorithm to compute the same result. To make this more precise, we give the following definition.

Definition 1.97 *Let $\{p_n\}_{n\in\mathbb{N}_0,n\ge\mu}$ with $\mu\in\mathbb{N}_0$ be a set of functions, orthogonal over an interval $[a,b]$. Furthermore, let $\hat{f}_n$ for $n=\mu,\mu+1,\ldots,N$, with $N\in\mathbb{N}_0$, be given coefficients, possibly complex, and let x_i for $i=1,2,\ldots,I$ be points in $[a,b]$. Then the* *evaluation of the sums*

$$f(x_i) = \sum_{j=\mu}^{N} \hat{f}_j p_j(x_i), \quad i = 1, 2, \ldots, I, \tag{1.58}$$

is called a discrete polynomial transform. *An algorithm to compute the same result up to a given accuracy ε (in some sensible measure) with no more than $\mathcal{O}\big((N \log N + I) \log(1/\varepsilon)\big)$ arithmetic operations is called a* fast polynomial transform.

Typically, algorithms to evaluate the sums (1.58) need $\mathcal{O}(NI)$ arithmetic operations. Since the number of nodes I is usually comparable to N this would typically need $\mathcal{O}(N^2)$ arithmetic operations. For large N this must often be considered too expensive. A fast polynomial transform, under the same circumstances, can calculate, or at least approximate, the same result with asymptotically much less operations. The methods developed in the following two chapters provide a way to achieve this favorable cost by allowing one to efficiently replace the sums (1.58) with an equivalent form

$$f(x_i) = \sum_{j=\mu}^{N} \hat{c}_j \mathrm{e}^{\mathrm{i}j \arccos(x_i)}, \quad i = 1, 2, \ldots, I,$$

with new coefficients $\hat{c}_j$. This is a plain Fourier sum that can be evaluated using either the fast Fourier transform (FFT) or its non-equispaced variant (NFFT) with not more than $\mathcal{O}\big(N \log N + I \log(1/\varepsilon)\big)$ operations. If the first part, that is, replacing the coefficients $\hat{f}_j$ with the coefficients $\hat{c}_j$, can be carried out efficiently enough, the result is a fast polynomial transform in the sense of Definition 1.97. In the following two chapters, we will concentrate on this first step. Two applications that demonstrate these principles are given in Chapter 4.

2 – Techniques based on semiseparable matrices

In this chapter, methods are developed for the efficient conversion between expansions in different sequences of classical orthogonal polynomials or classical associated functions, respectively. The algorithms introduced here have in common that they rely on methods to efficiently compute the eigendecomposition of certain semiseparable matrices. Before the actual algorithms are derived, a brief introduction to semiseparable matrices, the closely related banded matrices, and so-called checkerboard-like matrices is given.
Generally, for any matrix with structure a number of elementary operations, like matrix-vector multiplication or matrix inversion, can be carried out efficiently by exploiting the structure found in the matrix. Perhaps the most intuitive example are banded matrices which have their non-vanishing components confined to a banded pattern. Semiseparable matrices, on the other hand, are usually fully populated, but similarly have a reduced number of degrees of freedom. And there is another variation of structured matrices that will be important, that is, the class of checkerboard-like matrices. A matrix is said to be checkerboard-like if the components that correspond to the white fields on a checkerboard vanish.
Sections 2.1 and 2.2 provide a brief introduction to basic notational conventions, as well as the classes of banded, semiseparable, and checkerboard-like matrices. Section 2.3 introduces an algorithm to efficiently compute the eigendecomposition of triangular generator representable semiseparable matrices. In Section 2.4 we show how this can be used to obtain fast algorithms to apply connection matrices between different sequences of classical orthogonal polynomials. A number of numerical results is given. A similar algorithm for the eigendecomposition of symmetric extended generator representable semiseparable matrices is given in Section 2.5. It is used in Section 2.6 to obtain an efficient method to apply connection matrices between classical associated functions of different orders.

2.1 Notation

Analyzing the properties of structured matrices relies on a succinct regime for notation. A convenient way is to use a MATLAB[1]-like syntax. The following definition introduces a couple of symbols to denote different parts of a matrix.

Definition 2.1 *Let* $\mathbf{A}$ *be an* $n \times n$ *matrix and let* k *enumerate all diagonals of* $\mathbf{A}$*, from the* $(n-1)$*st subdiagonal (where* $k = -n+1$*) up to the* $(n-1)$*st super-diagonal (where* $k = n-1$*). Furthermore, denote by* $\mathbf{d}$ *an arbitrary column vector of length* $n - |k|$*. Then we denote by*

(i) $\mathrm{diag}(\mathbf{A}, k)$ *the column vector that contains the* k*th diagonal of* $\mathbf{A}$*,*
(ii) $\mathrm{diag}(\mathbf{d}, k)$ *the matrix whose* k*th diagonal contains the entries of* $\mathbf{d}$*,*
(iii) $\mathrm{triu}(\mathbf{A}, k)$ *the matrix with the part on and above the* k*th diagonal of* $\mathbf{A}$*,*
(iv) $\mathrm{tril}(\mathbf{A}, k)$ *the matrix with the part on and below the* k*th diagonal of* $\mathbf{A}$*.*

For example, $\mathrm{triu}(\mathbf{A}, 1)$ is the matrix that has been obtained from the matrix $\mathbf{A}$ by annihilating all components that are not contained in the strictly upper triangular part. To further simplify the notation, we define the shortcuts

$$\mathrm{diag}(\mathbf{A}) := \mathrm{diag}(\mathbf{A}, 0), \qquad \mathrm{triu}(\mathbf{A}) := \mathrm{triu}(\mathbf{A}, 0), \qquad \mathrm{tril}(\mathbf{A}) := \mathrm{tril}(\mathbf{A}, 0).$$

Vectors and matrices that contain only zeros or ones are denoted $\mathbf{0}$ and $\mathbf{1}$, respectively. It will be clear from the context whether this refers to a vector or a matrix. To denote

[1] MATLAB is a registered trademark of The MathWorks, Inc.

sub-matrices we use MATLAB-style notation. For example, $\mathbf{A}(i:j,k:l)$ denotes the sub-matrix formed by rows i to j and columns k to l of the matrix $\mathbf{A}$.

2.2 Structured matrices

2.2.1 Banded matrices. In a banded matrix, the entries are confined to a band of contiguous diagonals, usually comprising the main diagonal and a number of diagonals on either side. These matrices are special cases of *Hessenberg* matrices which can be regarded as "almost" triangular matrices. Let us make the definition precise.

Definition 2.2 *An $n \times n$ matrix $\mathbf{A}$ is called*

(i) (strictly) upper (p)-Hessenberg, *if* $\mathrm{tril}(\mathbf{A}, -p-1) = \mathbf{0}$ *(and* $\mathrm{diag}(\mathbf{A}, -p) \neq \mathbf{0}$*),*
(ii) (strictly) lower (q)-Hessenberg, *if* $\mathrm{triu}(\mathbf{A}, q+1) = \mathbf{0}$ *(and* $\mathrm{diag}(\mathbf{A}, q) \neq \mathbf{0}$*),*
(iii) (strictly) (p,q)-banded, *if $\mathbf{A}$ is simultaneously (strictly) lower (q)-Hessenberg and (strictly) upper (p)-Hessenberg,*

with $0 \leq p, q$.

Note that the strict variants require that all entries on the diagonals adjacent to the vanishing parts of $\mathbf{A}$ be non-zero. Noteworthy cases of banded matrices are those with just one, two, or at most three non-vanishing diagonals. These are defined as follows.

Definition 2.3 *An $n \times n$ matrix $\mathbf{A}$ is called*

(i) diagonal, *if $\mathbf{A}$ is $(0,0)$-banded or, equivalently, if* $\mathbf{A} = \mathrm{diag}(\mathrm{diag}(\mathbf{A}))$,
(ii) lower bidiagonal, *if $\mathbf{A}$ is $(1,0)$-banded,*
(iii) upper bidiagonal, *if $\mathbf{A}$ is $(0,1)$-banded,*
(iv) tridiagonal, *if $\mathbf{A}$ is $(1,1)$-banded,*
(v) lower triangular, *if $\mathbf{A}$ is $(n-1,0)$-banded,*
(vi) upper triangular, *if $\mathbf{A}$ is $(0,n-1)$-banded.*

The matrix classes introduced in Definitions 2.2 and 2.3 are illustrated in Figure 2.1.

2.2.2 Semiseparable matrices. While banded matrices have been well-known for some time, the class of semiseparable matrices has only recently become more popular. A comprehensive introduction to the topic can be found in [78, 79]. Historically, semiseparable matrices had been investigated independently in a number of different fields, e.g., in integral equations and statistics; see [77, 78, Chapter 3, p. 109]. In some areas, semiseparable matrices had been known before the term *semiseparable* was eventually coined. But, slight differences in the conventions used over time have also been a cause for misunderstandings. Therefore, one has to be careful with the definitions used, especially when referring to the literature. This text follows [78, 79]. We will take a number of important results from there.

In the following, a number of definitions surrounding semiseparable matrices and other similar classes of matrices will be given. We remind the reader that it is important to note all subtle details found in the different definitions to avoid confusion.

Definition 2.4 *An $n \times n$ matrix $\mathbf{A}$ is called* (p,q)-generator representable semiseparable, *with $0 \leq p, q$, if the following two conditions are satisfied:*

$$\mathrm{tril}(\mathbf{A}, p-1) = \mathrm{tril}\left(\mathbf{U}\,\mathbf{V}^{\mathrm{T}}, p-1\right),$$
$$\mathrm{triu}(\mathbf{A}, -q+1) = \mathrm{triu}\left(\mathbf{W}\,\mathbf{Z}^{\mathrm{T}}, -q+1\right),$$

with matrices $\mathbf{U}$ and $\mathbf{V}$ of size $n \times p$ and matrices $\mathbf{W}$ and $\mathbf{Z}$ of size $n \times q$. This means that the lower triangular part of the matrix $\mathbf{A}$, up to and including the $(p-1)$st superdiagonal, stems from a rank-p matrix. Similarly, the upper triangular part of $\mathbf{A}$, starting from the

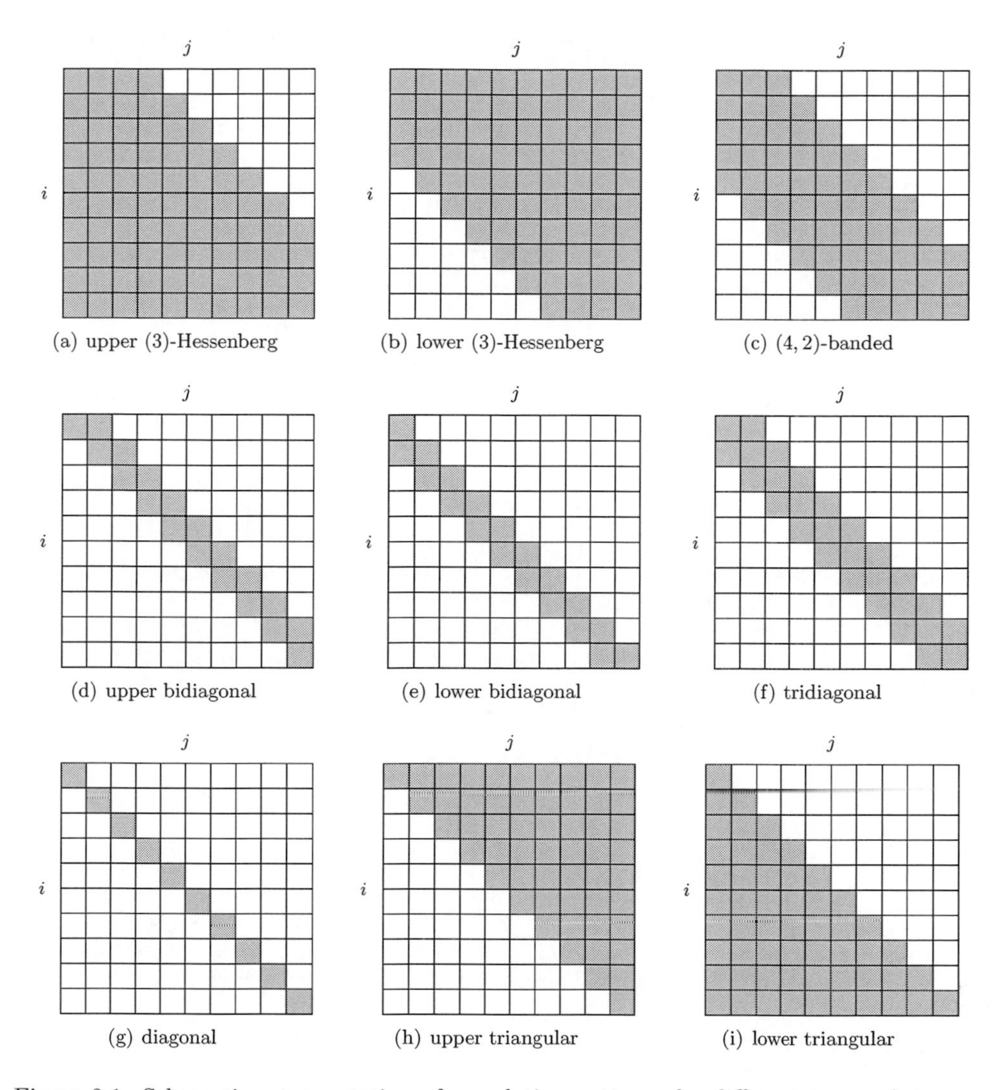

Figure 2.1: Schematic representation of population patterns for different types of structured matrices. The non-shaded boxes represent vanishing entries.

$(q-1)$st subdiagonal, has been taken from a rank-q matrix. Thus, the matrix $\mathbf{A}$ might be represented as

$$\mathbf{A} = \operatorname{tril}\left(\mathbf{U}\,\mathbf{V}^{\mathrm{T}}, p-1\right) + \operatorname{triu}\left(\mathbf{W}\,\mathbf{Z}^{\mathrm{T}}, p\right).$$

Note that the definition requires that

$$\operatorname{triu}\left(\operatorname{tril}\left(\mathbf{U}\,\mathbf{V}^{\mathrm{T}}, p-1\right), -q+1\right) = \operatorname{triu}\left(\operatorname{tril}\left(\mathbf{W}\,\mathbf{Z}^{\mathrm{T}}, p-1\right), -q+1\right),$$

so the overlapping parts must, of course, be identical. The matrices $\mathbf{U}$, $\mathbf{V}$, $\mathbf{W}$, and $\mathbf{Z}$ are called the *generators* of $\mathbf{A}$, hence the name *generator representable semiseparable*. There is a slightly different definition that describes the more general class of *extended generator representable semiseparable matrices.*

Definition 2.5 *An $n \times n$ matrix $\mathbf{A}$ is called* extended (p,q)-generator representable semiseparable, *with $0 \leq p, q$, if the following two conditions are satisfied:*

$$\operatorname{tril}(\mathbf{A}) = \operatorname{tril}\left(\mathbf{U}\,\mathbf{V}^{\mathrm{T}}\right),$$
$$\operatorname{triu}(\mathbf{A}) = \operatorname{triu}\left(\mathbf{W}\,\mathbf{Z}^{\mathrm{T}}\right),$$

with matrices $\mathbf{U}$ and $\mathbf{V}$ of size $n \times p$ and matrices $\mathbf{W}$ and $\mathbf{Z}$ of size $n \times q$. This means that the lower triangular part of the matrix $\mathbf{A}$ is coming from a rank-p matrix. Similarly, the upper triangular part of $\mathbf{A}$ is coming from a rank-q matrix. Thus, the matrix $\mathbf{A}$ might be represented as

$$\mathbf{A} = \operatorname{tril}\left(\mathbf{U}\,\mathbf{V}^{\mathrm{T}}\right) + \operatorname{triu}\left(\mathbf{W}\,\mathbf{Z}^{\mathrm{T}}, 1\right).$$

The definition implies

$$\operatorname{diag}\left(\mathbf{U}\,\mathbf{V}^{\mathrm{T}}\right) = \operatorname{diag}\left(\mathbf{W}\,\mathbf{Z}^{\mathrm{T}}\right).$$

To the casual reader, extended (p,q)-generator representable semiseparable matrices might appear as the more natural generalization of $(1,1)$-generator representable semiseparable matrices. But, the class of (p,q)-generator representable semiseparable matrices as defined in Definition 2.4 is important for a number of reasons. Unlike the extended variant, for example, it is closed under multiplication; see [78, p. 432, Corollary 9.59].

So far, the definitions of semiseparable matrices have included the diagonal into the upper and lower triangular rank structure. This is problematic for the handling of triangular matrices with a semiseparable structure. For example, of the following two matrices, the first one is (extended) $(1,1)$-generator representable semiseparable, but the second one is not:

$$\begin{pmatrix} 1 & 1 & 1 \\ 1 & 1 & 1 \\ 1 & 1 & 1 \end{pmatrix}, \qquad \begin{pmatrix} 1 & 1 & 1 \\ 0 & 1 & 1 \\ 0 & 0 & 1 \end{pmatrix}. \tag{2.1}$$

A solution to this issue would be to define more general classes of semiseparable matrices by dropping the requirement of a generator representation. This gives the class of (extended) (p,q)-semiseparable matrices; see [78, p. 300]. For our purposes, however, we would like to keep the generator representation but at the same time allow for triangular matrices with a semiseparable structure like the second matrix in (2.1). The following definition is to provide that.

Definition 2.6 *An $n \times n$ matrix $\mathbf{A}$ is called* upper (p)-generator representable semiseparable, *with $0 \leq p$, if it satisfies*

$$\mathbf{A} = \operatorname{triu}\left(\mathbf{U}\,\mathbf{V}^{\mathrm{T}}\right),$$

with matrices $\mathbf{U}$ and $\mathbf{V}$ of size $n \times p$. This means that the upper triangular part of the matrix $\mathbf{A}$ stems from a rank-p matrix. Similarly, a lower (p)-generator representable semiseparable *matrix is given by*

$$\mathbf{A} = \operatorname{tril}\left(\mathbf{U}\,\mathbf{V}^{\mathrm{T}}\right).$$

We have seen upper (1)- and upper (2)-generator representable semiseparable matrices in Chapter 1 as certain special cases of connection matrices between the classical orthogonal polynomials; see Section 1.4.2. The following result is a special case of Theorem 9.56 in [78, p. 431] and was needed in Chapter 1 to assert the semiseparable structure of some connection matrices for larger semiseparability rank $p > 1$; see Corollary 1.46, Corollary 1.55, Corollary 1.63, and Corollary 1.70.

Theorem 2.7 *Suppose that* $\mathbf{A}_1$ *and* $\mathbf{A}_2$ *are, respectively, upper* (p_1)*-generator representable semiseparable and upper* (p_2)*-generator representable semiseparable. Then the matrix* $\mathbf{A} = \mathbf{A}_1\mathbf{A}_2$ *is upper* $(p_1 + p_2)$*-generator representable semiseparable. The analogous result holds for lower generator representable semiseparable matrices.*

There are many classes of matrices that are closely related to semiseparable matrices and one can easily change above definitions to come to slightly altered versions. One such generalization is to allow for the addition of a diagonal free of choice.

Definition 2.8 *A matrix* $\mathbf{A}$ *is called* diagonal plus (extended) (p, q)-generator representable semiseparable *if it can be written as the sum of a diagonal matrix* $\mathbf{D}$ *and an (extended)* (p, q)*-generator representable semiseparable matrix* $\mathbf{B}$,

$$\mathbf{A} = \mathbf{D} + \mathbf{B}.$$

Similarly, a diagonal plus upper (lower) (p)-generator representable semiseparable *matrix can be represented as the sum of a diagonal matrix and an upper (lower)* (p)*-generator representable semiseparable matrix.*

2.2.3 Checkerboard-like matrices. The interest for structured matrices in this work is driven by their importance for algorithms related to the connection problem for classical orthogonal polynomials and their associated functions. If these are orthogonal with respect to a symmetric measure, then it is easily verified that the connection matrix between two such sequences of polynomials or associated functions has a checkerboard-like population pattern: in each row, every second element vanishes due to orthogonality. This behavior is made precise by the following definition.

Definition 2.9 *A matrix* $\mathbf{A} = (a_{i,j})$ *is called* checkerboard-like *if the following condition is satisfied:*

$$a_{i,j} = 0, \quad \text{if } i + j \text{ odd.} \tag{2.2}$$

It is possible to impose this condition on every type of matrix discussed so far to obtain their checkerboard-like counterparts. For example, Figure 2.2 shows checkerboard-like versions of Hessenberg and banded matrices. Please note that these are already contained in the original definition of Hessenberg and banded matrices used so far, since the checkerboard-like structure here simply implies that certain diagonals must vanish.
For semiseparable matrices the situation is more difficult, as there is the valid question how the definitions from the previous section should be modified accordingly. It is easy to see that simply removing all entries that vanish owing to (2.2) is incompatible with the original definition of semiseparability. Therefore, it is better to take a different viewpoint and first investigate how checkerboard-like matrices with a semiseparable structure can arise from semiseparable matrices that lack the pattern. The key is the following observation: Take, for example, an upper (2)-generator representable semiseparable matrix $\mathbf{A} = (a_{i,j}) = \operatorname{triu}\left(\mathbf{U}\,\mathbf{V}^{\mathrm{T}}\right)$, with $n \times 2$ matrices $\mathbf{U} = (u_{i,j})$ and $\mathbf{V} = (v_{i,j})$. Now, suppose that the entries of the matrices $\mathbf{U}$ and $\mathbf{V}$ are related by $u_{i,2} = (-1)^i u_{i,1}$ and $v_{i,2} = (-1)^i v_{i,1}$. Then we have

$$a_{i,j} = u_{i,1}v_{j,1} + u_{i,2}v_{j,2} = (1 + (-1)^{i+j})u_{i,1}v_{j,1}, \qquad \text{with } i \leq j.$$

This evaluates to zero whenever $i + j$ is odd, hence the matrix $\mathbf{A}$ has a checkerboard-like structure. Moreover, the matrix clearly looks like one that had been obtained from an upper (1)-generator representable semiseparable matrix by imposing the checkerboard structure afterwards. The last interpretation is useful to understand how this type of matrices typically arises in the first place. For a convenient notation, we introduce the symbols $\operatorname{trilc}(\cdot)$ and $\operatorname{triuc}(\cdot)$ which are similar to $\operatorname{tril}(\cdot)$ and $\operatorname{triu}(\cdot)$, respectively, but also incorporate the checkerboard pattern.

Definition 2.10 *Let* $\mathbf{U}$ *and* $\mathbf{V}$ *be two* $n \times p$ *matrices, with* $0 \leq p$. *Then the matrix* $\operatorname{trilc}(\mathbf{U}\mathbf{V}^{\mathrm{T}}, k)$ *is defined by*

$$\operatorname{trilc}(\mathbf{U}\mathbf{V}^{\mathrm{T}}, k) = \operatorname{tril}(\tilde{\mathbf{U}}\tilde{\mathbf{V}}^{\mathrm{T}}, k),$$

where the $n \times 2p$ *matrices* $\tilde{\mathbf{U}} = (\tilde{u}_{i,j})$ *and* $\tilde{\mathbf{V}} = (\tilde{v}_{i,j})$ *are defined by*

$$\tilde{u}_{i,2j} = (-1)^i \tilde{u}_{i,2j+1} = \frac{u_{i,j}}{2}, \qquad \tilde{v}_{i,2j} = (-1)^i \tilde{v}_{i,2j+1} = \frac{v_{i,j}}{2},$$

for $i = 1, 2, \ldots, n$ *and* $j = 1, 2, \ldots, p$. *Similarly, the matrix* $\operatorname{triuc}(\mathbf{U}\mathbf{V}^{\mathrm{T}}, k)$ *is defined by*

$$\operatorname{triuc}(\mathbf{U}\mathbf{V}^{\mathrm{T}}, k) = \operatorname{triu}(\tilde{\mathbf{U}}\tilde{\mathbf{V}}^{\mathrm{T}}, k).$$

The last definition is equivalent to defining the matrix $\operatorname{trilc}(\mathbf{U}\mathbf{V}^{\mathrm{T}}, k)$ and the matrix $\operatorname{triuc}(\mathbf{U}\mathbf{V}^{\mathrm{T}}, k)$ as the matrices $\operatorname{tril}(\mathbf{U}\mathbf{V}^{\mathrm{T}}, k)$ and $\operatorname{triu}(\mathbf{U}\mathbf{V}^{\mathrm{T}}, k)$, respectively, with the addition of the checkerboard pattern.
There is yet another interpretation of checkerboard-like matrices that is particularly useful from a computational point of view: every matrix with checkerboard structure can be decomposed into two independent smaller matrices of the same type that lack the checkerboard pattern. This can be achieved by appropriate permutations of rows and columns. Figure 2.3 illustrates this procedure. The upshot is that with these equivalent viewpoints, one and the same matrix can be seen as, for example, a particular upper (2)-generator representable semiseparable matrix, as an upper (1)-generator representable semiseparable matrix with checkerboard structure, or as two independent upper (1)-generator representable semiseparable matrices without the checkerboard structure. One is free to pick the most convenient representation for a given purpose. This, of course, extends to all other types of semiseparable matrices mentioned.

2.3 The eigendecomposition of triangular semiseparable matrices

This section is to introduce a method to compute the eigendecomposition of diagonal plus upper or lower generator representable semiseparable matrices. The procedure has first been published by the author in [38]. It is similar to an earlier method for diagonal plus symmetric $(1,1)$-generator representable semiseparable matrices from [9], but has subtle differences that should be noted. There are other methods, for example the recursive expressions of Eidelman, Gohberg, and Olshevsky [20] for the characteristic polynomial and the eigenvectors of quasiseparable matrices. The algorithm presented in the following has the blessing property that it not only allows one to compute eigenvalues and eigenvectors explicitly. It leads to an efficient approximate method to compute the product of the eigenvector matrix, or its inverse or transpose, with any vector. This way, the cost of setting up the full eigenvector matrix can be avoided. A restriction that we must acknowledge, is that the matrix at hand must have simple eigenvalues. This condition will later always be satisfied when we employ the method for our purposes in Section 2.4.
We start with matrices that have semiseparability rank $p = 1$. Recall that diagonal plus lower or upper (1)-generator representable semiseparable matrices have, respectively, the form

$$\mathbf{A} = \operatorname{diag}(\mathbf{d}) + \operatorname{triu}(\mathbf{u}\,\mathbf{v}^{\mathrm{T}}), \quad \text{or} \quad \mathbf{B} = \operatorname{diag}(\mathbf{d}) + \operatorname{tril}(\mathbf{u}\,\mathbf{v}^{\mathrm{T}}). \tag{2.3}$$

We will only treat upper generator representable semiseparable matrices. The lower triangular case can approached analogously. Moreover, we will represent a diagonal plus upper (1)-generator representable semiseparable matrix $\mathbf{A}$ as

$$\mathbf{A} = \operatorname{diag}(\mathbf{d}) + \operatorname{triu}(\mathbf{u}\,\mathbf{v}^{\mathrm{T}}, 1),$$

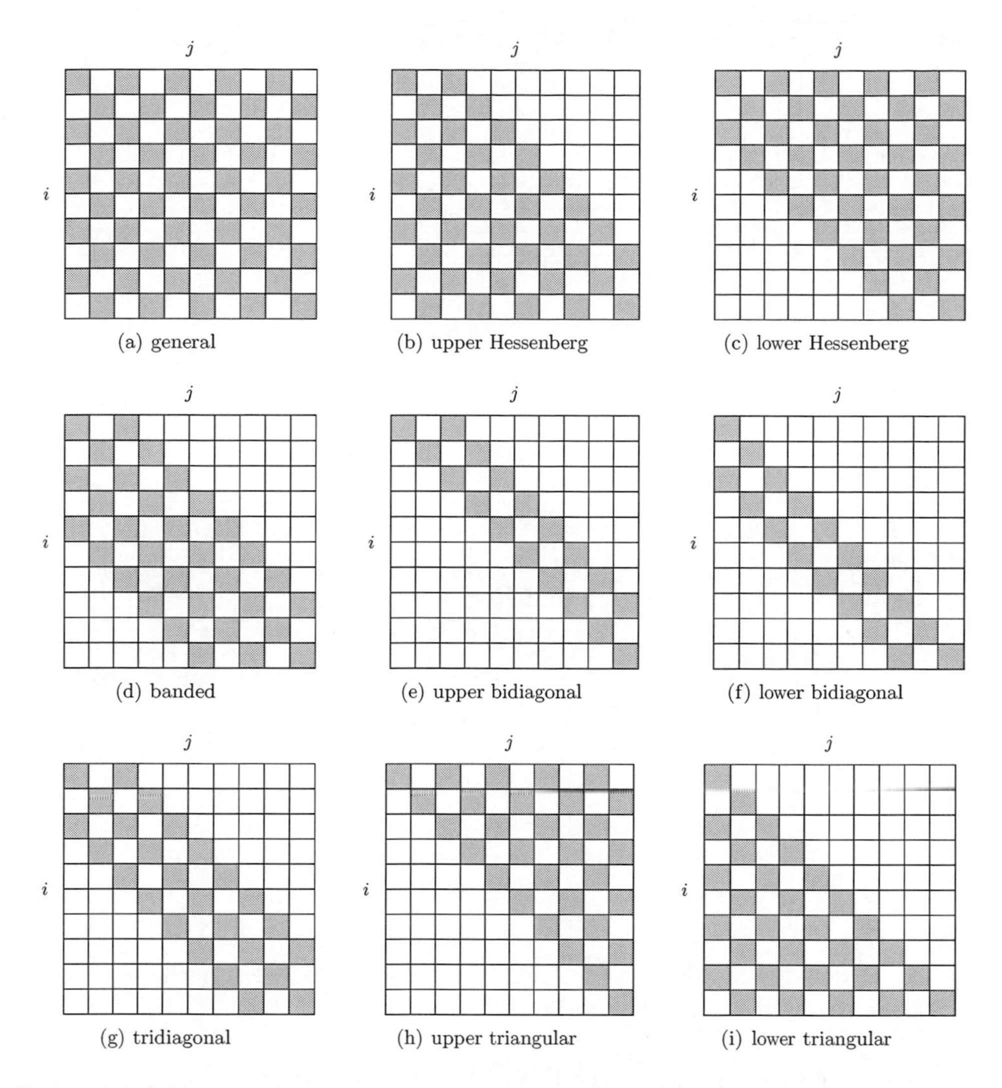

Figure 2.2: Schematic representation of population patterns for different types of structured matrices with checkerboard-like structure. The non-shaded boxes represent vanishing entries.

by absorbing the $\mathrm{diag}(\mathbf{u}\,\mathbf{v}^{\mathrm{T}})$ part in (2.3) into the vector $\mathbf{d}$. This is solely for notational convenience. An upper (1)-generator representable semiseparable matrix $\mathbf{A}$ has an eigendecomposition of the form

$$\mathbf{A} = \mathbf{Q}\,\mathbf{D}\,\mathbf{Q}^{-1},$$

with an invertible upper triangular eigenvector matrix $\mathbf{Q}$ and a diagonal eigenvalue matrix $\mathbf{D} = \mathrm{diag}(\mathbf{d})$. The method we propose to compute the eigendecomposition of such matrix follows a divide-and-conquer approach. It is based on that the matrix $\mathbf{A}$ can be divided

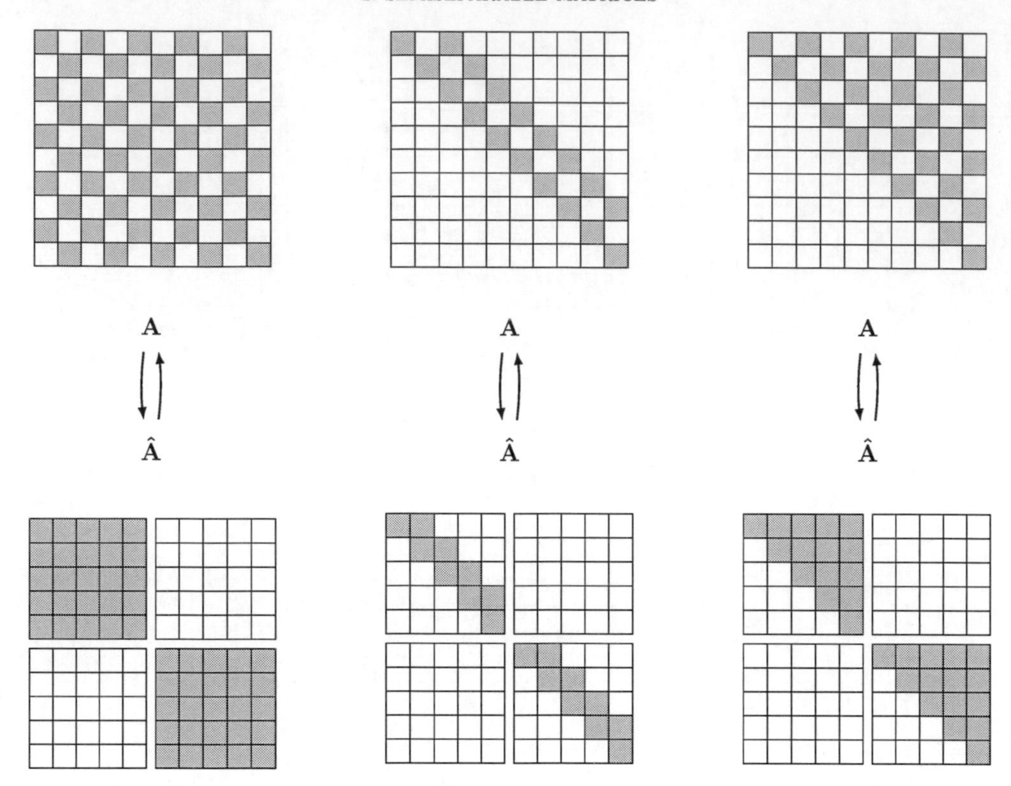

Figure 2.3: Examples of checkerboard-like matrices $\mathbf{A}$ and how these can be rearranged into two smaller counterparts without the checkerboard structure. These are contained within a permuted matrix $\hat{\mathbf{A}}$.

into smaller matrices of the same type, say $\mathbf{A}_1$ and $\mathbf{A}_2$, and that the eigendecompositions of these smaller matrices can be efficiently combined to the eigendecomposition of $\mathbf{A}$. This is established in the following.

2.3.1 Divide-and-conquer method. The method is split into two phases. In the *divide phase*, the matrix $\mathbf{A}$ is recursively divided into smaller matrices until these are sufficiently small such that their eigendecompositions can be computed with a standard algorithm free of choice. In the *conquer phase*, we seek a method to efficiently combine these eigendecompositions.

Divide phase

We start with a given a diagonal plus upper (1)-generator representable semiseparable matrix $\mathbf{A} = \operatorname{diag}(\mathbf{d}) + \operatorname{triu}(\mathbf{u}\,\mathbf{v}^{\mathrm{T}}, 1)$ of size $n \times n$ which we would like to write using two smaller matrices of same type. This can be done as follows. Split each of the vectors $\mathbf{d}$, $\mathbf{u}$, $\mathbf{v}$ into two vectors, with the first $\lfloor n/2 \rfloor$ components in the first vector, and the remaining components in the second vector. That is, define $\mathbf{d}_1$, $\mathbf{d}_2$, $\mathbf{u}_1$, $\mathbf{u}_2$, $\mathbf{v}_1$, and $\mathbf{v}_2$ such that

$$\mathbf{d} = \begin{pmatrix} \mathbf{d}_1 \\ \mathbf{d}_2 \end{pmatrix}, \qquad \mathbf{u} = \begin{pmatrix} \mathbf{u}_1 \\ \mathbf{u}_2 \end{pmatrix}, \qquad \mathbf{v} = \begin{pmatrix} \mathbf{v}_1 \\ \mathbf{v}_2 \end{pmatrix}.$$

Also, define vectors $\hat{\mathbf{u}}$ and $\hat{\mathbf{v}}$ as

$$\hat{\mathbf{u}} = \begin{pmatrix} \mathbf{u}_1 \\ \mathbf{0} \end{pmatrix}, \qquad \hat{\mathbf{v}} = \begin{pmatrix} \mathbf{0} \\ \mathbf{v}_2 \end{pmatrix}.$$

The matrix $\mathbf{A}$ may then be decomposed as

$$\mathbf{A} = \begin{pmatrix} \mathbf{A}_1 & \mathbf{0} \\ \mathbf{0} & \mathbf{A}_2 \end{pmatrix} + \hat{\mathbf{u}}\,\hat{\mathbf{v}}^{\mathrm{T}},$$

where $\mathbf{A}_1$ and $\mathbf{A}_2$ are defined to the diagonal plus upper (1)-generator representable semiseparable matrices

$$\mathbf{A}_1 = \operatorname{diag}(\mathbf{d}_1) + \operatorname{triu}(\mathbf{u}_1\,\mathbf{v}_1^{\mathrm{T}}, 1), \qquad \mathbf{A}_2 = \operatorname{diag}(\mathbf{d}_2) + \operatorname{triu}(\mathbf{u}_2\,\mathbf{v}_2^{\mathrm{T}}, 1).$$

These are submatrices of the original matrix $\mathbf{A}$ that may be decomposed in a similar manner. Note that the vector $\mathbf{d}$ contains the eigenvalues of $\mathbf{A}$ and that the components of the vectors $\mathbf{d}_1$ and $\mathbf{d}_2$ are the eigenvalues of $\mathbf{A}_1$ and $\mathbf{A}_2$, respectively. The rank-one modification $\hat{\mathbf{u}}\,\hat{\mathbf{v}}^{\mathrm{T}}$ is non-symmetric, but of a particular type, since it modifies only a certain block in the strictly upper triangular part. Note that the decomposition does not require the computation of any new quantities.

Conquer phase

Suppose that two diagonal plus upper (1)-generator representable semiseparable matrices $\mathbf{A}_1$ and $\mathbf{A}_2$ have the eigendecomposition

$$\mathbf{A}_1 = \mathbf{Q}_1\,\mathbf{D}_1\,\mathbf{Q}_1^{-1}, \qquad \mathbf{A}_2 = \mathbf{Q}_2\,\mathbf{D}_2\,\mathbf{Q}_2^{-1},$$

with diagonal eigenvalue matrices $\mathbf{D}_1 = \operatorname{diag}(\mathbf{d}_1)$, $\mathbf{D}_2 = \operatorname{diag}(\mathbf{d}_2)$ and invertible upper triangular eigenvector matrices $\mathbf{Q}_1$ and $\mathbf{Q}_2$. Then this implies the representation

$$\mathbf{A} = \begin{pmatrix} \mathbf{Q}_1 & \mathbf{0} \\ \mathbf{0} & \mathbf{Q}_2 \end{pmatrix} (\mathbf{D} + \mathbf{w}\,\mathbf{z}^{\mathrm{T}}) \begin{pmatrix} \mathbf{Q}_1 & \mathbf{0} \\ \mathbf{0} & \mathbf{Q}_2 \end{pmatrix}^{-1},$$

where $\mathbf{w}$ and $\mathbf{z}$ are vectors defined by

$$\mathbf{w} = \begin{pmatrix} \mathbf{Q}_1 & \mathbf{0} \\ \mathbf{0} & \mathbf{Q}_2 \end{pmatrix}^{-1} \hat{\mathbf{u}} = \begin{pmatrix} \mathbf{Q}_1^{-1}\,\mathbf{u}_1 \\ \mathbf{0} \end{pmatrix}, \qquad \mathbf{z} = \begin{pmatrix} \mathbf{Q}_1 & \mathbf{0} \\ \mathbf{0} & \mathbf{Q}_2 \end{pmatrix}^{\mathrm{T}} \hat{\mathbf{v}} = \begin{pmatrix} \mathbf{0} \\ \mathbf{Q}_2^{\mathrm{T}}\,\mathbf{v}_2 \end{pmatrix}.$$

Suppose that we can compute the eigendecomposition of the non-symmetric rank-one modified diagonal matrix $\mathbf{D} + \mathbf{w}\,\mathbf{z}^{\mathrm{T}}$. This problem will be dealt with in the next section. Let this eigendecomposition be written as

$$\mathbf{D} + \mathbf{w}\,\mathbf{z}^{\mathrm{T}} = \mathbf{P}\,\mathbf{D}\,\mathbf{P}^{-1}. \tag{2.4}$$

We can then write the eigendecomposition of $\mathbf{A}$ as

$$\mathbf{A} = \begin{pmatrix} \mathbf{Q}_1 & \mathbf{0} \\ \mathbf{0} & \mathbf{Q}_2 \end{pmatrix} \mathbf{P}\,\mathbf{D}\,\mathbf{P}^{-1} \begin{pmatrix} \mathbf{Q}_1 & \mathbf{0} \\ \mathbf{0} & \mathbf{Q}_2 \end{pmatrix}^{-1}$$

and

$$\begin{pmatrix} \mathbf{Q}_1 & \mathbf{0} \\ \mathbf{0} & \mathbf{Q}_2 \end{pmatrix} \mathbf{P}$$

is the desired eigenvector matrix of $\mathbf{A}$. Two problems remain to be solved. First, the eigendecomposition of the rank-one modified matrix $\mathbf{D} + \mathbf{w}\,\mathbf{z}^{\mathrm{T}}$ needs to be obtained efficiently. Second, the representation of the eigenvector matrix $\mathbf{Q}$ of the matrix $\mathbf{A}$ via

$$\mathbf{Q} = \begin{pmatrix} \mathbf{Q}_1 & \mathbf{0} \\ \mathbf{0} & \mathbf{Q}_2 \end{pmatrix} \mathbf{P} \tag{2.5}$$

does not instantly constitute a fast algorithm to apply the matrix $\mathbf{Q}$ to a vector. Both problems are addressed in the following two sections.

2.3.2 Non-symmetric rank-one modified eigenproblem. We need to study the eigendecomposition of a diagonal matrix with a particular rank-one modification. The matrix has the form

$$\mathbf{D} + \mathbf{w}\,\mathbf{z}^{\mathrm{T}},$$

with a diagonal matrix $\mathbf{D} = \operatorname{diag}(\mathbf{d})$ and vectors $\mathbf{d}$, $\mathbf{w}$, and $\mathbf{z}$ given by

$$\begin{aligned} \mathbf{d} &= (d_1, d_2, \ldots, d_n)^{\mathrm{T}}, \\ \mathbf{w} &= (w_1, w_2, \ldots, w_k, 0, 0, \ldots, 0)^{\mathrm{T}}, \\ \mathbf{z} &= (0, 0, \ldots, 0, z_{k+1}, z_{k+2}, \ldots, z_n)^{\mathrm{T}}, \end{aligned}$$

with $1 \leq k \leq n$. The rank-one modification $\mathbf{w}\,\mathbf{z}^{\mathrm{T}}$ is non-symmetric, but has a particular form since the last $n-k$ entries of $\mathbf{w}$ and the first k entries of $\mathbf{z}$ are zero. This implies that the matrix $\mathbf{D} + \mathbf{w}\,\mathbf{z}^{\mathrm{T}}$ can be written as

$$\mathbf{D} + \mathbf{w}\,\mathbf{z}^{\mathrm{T}} = \begin{pmatrix} d_1 & & & & w_1 z_{k+1} & w_1 z_{k+2} & \cdots & w_1 z_n \\ & d_2 & & & w_2 z_{k+1} & w_2 z_{k+2} & & w_2 z_n \\ & & \ddots & & \vdots & \vdots & \ddots & \vdots \\ & & & d_k & w_k z_{k+1} & w_k z_{k+2} & \cdots & w_k z_n \\ & & & & d_{k+1} & & & \\ & & & & & d_{k+2} & & \\ & & & & & & \ddots & \\ & & & & & & & d_n \end{pmatrix}.$$

Before the eigendecomposition of above matrix can be studied in more detail, the following definition is needed.

Definition 2.11 *A matrix $\mathbf{C} = (c_{i,j})_{i,j=1}^{n,m}$ is called a* Cauchy-like matrix, *if the following condition is satisfied:*

$$c_{i,j} = \frac{w_i z_j}{y_i - x_j},$$

with certain numbers y_i, w_i for $i = 1, 2, \ldots, n$, and x_j, z_j for $j = 1, 2, \ldots, m$.

The following theorem, first established in [38], shows the detailed structure of the eigendecomposition of the matrix $\mathbf{D} + \mathbf{w}\,\mathbf{z}^{\mathrm{T}}$.

Theorem 2.12 *Let $\mathbf{D}$ be an $n \times n$ diagonal matrix with pairwise distinct diagonal entries $d_1, d_2, \ldots, d_n$, and $\mathbf{w}$ and $\mathbf{z}$ vectors defined by*

$$\mathbf{w} = (w_1, w_2, \ldots, w_k, 0, 0, \ldots, 0)^{\mathrm{T}}, \qquad \mathbf{z} = (0, 0, \ldots, 0, z_{k+1}, z_{k+2}, \ldots, z_n)^{\mathrm{T}},$$

with $1 \leq k \leq n$. Then, for the matrix $\mathbf{B} = \mathbf{D} + \mathbf{w}\,\mathbf{z}^{\mathrm{T}}$ the following statements hold:

(i) *The eigenvalues of the matrices $\mathbf{D}$ and $\mathbf{B}$ are the numbers $d_1, d_2, \ldots, d_n$.*

(ii) *The matrix $\mathbf{B}$ has the eigendecomposition*

$$\mathbf{B} = \mathbf{P}\,\mathbf{D}\,\mathbf{P}^{-1},$$

where an eigenvector matrix $\mathbf{P}$ that contains $\|\cdot\|_2$-normalized eigenvectors of $\mathbf{B}$ and its inverse $\mathbf{P}^{-1}$ have the form

$$\mathbf{P} = \begin{pmatrix} \mathbf{I} & \mathbf{C}\hat{\mathbf{D}} \\ \mathbf{0} & \hat{\mathbf{D}} \end{pmatrix}, \qquad \mathbf{P}^{-1} = \begin{pmatrix} \mathbf{I} & -\mathbf{C} \\ \mathbf{0} & \hat{\mathbf{D}}^{-1} \end{pmatrix}. \tag{2.6}$$

Here, $\mathbf{I}$ *denotes the* $k \times k$ *identity matrix,* $\hat{\mathbf{D}}$ *is an* $(n-k) \times (n-k)$ *diagonal matrix with non-zero entries* $\hat{d}_j$*,* $j = k+1, k+2, \ldots, n$*, and* $\mathbf{C}$ *is a certain* $k \times (n-k)$ *matrix.*

(iii) *The entries of the diagonal matrix* $\hat{\mathbf{D}} = (\hat{d}_j)_{j=k+1}^{n}$ *are given by*

$$\hat{d}_j = \pm \left(1 + \sum_{i=1}^{k} \frac{w_i^2\, z_j^2}{(d_i - d_j)^2} \right)^{-1/2},$$

and the Cauchy-like matrix $\mathbf{C} = (c_{i,j})_{i=1,\, j=k+1}^{k,n}$ *has its entries defined by*

$$c_{i,j} = -\frac{w_i\, z_j \hat{d}_j}{d_i - d_j}.$$

PROOF.

(i) It is clear that the diagonal entries $d_1, d_2, \ldots, d_n$ of the diagonal matrix $\mathbf{D}$ coincide with its eigenvalues. Since $\mathbf{B} = \mathbf{D} + \mathbf{w}\,\mathbf{z}^{\mathrm{T}}$ is an upper triangular matrix with the same main diagonal as $\mathbf{D}$, this also holds for $\mathbf{B}$.

(ii) The matrix $\mathbf{B}$ clearly has real eigenvalues and the eigendecomposition of $\mathbf{B}$ reads $\mathbf{B} = \mathbf{P}\,\mathbf{D}\,\mathbf{P}^{-1}$ with the invertible eigenvector matrix $\mathbf{P}$. To show that $\mathbf{P}$ has the form

$$\mathbf{P} = \begin{pmatrix} \mathbf{I} & \mathbf{C}\hat{\mathbf{D}} \\ \mathbf{0} & \hat{\mathbf{D}} \end{pmatrix},$$

we first prove that the coordinate vectors $\mathbf{e}_i$, $i = 1, 2, \ldots, k$, are eigenvectors of $\mathbf{B}$ to the respective eigenvalues d_i. For this, note that $\mathbf{z}^{\mathrm{T}}\,\mathbf{e}_i = 0$ which implies

$$\mathbf{B}\,\mathbf{e}_i = \mathbf{D}\,\mathbf{e}_i + \mathbf{w}\,\mathbf{z}^{\mathrm{T}}\,\mathbf{e}_i = d_i\,\mathbf{e}_i, \qquad \text{with } 1 \le i \le k.$$

Thus, the first k columns of $\mathbf{P}$ may be taken as

$$\mathbf{P}\,(1:n, 1:k) = \begin{pmatrix} \mathbf{I} \\ \mathbf{0} \end{pmatrix}.$$

It remains to prove that

$$\mathbf{P}\,(1:n, k+1:n) = \begin{pmatrix} \mathbf{C}\,\hat{\mathbf{D}} \\ \hat{\mathbf{D}} \end{pmatrix}.$$

Now, assume that $k+1 \le i, j \le n$. It suffices to show that the ith component of the jth column of $\mathbf{P}$, denoted $(\mathbf{p}_j)_i$, satisfies

$$(\mathbf{p}_j)_i = \delta_{i,j}\hat{d}_i, \qquad \text{with } \hat{d}_i \neq 0.$$

To verify this, fix an arbitrary index j. Since the vector $\mathbf{p}_j$ is an eigenvector of $\mathbf{B}$, we have

$$\mathbf{B}\,\mathbf{p}_j = \left(\mathbf{D} + \mathbf{w}\,\mathbf{z}^{\mathrm{T}}\right)\mathbf{p}_j = d_j\,\mathbf{p}_j.$$

Note that the last $n-k$ rows of the matrix $\mathbf{w}\,\mathbf{z}^{\mathrm{T}}$ vanish. This implies

$$d_j\,(\mathbf{p}_j)_i = d_i\,(\mathbf{p}_j)_i\,.$$

Since $d_i \neq d_j$ while $i \neq j$ by assumption, we have

$$(\mathbf{p}_j)_i = \delta_{i,j}\hat{d}_i,$$

with a certain number $\hat{d}_i$. This number cannot vanish since this would render the eigenvector matrix $\mathbf{P}$ singular. The form of the inverse matrix $\mathbf{P}^{-1}$ can be obtained from the well-known block matrix inversion formula; see [35, p. 18].

(iii) We first show that the upper right block of the eigenvector matrix $\mathbf{P}$ has the representation $\mathbf{C}\hat{\mathbf{D}}$ with a Cauchy-like matrix $\mathbf{C} = (c_{i,j})_{i=1,\, j=k+1}^{k,n}$ whose components are

$$c_{i,j} = -\frac{w_i\, z_j \hat{d}_j}{d_i - d_j}.$$

Assume that $1 \le i \le k$ and $k+1 \le j \le n$. The vector $\mathbf{p}_j$ and the diagonal entry d_j form an eigenpair of $\mathbf{B}$, i.e., $\mathbf{B}\,\mathbf{p}_j = \left(\mathbf{D} + \mathbf{w}\mathbf{z}^{\mathrm{T}}\right)\,\mathbf{p}_j = d_j\,\mathbf{p}_j$. By subtracting $d_j\,\mathbf{p}_j$ and $\mathbf{w}\,\mathbf{z}^{\mathrm{T}}\,\mathbf{p}_j$ on each side and after rearranging terms, we obtain

$$(\mathbf{D} - d_j\mathbf{I})\,\mathbf{p}_j = -\mathbf{w}\mathbf{p}_j^{\mathrm{T}}\mathbf{z}.$$

For the ith component in this equation we can verify that

$$\left((\mathbf{D} - d_j\mathbf{I})\mathbf{p}_j\right)_i = (d_i - d_j)\,(\mathbf{p}_j)_i = -\left(\mathbf{w}\mathbf{p}_j^{\mathrm{T}}\mathbf{z}\right)_i.$$

Since the first k components of the vector $\mathbf{z}$ and the last $(n-k)$ components of the vector $\mathbf{p}_j$, except for the entry $(\mathbf{p}_j)_j = \hat{d}_j$, are zero, it is verified that $\mathbf{p}_j^{\mathrm{T}}\mathbf{z} = \hat{d}_j z_j$. This implies $(d_i - d_j)\,(\mathbf{p}_j)_i = -w_i\hat{d}_j z_j$ and finally

$$(\mathbf{p}_j)_i = -\frac{w_i\, z_j \hat{d}_j}{d_i - d_j}.$$

We have thus proved the desired representation for the upper right part of $\mathbf{P}$,

$$\mathbf{P}(1:k, k+1:n) = \mathbf{C}\,\hat{\mathbf{D}}.$$

To obtain the explicit expression for the values $\hat{d}_j$, recall that each eigenvector $\mathbf{p}_j$ is assumed to be normalized, i.e.,

$$\|\mathbf{u}_j\|_2^2 = \sum_{i=1}^{k}\left(\frac{w_i\, z_j}{d_i - d_j}\hat{d}_j\right)^2 + \hat{d}_j^2 = \left(\sum_{i=1}^{k}\left(\frac{w_i\, z_j}{d_i - d_j}\right)^2 + 1\right)\hat{d}_j^2 = 1.$$

This implies

$$\hat{d}_j = \pm\left(1 + \sum_{i=1}^{k}\frac{w_i^2\, z_j^2}{(d_i - d_j)^2}\right)^{-1/2}. \tag{2.7}$$

□

The last theorem allows us to calculate the eigenvector matrix of the rank-one modified diagonal matrix $\mathbf{D}+\mathbf{w}\,\mathbf{z}^{\mathrm{T}}$. If one assumes that the eigenvectors are normalized, then these are uniquely defined up to signedness; cf. (2.7). Note that it would also be allowed to change the sign of individual columns in the identity matrix $\mathbf{I}$ in (2.6).

Remark 2.13 In some cases that will be observed later, it might be favorable to use a different normalization of the eigenvectors by requiring that the diagonal entries of the eigenvector matrix $\mathbf{P}$ be equal to one. This implies $\hat{d}_j = 1$ for $j = k+1, \ldots, n$. Then the matrix $\mathbf{P}$ and its inverse $\mathbf{P}^{-1}$ take the form

$$\mathbf{P} = \begin{pmatrix} \mathbf{I} & \mathbf{C} \\ \mathbf{0} & \mathbf{I} \end{pmatrix}, \qquad \mathbf{P}^{-1} = \begin{pmatrix} \mathbf{I} & -\mathbf{C} \\ \mathbf{0} & \mathbf{I} \end{pmatrix}.$$

In the divide-and-conquer method, this can be used to ensure that the final eigenvector matrix $\mathbf{Q}$ will also have a unit diagonal; cf. (2.5).

2.3.3 Efficient application of the eigenvector matrix. An efficient method is needed to apply the matrices $\mathbf{P}$ and $\mathbf{P}^{-1}$, obtained from the rank-one modified system $\mathbf{D}+\mathbf{wz}^{\mathrm{T}}$, to an arbitrary vector. As we have seen, these have the representation (2.6) with a Cauchy-like matrix $\mathbf{C}$. While the matrices $\mathbf{I}$, $\hat{\mathbf{D}}$, and $\hat{\mathbf{D}}^{-1}$ are easily applied to a vector, one also needs an efficient method to apply the Cauchy-like matrix $\mathbf{C}$. More precisely, calculating the matrix-vector product $\mathbf{y} = \mathbf{Cx}$, with vectors $\mathbf{x} = (x_j)_{j=k+1}^{n}$ and $\mathbf{y} = (y_i)_{i=1}^{k}$, is equivalent to calculating the sums

$$y_i = -\sum_{j=k+1}^{n} \frac{w_i z_j}{d_i - d_j} x_j, \qquad \text{with } i = 1, 2, \ldots, k.$$

Methods to efficiently compute such sums are available and commonly subsumed under the name *fast multipole method (FMM)*. The original algorithm was introduced by Rokhlin and Greengard [28]. Variations are also described, for example, in [18]. The principle common to all methods is that a Cauchy-like matrix can be decomposed into tiles that are well approximated by low-rank matrices. This is done by imposing a binary tree structure on the nodes d_i. An efficient organization of the computation leads to an algorithm with $\mathcal{O}(n\log(1/\varepsilon))$ arithmetic operations for an $n \times n$ matrix, where ε is the desired accuracy. The amount of memory needed is in the same order.
It should be noted that direct application of the FMM requires that the nodes d_i be ordered increasingly. This can be arranged for by column and row permutations to the matrix $\mathbf{D}+\mathbf{wz}^{\mathrm{T}}$, if necessary. But this will not be needed for our purposes in Section 2.4.
Typically, the FMM is divided into two stages. In the first, all necessary information that depends only on the matrix to be applied is pre-computed and stored. Then, in the second stage, any number of matrix-vector products with any vector can be carried out efficiently using the pre-computed information. Usually, both stages need $\mathcal{O}(n\log(1/\varepsilon))$ arithmetic operations, but the former one with a much larger constant hidden in the asymptotic notation. This justifies to pre-compute as much information as possible in the first stage.

2.3.4 Complexity of the divide-and-conquer algorithm. We have seen that each eigenvector matrix $\mathbf{P}$ and its inverse $\mathbf{P}^{-1}$, obtained from a rank-one modified diagonal matrix in the divide-and-conquer method, can be applied efficiently to any vector. Moreover, (2.5) explains how this can be used to apply the eigenvector matrix $\mathbf{Q}$ (or the inverse) of an upper (1)-generator representable semiseparable matrix $\mathbf{A}$ to a vector. As a result, a factorization of the matrix $\mathbf{Q}$ into a product of block-diagonal matrices is obtained, for example,

$$\mathbf{Q} = \begin{pmatrix} \mathbf{Q}_{000} & & & \\ & \mathbf{Q}_{001} & & \\ & & \ddots & \\ & & & \mathbf{Q}_{111} \end{pmatrix} \begin{pmatrix} \mathbf{P}_{00} & & & \\ & \mathbf{P}_{01} & & \\ & & \mathbf{P}_{10} & \\ & & & \mathbf{P}_{11} \end{pmatrix} \begin{pmatrix} \mathbf{P}_{0} & \\ & \mathbf{P}_{1} \end{pmatrix} \mathbf{P}, \qquad (2.8)$$

which corresponds to the exemplary decomposition shown in Figure 2.4. Here, the original matrix has been decomposed to four levels. Using plain matrix-vector multiplications, the divide-and-conquer method needs $O(n^2)$ arithmetic operations and $O(n^2)$ of memory to explicitly compute the full eigendecomposition of the $n \times n$ matrix $\mathbf{A}$. This is asymptotically better than standard methods which have an arithmetic cost of $\mathcal{O}(n^3)$ to compute the same result; see [26, Chapter 7]. To apply the matrix $\mathbf{Q}$ to a vector clearly takes $O(n^2)$ operations. Recall that our goal is not to compute the eigendecomposition explicitly, but only to apply the eigenvector matrix $\mathbf{Q}$ to a vector. If we use the FMM to accelerate the calculation of matrix-vector products, then (2.8) makes clear that the matrix $\mathbf{Q}$ can

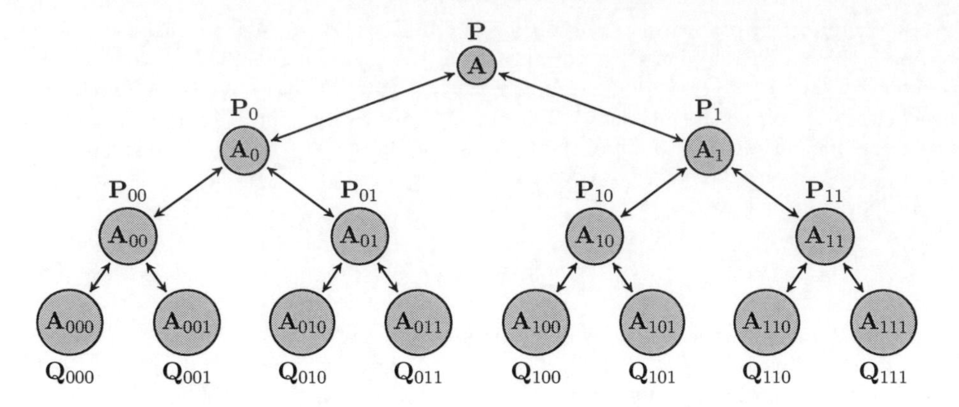

Figure 2.4: Schematic representation of a decomposition in the divide-and-conquer method for an upper or lower (1)-generator representable semiseparable matrix. The initial matrix $\mathbf{A}$ is recursively divided into smaller matrices $\mathbf{A}_0$, $\mathbf{A}_1$ and so forth. The eigenvector matrices $\mathbf{Q}_{000}, \mathbf{Q}_{001}, \ldots, \mathbf{Q}_{111}$ corresponding to the smallest matrices $\mathbf{A}_{000}, \mathbf{A}_{001}, \ldots, \mathbf{A}_{111}$ are computed explicitly. Then, these are combined to eigenvector matrices for larger matrices.

be multiplied with a vector using only $\mathcal{O}\big(n \log n \log(1/\varepsilon)\big)$ arithmetic operations. Moreover, one never needs to set up the matrix $\mathbf{Q}$ explicitly. For example, in (2.8) only the information needed to define the Cauchy-like matrices $\mathbf{P}$, $\mathbf{P}_0$, $\mathbf{P}_1$, $\mathbf{P}_{00}$, $\mathbf{P}_{01}$, $\mathbf{P}_{10}$, $\mathbf{P}_{11}$, and the full eigenvector matrices $\mathbf{Q}_{000}, \mathbf{Q}_{001}, \ldots, \mathbf{Q}_{111}$ need to be stored. Therefore, only $\mathcal{O}\big(n \log n \log(1/\varepsilon)\big)$ memory is required to store all needed information.

2.3.5 Generalization to higher semiseparability ranks. We can modify the divide-and-conquer method to handle diagonal plus upper (p)-generator representable semiseparable matrices for arbitrary semiseparability rank $p \geq 1$. The matrix $\mathbf{A}$ can then be decomposed into

$$\mathbf{A} = \operatorname{diag}(\mathbf{d}) + \operatorname{triu}(\mathbf{U}\mathbf{V}^{\mathrm{T}}, 1),$$

with $n \times p$ matrices $\mathbf{U}$ and $\mathbf{V}$. In the divide phase, this may be written as

$$\mathbf{A} = \begin{pmatrix} \mathbf{A}_1 & \mathbf{0} \\ \mathbf{0} & \mathbf{A}_2 \end{pmatrix} + \hat{\mathbf{U}}\hat{\mathbf{V}}^{\mathrm{T}},$$

with the matrics $\hat{\mathbf{U}}$ and $\hat{\mathbf{V}}$ defined by

$$\hat{\mathbf{U}} = \begin{pmatrix} \mathbf{U}(1:k,:) \\ 0 \end{pmatrix}, \qquad \hat{\mathbf{V}} = \begin{pmatrix} 0 \\ \mathbf{V}(k+1:n,:) \end{pmatrix}.$$

For the conquer phase, assume that the eigendecompositions of $\mathbf{A}_1$ and $\mathbf{A}_2$ are

$$\mathbf{A}_1 = \mathbf{Q}_1\mathbf{D}_1\mathbf{Q}_1^{-1}, \qquad \mathbf{A}_2 = \mathbf{Q}_2\mathbf{D}_2\mathbf{Q}_2^{-1}.$$

This implies the representation

$$\mathbf{A} = \begin{pmatrix} \mathbf{Q}_1 & \mathbf{0} \\ \mathbf{0} & \mathbf{Q}_2 \end{pmatrix} (\mathbf{D} + \mathbf{W}\,\mathbf{Z}^{\mathrm{T}}) \begin{pmatrix} \mathbf{Q}_1 & \mathbf{0} \\ \mathbf{0} & \mathbf{Q}_2 \end{pmatrix}^{-1},$$

where the matrices $\mathbf{W}$ and $\mathbf{Z}$ are defined by

$$\mathbf{W} = \begin{pmatrix} \mathbf{Q}_1 & \mathbf{0} \\ \mathbf{0} & \mathbf{Q}_2 \end{pmatrix}^{-1} \hat{\mathbf{U}} = \begin{pmatrix} \mathbf{Q}_1^{-1}\mathbf{U}(1:k,:) \\ \mathbf{0} \end{pmatrix},$$

$$\mathbf{Z} = \begin{pmatrix} \mathbf{Q}_1 & \mathbf{0} \\ \mathbf{0} & \mathbf{Q}_2 \end{pmatrix}^{\mathrm{T}} \hat{\mathbf{V}} = \begin{pmatrix} \mathbf{0} \\ \mathbf{Q}_2^{\mathrm{T}}\mathbf{V}(k+1:n,:) \end{pmatrix}.$$

Let the eigendecomposition of the upper triangular rank-p modified diagonal matrix $\mathbf{D} + \mathbf{W}\,\mathbf{Z}^{\mathrm{T}}$ be written as

$$\mathbf{D} + \mathbf{W}\,\mathbf{Z}^{\mathrm{T}} = \mathbf{P}\,\mathbf{D}\,\mathbf{P}^{-1}. \tag{2.9}$$

Then the eigendecomposition of the matrix $\mathbf{A}$ can be written as

$$\mathbf{A} = \begin{pmatrix} \mathbf{Q}_1 & \mathbf{0} \\ \mathbf{0} & \mathbf{Q}_2 \end{pmatrix} \mathbf{P}\,\mathbf{D}\,\mathbf{P}^{-1} \begin{pmatrix} \mathbf{Q}_1 & \mathbf{0} \\ \mathbf{0} & \mathbf{Q}_2 \end{pmatrix}^{-1}.$$

The generalization to higher semiseparability ranks is straightforward but it remains to efficiently compute the eigendecomposition of the upper triangular rank-p modified diagonal matrix $\mathbf{D} + \mathbf{W}\,\mathbf{Z}^{\mathrm{T}}$. This can be reduced to the case $p = 1$ as follows. We write the matrix in the form

$$\begin{aligned} \mathbf{D} + \mathbf{W}\,\mathbf{Z}^{\mathrm{T}} &= \mathbf{D} + (\mathbf{w}_1, \mathbf{w}_2, \ldots, \mathbf{w}_p)\,(\mathbf{z}_1, \mathbf{z}_2, \cdots, \mathbf{z}_p)^{\mathrm{T}} \\ &= \mathbf{D} + \mathbf{w}_1\,\mathbf{z}_1^{\mathrm{T}} + \mathbf{w}_2\,\mathbf{z}_2^{\mathrm{T}} + \cdots + \mathbf{w}_p\,\mathbf{z}_p^{\mathrm{T}}. \end{aligned} \tag{2.10}$$

Now we invoke Theorem 2.12 to cheaply obtain the eigendecomposition of the matrix $\mathbf{D} + \mathbf{w}_1\,\mathbf{z}_1^{\mathrm{T}}$, i.e.,

$$\mathbf{D} + \mathbf{w}_1\,\mathbf{z}_1^{\mathrm{T}} = \mathbf{P}_1\,\mathbf{D}\,\mathbf{P}_1^{-1}.$$

Then, (2.10) may be written as

$$\mathbf{D} + \mathbf{w}_1\,\mathbf{z}_1^{\mathrm{T}} + \mathbf{w}_2\,\mathbf{z}_2^{\mathrm{T}} + \cdots + \mathbf{w}_p\,\mathbf{z}_p^{\mathrm{T}} = \mathbf{P}_1\left(\mathbf{D} + \tilde{\mathbf{w}}_2\,\tilde{\mathbf{z}}_2^{\mathrm{T}} + \tilde{\mathbf{w}}_3\,\tilde{\mathbf{z}}_3^{\mathrm{T}} + \ldots + \tilde{\mathbf{w}}_p\,\tilde{\mathbf{z}}_p^{\mathrm{T}}\right)\mathbf{P}_1^{-1},$$

where the vectors $\tilde{\mathbf{w}}_2, \tilde{\mathbf{w}}_3, \ldots, \tilde{\mathbf{w}}_p$ and $\tilde{\mathbf{z}}_2, \tilde{\mathbf{z}}_3, \ldots, \tilde{\mathbf{z}}_p$ are defined by

$$\tilde{\mathbf{w}}_i = \mathbf{P}_1^{-1}\mathbf{w}_i, \quad \tilde{\mathbf{z}}_i = \mathbf{P}_1^{\mathrm{T}}\mathbf{z}_i, \qquad \text{with } i = 2, 3, \ldots, p.$$

Thus, it remains to compute the eigendecomposition of the rank-$(p-1)$ modified diagonal matrix $\mathbf{D} + \tilde{\mathbf{w}}_2\,\tilde{\mathbf{z}}_2^{\mathrm{T}} + \tilde{\mathbf{w}}_3\,\tilde{\mathbf{z}}_3^{\mathrm{T}} + \ldots + \tilde{\mathbf{w}}_p\,\tilde{\mathbf{z}}_p^{\mathrm{T}}$ which can be dealt with similarly. Finally, the eigendecomposition

$$\mathbf{D} + \mathbf{W}\,\mathbf{Z}^{\mathrm{T}} = \mathbf{P}\,\mathbf{D}\,\mathbf{P}^{-1} = \mathbf{P}_1 \cdot \mathbf{P}_2 \cdot \cdots \cdot \mathbf{P}_p \cdot \mathbf{D} \cdot \mathbf{P}_p^{-1} \cdot \mathbf{P}_{p-1}^{-1} \cdot \cdots \cdot \mathbf{P}_1^{-1}$$

is obtained, where the matrices $\mathbf{P}_i$ for $i = 1, 2, \ldots, p$, all stem from the eigendecomposition of certain upper triangular rank-one modified diagonal matrices according to Theorem 2.12. Each of these matrices can be applied efficiently with the FMM.

2.4 Classical orthogonal polynomials and semiseparable matrices

In this section, we will make use of the algorithms for the eigendecomposition of diagonal plus upper (p)-generator representable semiseparable matrices from the last section. This will allow us to obtain an efficient algorithm for applying connection matrices between sequences of classical orthogonal polynomials. To this end, we will identify the corresponding connection matrices with the eigenvector matrices of certain explicitly constructed upper generator representable semiseparable matrices.

2.4.1 Differentiated expressions of classical orthogonal polynomials. We start with the expansion of differentiated classical orthogonal polynomials back into the same basis. The following result shows that derivatives of monic Laguerre polynomials $\bar{L}_n^{(\alpha)}$ can be linked to an upper (1)-generator representable semiseparable matrix.

Lemma 2.14 *Let* $\{\bar{L}_n^{(\alpha)}\}_{n\in\mathbb{N}_0}$ *be the monic Laguerre polynomials which are orthogonal with respect to an inner product* $\langle\,\cdot\,,\,\cdot\,\rangle$ *and which satisfy the derivative identity*

$$\frac{\mathrm{d}}{\mathrm{d}x}\bar{L}_n^{(\alpha)}(x) = \bar{A}_n \sum_{i=0}^{n-1} \bar{B}_i \bar{L}_i^{(\alpha)}(x),$$

with certain numbers $\bar{A}_n$ *and* $\bar{B}_i$*; cf. Theorem* 1.75*. Then the matrix* $\mathbf{G} = (g_{i,j})_{i,j=0}^n$*, defined by*

$$g_{i,j} = \frac{\langle \bar{L}_i^{(\alpha)}, \frac{\mathrm{d}}{\mathrm{d}x}\bar{L}_j^{(\alpha)}\rangle}{\langle \bar{L}_i^{(\alpha)}, \bar{L}_i^{(\alpha)}\rangle}, \qquad \text{with } i,j = 0,1,\dots,n,$$

is diagonal plus upper (1)*-generator representable semiseparable and may be written as*

$$\mathbf{G} = \mathrm{diag}(\mathbf{0}) + \mathrm{triu}(\mathbf{u}\,\mathbf{v}^{\mathrm{T}}, 1),$$

with the vectors $\mathbf{u}$ *and* $\mathbf{v}$ *given by*

$$\mathbf{u} = (\bar{B}_0, \bar{B}_1, \dots, \bar{B}_n)^{\mathrm{T}}, \qquad \mathbf{v} = (\bar{A}_0, \bar{A}_1, \dots, \bar{A}_n)^{\mathrm{T}}.$$

PROOF. By orthogonality, we verify that $\langle \bar{L}_i^{(\alpha)}, \frac{\mathrm{d}}{\mathrm{d}x}\bar{L}_j^{(\alpha)}\rangle = 0$ whenever $j < i+1$. This implies that $\mathbf{G}$ must be a strictly upper triangular matrix. For $j \geq i+1$, we have

$$g_{i,j} = \frac{\langle \bar{L}_i^{(\alpha)}, \frac{\mathrm{d}}{\mathrm{d}x}\bar{L}_j^{(\alpha)}\rangle}{\langle \bar{L}_i^{(\alpha)}, \bar{L}_i^{(\alpha)}\rangle} = \frac{\bar{A}_j \sum_{k=0}^{j-1} \bar{B}_k \langle \bar{L}_i^{(\alpha)}, \bar{L}_k^{(\alpha)}\rangle}{\langle \bar{L}_i^{(\alpha)}, \bar{L}_i^{(\alpha)}\rangle} = \bar{B}_i \bar{A}_j \frac{\langle \bar{L}_i^{(\alpha)}, \bar{L}_i^{(\alpha)}\rangle}{\langle \bar{L}_i^{(\alpha)}, \bar{L}_i^{(\alpha)}\rangle} = \bar{B}_i \bar{A}_j.$$

□

The following result establishes a similar link to semiseparable matrices for Jacobi polynomials. After that, it is observed that the semiseparable structure is preserved if the derivatives of Jacobi polynomials are multiplied by a linear term. We omit the proof for the first result since it is very similar to that of the last.

Lemma 2.15 *Let* $\{\bar{P}_n^{(\alpha,\beta)}\}_{n\in\mathbb{N}_0}$ *be the monic Jacobi polynomials which are orthogonal with respect to an inner product* $\langle\,\cdot\,,\,\cdot\,\rangle$*, and which satisfy the derivative identity*

$$\frac{\mathrm{d}}{\mathrm{d}x}\bar{P}_n^{(\alpha,\beta)}(x) = \bar{A}_n \sum_{i=0}^{n-1} \bar{B}_i \bar{P}_i^{(\alpha,\beta)}(x) + \bar{C}_n \sum_{i=0}^{n-1} \bar{D}_i \bar{P}_i^{(\alpha,\beta)}(x), \tag{2.11}$$

with certain numbers $\bar{A}_n$*,* $\bar{B}_i$*,* $\bar{C}_n$*, and* $\bar{D}_i$*; cf. Theorem* 1.75*. Then the matrix* $\mathbf{G} = (g_{i,j})_{i,j=0}^n$*, defined by*

$$g_{i,j} = \frac{\langle \bar{P}_i^{(\alpha,\beta)}, \frac{\mathrm{d}}{\mathrm{d}x}\bar{P}_j^{(\alpha,\beta)}\rangle}{\langle \bar{P}_i^{(\alpha,\beta)}, \bar{P}_i^{(\alpha,\beta)}\rangle}, \qquad \text{with } i,j = 0,1,\dots,n,$$

is diagonal plus upper (2)*-generator representable semiseparable and may be written as*

$$\mathbf{G} = \mathrm{diag}(\mathbf{0}) + \mathrm{triu}(\mathbf{u}_1\,\mathbf{v}_1^{\mathrm{T}}, 1) + \mathrm{triu}(\mathbf{u}_2\,\mathbf{v}_2^{\mathrm{T}}, 1),$$

with the vectors $\mathbf{u}_1$*,* $\mathbf{v}_1$*,* $\mathbf{u}_2$*, and* $\mathbf{v}_2$ *given by*

$$\mathbf{u}_1 = (\bar{B}_0, \bar{B}_1, \dots, \bar{B}_n)^{\mathrm{T}}, \qquad \mathbf{v}_1 = (\bar{A}_0, \bar{A}_1, \dots, \bar{A}_n)^{\mathrm{T}}.$$

$$\mathbf{u}_2 = (\bar{D}_0, \bar{D}_1, \dots, \bar{D}_n)^{\mathrm{T}}, \qquad \mathbf{v}_2 = (\bar{C}_0, \bar{C}_1, \dots, \bar{C}_n)^{\mathrm{T}}.$$

Lemma 2.16 *Let $\{\bar{P}_n^{(\alpha,\beta)}\}_{n\in\mathbb{N}_0}$ be the monic Jacobi polynomials which are orthogonal with respect to an inner product $\langle\cdot,\cdot\rangle$, which satisfy the derivative identity* (2.11) *and moreover have a three-term recurrence formula of the form*

$$x\bar{P}_n^{(\alpha,\beta)}(x) = \bar{a}'_n \bar{P}_{n+1}^{(\alpha,\beta)} + \bar{b}'_n \bar{P}_n^{(\alpha,\beta)}(x) + \bar{c}'_n \bar{P}_{n-1}^{(\alpha,\beta)}(x).$$

Then the matrix $\mathbf{G} = (g_{i,j})_{i,j=0}^n$, defined by

$$g_{i,j} = \frac{\langle \bar{P}_i^{(\alpha,\beta)}, (rx+s)\frac{\mathrm{d}}{\mathrm{d}x}\bar{P}_j^{(\alpha,\beta)}\rangle}{\langle \bar{P}_i^{(\alpha,\beta)}, \bar{P}_i^{(\alpha,\beta)}\rangle}, \qquad \text{with } r,s\in\mathbb{R} \text{ and } i,j=0,1,\ldots,n,$$

is diagonal plus upper (2)*-generator representable semiseparable and may be written as*

$$\mathbf{G} = \operatorname{diag}(\mathbf{d}) + \operatorname{triu}(\mathbf{u}_1\mathbf{v}_1^{\mathrm{T}}, 1) + \operatorname{triu}(\mathbf{u}_2\mathbf{v}_2^{\mathrm{T}}, 1),$$

with the vectors $\mathbf{d}$, $\mathbf{u}_1$, $\mathbf{v}_1$, $\mathbf{u}_2$, *and* $\mathbf{v}_2$ *given by*

$$\begin{aligned}
\mathbf{d} &= \big(r(\bar{A}_j\bar{B}_{j-1}\bar{a}'_{j-1} + \bar{C}_j\bar{D}_{j-1}\bar{a}'_{j-1})\big)_{j=0}^n,\\
\mathbf{u}_1 &= \big(r(\bar{B}_{i-1}\bar{a}'_{i-1} + \bar{B}_i\bar{b}'_i + \bar{B}_{i+1}\bar{c}'_{i+1}) + s\bar{B}_i\big)_{i=0}^n, & \mathbf{v}_1 &= (\bar{A}_j)_{j=0}^n,\\
\mathbf{u}_2 &= \big(r(\bar{D}_{i-1}\bar{a}'_{i-1} + \bar{D}_i\bar{b}'_i + \bar{D}_{i+1}\bar{c}'_{i+1}) + s\bar{D}_i\big)_{i=0}^n, & \mathbf{v}_2 &= (\bar{C}_j)_{j=0}^n.
\end{aligned}$$

PROOF. By orthogonality, we verify that $\langle \bar{P}_i^{(\alpha,\beta)}, (rx+s)\frac{\mathrm{d}}{\mathrm{d}x}\bar{P}_j^{(\alpha,\beta)}\rangle = 0$ whenever $j<i$. This implies that $\mathbf{G}$ must be an upper triangular matrix. For $j\geq i$, we have

$$\begin{aligned}
g_{i,j} &= \frac{\langle \bar{P}_i^{(\alpha,\beta)}, (rx+s)\frac{\mathrm{d}}{\mathrm{d}x}\bar{P}_j^{(\alpha,\beta)}\rangle}{\langle \bar{P}_i^{(\alpha,\beta)}, \bar{P}_i^{(\alpha,\beta)}\rangle}\\
&= \frac{\bar{A}_j\sum_{k=0}^{j-1}\bar{B}_k\langle \bar{P}_i^{(\alpha,\beta)}, (rx+s)\bar{P}_k^{(\alpha,\beta)}\rangle + \bar{C}_j\sum_{k=0}^{j-1}\bar{D}_k\langle \bar{P}_i^{(\alpha,\beta)}, (rx+s)\bar{P}_k^{(\alpha,\beta)}\rangle}{\langle \bar{P}_i^{(\alpha,\beta)}, \bar{P}_i^{(\alpha,\beta)}\rangle}\\
&= \frac{\bar{A}_j\sum_{k=0}^{j-1}\bar{B}_k\langle \bar{P}_i^{(\alpha,\beta)}, r\bar{a}'_k\bar{P}_{k+1}^{(\alpha,\beta)} + (r\bar{b}'_k+s)\bar{P}_k^{(\alpha,\beta)} + r\bar{c}'_k\bar{P}_{k-1}^{(\alpha,\beta)})\rangle}{\langle \bar{P}_i^{(\alpha,\beta)}, \bar{P}_i^{(\alpha,\beta)}\rangle}\\
&\quad + \frac{\bar{C}_j\sum_{k=0}^{j-1}\bar{D}_k\langle \bar{P}_i^{(\alpha,\beta)}, r\bar{a}'_k\bar{P}_{k+1}^{(\alpha,\beta)} + (r\bar{b}'_k+s)\bar{P}_k^{(\alpha,\beta)} + r\bar{c}'_k\bar{P}_{k-1}^{(\alpha,\beta)})\rangle}{\langle \bar{P}_i^{(\alpha,\beta)}, \bar{P}_i^{(\alpha,\beta)}\rangle}\\
&= \begin{cases}
r(\bar{A}_j\bar{B}_{j-1}\bar{a}'_{j-1} + \bar{C}_j\bar{D}_{j-1}\bar{a}'_{j-1}), & \text{if } i=j,\\
\bar{A}_j\big(r(\bar{B}_{j-2}\bar{a}'_{j-2} + \bar{B}_{j-1}\bar{b}'_{j-1}) + s\bar{B}_{j-1}\big) & \\
\qquad +\bar{C}_j\big(r(\bar{D}_{j-2}\bar{a}'_{j-2} + \bar{D}_{j-1}\bar{b}'_{j-1}) + s\bar{D}_{j-1}\big), & \text{if } i=j-1,\\
\bar{A}_j\big(r(\bar{B}_{i-1}\bar{a}'_{i-1} + \bar{B}_i\bar{b}'_i + \bar{B}_{i+1}\bar{c}'_{i+1}) + s\bar{B}_i\big) & \\
\qquad +\bar{C}_j\big(r(\bar{D}_{i-1}\bar{a}'_{i-1} + \bar{D}_i\bar{b}'_i + \bar{D}_{i+1}\bar{c}'_{i+1}) + s\bar{D}_i\big), & \text{if } i<j-1.
\end{cases}
\end{aligned}$$

Observe that (1.39) implies

$$\bar{A}_j\bar{B}_j + \bar{C}_j\bar{D}_j = 0.$$

This allows us to absorb the second case into the third. Thus,

$$g_{i,j} = \begin{cases}
r(\bar{A}_j\bar{B}_{j-1}\bar{a}'_{j-1} + \bar{C}_j\bar{D}_{j-1}\bar{a}'_{j-1}), & \text{if } i=j,\\
\bar{A}_j\big(r(\bar{B}_{i-1}\bar{a}'_{i-1} + \bar{B}_i\bar{b}'_i + \bar{B}_{i+1}\bar{c}'_{i+1}) + s\bar{B}_i\big) & \\
\qquad +\bar{C}_j\big(r(\bar{D}_{i-1}\bar{a}'_{i-1} + \bar{D}_i\bar{b}'_i + \bar{D}_{i+1}\bar{c}'_{i+1}) + s\bar{D}_i\big), & \text{if } i<j,
\end{cases}$$

which proves the desired representation of the matrix $\mathbf{G}$. □

The following result is for Gegenbauer polynomials. It can be obtained from Lemma 2.16, but for the sake of completeness, the proof is also given.

Lemma 2.17 *Let $\{\bar{C}_n^{(\alpha)}\}_{n\in\mathbb{N}_0}$ be the monic Gegenbauer polynomials which are orthogonal with respect to an inner product $\langle\,\cdot\,,\,\cdot\,\rangle$ and which satisfy the derivative identity*

$$\frac{\mathrm{d}}{\mathrm{d}x}\bar{C}_n^{(\alpha)}(x) = \bar{A}_n \sum_{i=0}^{\lfloor (n-1)/2\rfloor} \bar{B}_{2i+\chi}\bar{C}_{2i+\chi}^{(\alpha)}(x),$$

with certain numbers $\bar{A}_n$ and $\bar{B}_{2i+\chi}$, cf. Corollary 1.76, where $\chi = \chi(n)$ is given by

$$\chi = \begin{cases} 1, & \text{if } n \text{ even},\\ 0, & \text{if } n \text{ odd}.\end{cases}$$

They also satisfy a three-term recurrence of the form

$$x\bar{C}_n^{(\alpha)}(x) = \bar{a}_n'\bar{C}_{n+1}^{(\alpha)} + \bar{c}_n'\bar{C}_{n-1}^{(\alpha)}(x).$$

Then the matrix $\mathbf{G} = (g_{i,j})_{i,j=0}^n$, defined by

$$g_{i,j} = \frac{\langle \bar{C}_i^{(\alpha)}, x\frac{\mathrm{d}}{\mathrm{d}x}\bar{C}_j^{(\alpha)}\rangle}{\langle \bar{C}_i^{(\alpha)}, \bar{C}_i^{(\alpha)}\rangle}, \qquad \text{with } i,j = 0,1,\ldots,n,$$

is checkerboard-like diagonal plus upper (1)-generator representable semiseparable and may be represented as

$$\mathbf{G} = \operatorname{diag}(\mathbf{d}) + \operatorname{triuc}(\mathbf{u}\,\mathbf{v}^{\mathrm{T}}, 1),$$

with the vectors $\mathbf{d}$, $\mathbf{u}$, and $\mathbf{v}$ given by

$$\mathbf{d} = \left(\bar{A}_j\bar{B}_{j-1}\bar{a}_{j-1}'\right)_{j=0}^n, \qquad \mathbf{u}_1 = \left(\bar{B}_{i-1}\bar{a}_{i-1}' + \bar{B}_{i+1}\bar{c}_{i+1}'\right)_{i=0}^n, \qquad \mathbf{v}_1 = (\bar{A}_j)_{j=0}^n.$$

Proof. By orthogonality and the fact that Gegenbauer polynomials are orthogonal with respect to a symmetric measure, we verify that $\langle \bar{C}_i^{(\alpha)}, x\frac{\mathrm{d}}{\mathrm{d}x}\bar{C}_j^{(\alpha)}\rangle = 0$ whenever $j < i$ or $j+i$ odd. This implies that $\mathbf{G}$ must be a checkerboard-like upper triangular matrix. If $j \geq i$ and $j+i$ even, we have

$$\begin{aligned} g_{i,j} &= \frac{\langle \bar{C}_i^{(\alpha)}, x\frac{\mathrm{d}}{\mathrm{d}x}\bar{C}_j^{(\alpha)}\rangle}{\langle \bar{C}_i^{(\alpha)}, \bar{C}_i^{(\alpha)}\rangle}\\ &= \frac{\bar{A}_j\sum_{k=0}^{\lfloor (j-1/2)\rfloor}\bar{B}_{2k+\chi}\langle \bar{C}_i^{(\alpha)}, x\bar{C}_{2k+\chi}^{(\alpha)}\rangle}{\langle \bar{C}_i^{(\alpha)}, \bar{C}_i^{(\alpha)}\rangle}\\ &= \frac{\bar{A}_j\sum_{k=0}^{\lfloor (j-1/2)\rfloor}\bar{B}_{2k+\chi}\langle \bar{C}_i^{(\alpha)}, \bar{a}_{2k+\chi}'\bar{C}_{2k+\chi+1}^{(\alpha)} + \bar{c}_{2k+\chi}'\bar{C}_{2k+\chi-1}^{(\alpha)}\rangle}{\langle \bar{C}_i^{(\alpha)}, \bar{C}_i^{(\alpha)}\rangle}\\ &= \begin{cases} \bar{A}_j\bar{B}_{j-1}\bar{a}_{j-1}', & \text{if } i = j,\\ \bar{A}_j\bar{B}_{i-1}\bar{a}_{i-1}' + \bar{A}_j\bar{B}_{i+1}\bar{c}_{i+1}', & \text{if } i < j.\end{cases} \end{aligned}$$

This can be matched with the desired representation of the matrix $\mathbf{G}$. □

2.4.2 Connection matrices and semiseparable matrices. We are now ready to turn to the conversion of expansion coefficients between different families of classical orthogonal polynomials. This will be restricted to polynomials of the same type, for example, from one sequence of Laguerre polynomials to another but not from Laguerre polynomials to Jacobi polynomials. Suppose that a degree-n polynomial f has been expanded into a sequence of orthogonal polynomials $\{p_n\}_{n\in\mathbb{N}_0}$,

$$f = \sum_{j=0}^{n} x_j p_j,$$

with known expansion coefficients x_j. Then we want to compute the coefficients y_j in the expansion

$$f = \sum_{j=0}^{n} y_j q_j,$$

where $\{q_n\}_{n\in\mathbb{N}_0}$ is another sequence of orthogonal polynomials different from the first one. If the coefficients x_j and y_j are collected in the two vectors $\mathbf{x}$ and $\mathbf{y}$, i.e.,

$$\mathbf{x} = (x_0, x_1, \ldots, x_n)^{\mathrm{T}}, \qquad \mathbf{y} = (y_0, y_1, \ldots, y_n)^{\mathrm{T}},$$

then $\mathbf{y}$ can be obtained from $\mathbf{x}$ via the matrix-vector product

$$\mathbf{y} = \mathbf{K}\,\mathbf{x},$$

where $\mathbf{K} = (\kappa_{i,j})_{i,j=0}^{n}$ is the connection matrix between the polynomial sequences $\{p_n\}_{n\in\mathbb{N}_0}$ and $\{q_n\}_{n\in\mathbb{N}_0}$; see Section 1.4. This can be verified by

$$\sum_{j=0}^{n} \kappa_{i,j} x_j = \sum_{j=0}^{n} \frac{\langle q_i, p_j\rangle}{\langle q_i, q_i\rangle} x_j = \frac{\langle q_i, \sum_{j=0}^{n} x_j p_j\rangle}{\langle q_i, q_i\rangle} = \frac{\langle q_i, f\rangle}{\langle q_i, Q_i\rangle} = y_i$$

with $i = 0, 1, \ldots, n$, where $\langle \cdot\,, \cdot\rangle$ is the inner product with respect to which the target polynomials $\{q_n\}_{n\in\mathbb{N}_0}$ are orthogonal.
To handle the conversion efficiently, we need to devise an efficient method to apply the connection matrix $\mathbf{K}$ to a vector. This will be done in two steps. First, a matrix $\mathbf{G}$ will be defined that has its eigenvector matrix $\mathbf{Q}$ identical to the desired connection matrix $\mathbf{K}$, provided that the columns of $\mathbf{Q}$ have been properly scaled. Second, it will be observed that the matrix $\mathbf{G}$ has a semiseparable structure that allows the application of the divide-and-conquer method from Section 2.3. To this end, we will calculate the entries of the matrix $\mathbf{G}$ explicitly. In total, we thus obtain a practical method to cheaply apply the connection matrix $\mathbf{K}$ to any vector with $\mathcal{O}\big(n \log n \log(1/\varepsilon)\big)$ arithmetic operations instead of the usual $\mathcal{O}(n^2)$, where ε is the desired accuracy.

Definition 2.18 *Let $\{p_n\}_{n\in\mathbb{N}_0}$ be a sequence of classical orthogonal polynomials which satisfy a differential equation of hypergeometric type,*

$$\sigma(x)p_n''(x) + \tau(x)p_n'(x) + \lambda_n p_n(x) = 0,$$

with the corresponding differential operator $\mathcal{D}$ given by

$$\mathcal{D} = -\sigma\frac{\mathrm{d}^2}{\mathrm{d}x^2} - \tau\frac{\mathrm{d}}{\mathrm{d}x}.$$

Let $\{q_n\}_{n\in\mathbb{N}_0}$ be a different sequence of classical polynomials, orthogonal with respect to the inner product $\langle \cdot\,, \cdot\rangle$. Then the matrix

$$\mathbf{G} = (g_{i,j})_{i,j=0}^{n}$$

is defined by

$$g_{i,j} = \frac{\langle q_i, \mathcal{D}(q_j)\rangle}{\langle q_i, q_i\rangle}. \tag{2.12}$$

Note that here, the operator $\mathcal{D}$ belongs to the polynomial sequence $\{p_n\}_{n\in\mathbb{N}_0}$, but is applied to the polynomials q_n. In the following, we will adopt the same notation for the matrix $\mathbf{G}$ that was used for the connection matrices $\mathbf{K}$ when it comes to concrete pairs of classical orthogonal polynomials. For example, the matrix $\mathbf{G}$ for two sequences of monic Laguerre polynomials, $\{L_n^{(\alpha)}\}_{n\in\mathbb{N}_0}$ and $\{L_n^{(\beta)}\}_{n\in\mathbb{N}_0}$, will be denoted $\bar{\mathbf{G}}^{\mathrm{L},(\alpha)\to(\beta)}$ or $\bar{\mathbf{G}}^{(\alpha)\to(\beta)}$, if it is clear from the context that we mean the Laguerre polynomials. With the definition of the matrix $\mathbf{G}$, we can now show that the connection matrix $\mathbf{K}$ contains its eigenvectors.

Lemma 2.19 *Let $\{p_n\}_{n\in\mathbb{N}_0}$ and $\{q_n\}_{n\in\mathbb{N}_0}$ be two sequences of classical orthogonal polynomials and let the matrix $\mathbf{G}$ be defined as before. Then for $j = 0, 2, \ldots, n$, any column $\boldsymbol{\kappa}_j$ of the connection matrix $\mathbf{K}$ between the two sequences is an eigenvector of the matrix $\mathbf{G}$ to the eigenvalue λ_j.*

Proof. Recall that we have $p_j = \sum_{i=0}^{j} \kappa_{i,j} q_i$. We denote by $(\mathbf{G}\,\boldsymbol{\kappa}_j)_i$ the $(i+1)$st component of the vector product $\mathbf{G}\boldsymbol{\kappa}_j$, where $\boldsymbol{\kappa}_j = (\kappa_{0,j}, \kappa_{1,j}, \ldots, \kappa_{n,j})^{\mathrm{T}}$ is the $(j+1)$st column of the connection matrix $\mathbf{K}$. Then we have

$$(\mathbf{G}\,\boldsymbol{\kappa}_j)_i = \sum_{k=0}^{n} g_{i,k}\kappa_{k,j} = \sum_{k=0}^{n} \frac{\langle q_i, \mathcal{D}(q_k)\rangle}{\langle q_i, q_i\rangle}\kappa_{k,j} = \frac{\langle q_i, \mathcal{D}\big(\sum_{k=0}^{j}\kappa_{k,j}q_k\big)\rangle}{\langle q_i, q_i\rangle} = \frac{\langle q_i, \mathcal{D}(p_j)\rangle}{\langle q_i, q_i\rangle}.$$

We know that the polynomial p_j is an eigenfunction of the differential operator $\mathcal{D}$ to the eigenvalue λ_j, that is, we can replace $\mathcal{D}(p_j) = \lambda_j p_j$. Then we work backward until we obtain

$$(\mathbf{G}\,\boldsymbol{\kappa}_j)_i = \lambda_j \sum_{k=0}^{n} g_{i,k}\kappa_{k,j} = \lambda_j\,(\boldsymbol{\kappa}_j)_i\,.$$

□

Remark 2.20 It is clear that the eigenvalues λ_j of the matrix $\mathbf{G}$ are simple, since these are distinct for all classical orthogonal polynomials. This was a requirement for the application of the divide-and-conquer algorithm from Section 2.3.

Knowing that the columns of the connection matrix $\mathbf{K}$ are eigenvectors of the matrix $\mathbf{G}$ does not readily yield a fast method to apply $\mathbf{K}$ to a vector. The idea to get such a method is to invoke the divide-and-conquer algorithm from Section 2.3 on the matrix $\mathbf{G}$. This allows to cheaply apply the eigenvector matrix of $\mathbf{G}$, that is, the connection matrix $\mathbf{K}$, to any vector.

There are two issues left that need to be resolved before this can be used. First, we need explicit expressions for the entries of the matrix $\mathbf{G}$ so that it can be fed to the divide-and-conquer algorithm. Second, we must work out how the eigenvectors of the matrix $\mathbf{G}$ should be scaled to make them coincide with the columns of the connection matrix $\mathbf{K}$. Of course, the entries of the matrix $\mathbf{G}$ depend on the parameters associated with each of the sequences of polynomials, $\{p_n\}_{n\in\mathbb{N}_0}$ and $\{q_n\}_{n\in\mathbb{N}_0}$. As we will see, the particular scaling of the target polynomials q_n is already encoded in the matrix $\mathbf{G}$. The scaling of the source polynomials p_n, however, is reflected in the appropriate scaling of the columns of the eigenvector matrix of $\mathbf{G}$.

2.4.3 Examples. In the following, explicit expressions for the entries of the matrix $\mathbf{G}$ for monic Laguerre, Jacobi, and Gegenbauer polynomials are given. For other normalizations, see the Appendix A.

Laguerre polynomials

Theorem 2.21 *Let* $\{\bar{L}_n^{(\alpha)}\}_{n\in\mathbb{N}_0}$ *and* $\{\bar{L}_n^{(\beta)}\}_{n\in\mathbb{N}_0}$ *with* $\alpha, \beta > -1$ *be two sequences of monic Laguerre polynomials. Then the corresponding matrix* $\bar{\mathbf{G}}^{(\alpha)\to(\beta)}$ *is diagonal plus upper* (1)*-generator representable semiseparable,*

$$\bar{\mathbf{G}}^{(\alpha)\to(\beta)} = \operatorname{diag}(\mathbf{d}) + (\beta-\alpha)\operatorname{triu}(\mathbf{u}\,\mathbf{v}^{\mathrm{T}}, 1),$$

with the vectors $\mathbf{d}$*,* $\mathbf{u}$*, and* $\mathbf{v}$ *given by*

$$\mathbf{d} = (j)_{j=0}^{n}, \qquad \mathbf{u} = \left(\frac{(-1)^j}{\Gamma(j+1)}\right)_{j=0}^{n}, \qquad \mathbf{v} = \left((-1)^{j+1}\Gamma(j+1)\right)_{j=0}^{n}.$$

PROOF. The polynomials $\bar{L}_n^{(\alpha)}$ satisfy the differential equation

$$xy''(x) + (1+\alpha-x)y'(x) + ny(x) = 0, \qquad \text{with } y = \bar{L}_n^{(\alpha)},$$

and similarly the polynomials $\bar{L}_n^{(\beta)}$ satisfy

$$xy''(x) + (1+\beta-x)y'(x) + ny(x) = 0, \qquad \text{with } y = \bar{L}_n^{(\beta)}.$$

Thus, the corresponding differential operators are given by

$$\mathcal{D}^{(\alpha)} = -x\frac{\mathrm{d}^2}{\mathrm{d}x^2} - (1+\alpha-x)\frac{\mathrm{d}}{\mathrm{d}x}, \qquad \mathcal{D}^{(\beta)} = -x\frac{\mathrm{d}^2}{\mathrm{d}x^2} - (1+\beta-x)\frac{\mathrm{d}}{\mathrm{d}x},$$

and we have

$$\mathcal{D}^{(\alpha)} - \mathcal{D}^{(\beta)} = (\beta-\alpha)\frac{\mathrm{d}}{\mathrm{d}x}.$$

This implies

$$\begin{aligned} g_{i,j} &= \frac{\langle L_i^{(\beta)}, \mathcal{D}^{(\alpha)}(L_j^{(\beta)})\rangle}{\langle L_i^{(\beta)}, L_i^{(\beta)}\rangle} \\ &= \frac{\langle L_i^{(\beta)}, \mathcal{D}^{(\beta)}(L_j^{(\beta)})\rangle}{\langle L_i^{(\beta)}, L_i^{(\beta)}\rangle} + \frac{\langle L_i^{(\beta)}, (\mathcal{D}^{(\alpha)} - \mathcal{D}^{(\beta)})(L_j^{(\beta)})\rangle}{\langle L_i^{(\beta)}, L_i^{(\beta)}\rangle}, \\ &= \frac{\langle L_i^{(\beta)}, \mathcal{D}^{(\beta)}(L_j^{(\beta)})\rangle}{\langle L_i^{(\beta)}, L_i^{(\beta)}\rangle} + (\beta-\alpha)\frac{\langle L_i^{(\beta)}, \frac{\mathrm{d}}{\mathrm{d}x}L_j^{(\beta)}\rangle}{\langle L_i^{(\beta)}, L_i^{(\beta)}\rangle}. \end{aligned}$$

Since the polynomial $\bar{L}_j^{(\beta)}$ is an eigenfunction of the operator $\mathcal{D}^{(\beta)}$ to the eigenvalue $\lambda_j = j$, we get for the first summand in the last expression

$$\frac{\langle L_i^{(\beta)}, \mathcal{D}^{(\beta)}(L_j^{(\beta)})\rangle}{\langle L_i^{(\beta)}, L_i^{(\beta)}\rangle} = j\frac{\langle L_i^{(\beta)}, L_j^{(\beta)}\rangle}{\langle L_i^{(\beta)}, L_i^{(\beta)}\rangle} = j\delta_{i,j}.$$

For the second summand, note that with the help of Lemma 2.14 and the derivative identity from Theorem 1.75, it is clear that

$$\frac{\langle L_i^{(\beta)}, \frac{\mathrm{d}}{\mathrm{d}x}L_j^{(\beta)}\rangle}{\langle L_i^{(\beta)}, L_i^{(\beta)}\rangle} = \begin{cases} (-1)^{i+j+1}\dfrac{\Gamma(j+1)}{\Gamma(i+1)}, & \text{if } i < j. \\ 0, & \text{else.} \end{cases}$$

This proves the desired representation of the matrix $\bar{\mathbf{G}}^{(\alpha)\to(\beta)}$. □

Jacobi polynomials

Theorem 2.22 *Let $\{\bar{P}_n^{(\alpha,\beta)}\}_{n\in\mathbb{N}_0}$ and $\{\bar{P}_n^{(\gamma,\delta)}\}_{n\in\mathbb{N}_0}$ with $\alpha,\beta,\gamma,\delta > -1$ be two sequences of monic Jacobi polynomials. Then the corresponding matrix $\bar{\mathbf{G}}^{(\alpha,\beta)\to(\gamma,\delta)}$ is diagonal plus upper (2)-generator representable semiseparable,*

$$\bar{\mathbf{G}}^{(\alpha,\beta)\to(\gamma,\delta)} = \operatorname{diag}(\mathbf{d}) + (\alpha-\gamma)\operatorname{triu}(\mathbf{u}_1\,\mathbf{v}_1^{\mathrm{T}},1) + (\beta-\delta)\operatorname{triu}(\mathbf{u}_2\,\mathbf{v}_2^{\mathrm{T}},1),$$

with the vectors $\mathbf{d}$, $\mathbf{u}_1$, $\mathbf{v}_1$, $\mathbf{u}_2$, and $\mathbf{v}_2$ given by

$$\mathbf{d} = \big(j(j+\alpha+\beta+1)\big)_{j=0}^{n},$$

$$\mathbf{u}_1 = \left(\frac{\Gamma(2j+\gamma+\delta+2)}{2^j\Gamma(j+1)\Gamma(j+\delta+1)}\right)_{j=0}^{n}, \qquad \mathbf{v}_1 = \left(\frac{2^j\Gamma(j+1)\Gamma(j+\delta+1)}{\Gamma(2j+\gamma+\delta+1)}\right)_{j=0}^{n},$$

$$\mathbf{u}_2 = \left(\frac{(-1)^j\Gamma(2j+\gamma+\delta+2)}{2^j\Gamma(j+1)\Gamma(j+\gamma+1)}\right)_{j=0}^{n}, \qquad \mathbf{v}_2 = \left(\frac{2^j\Gamma(j+1)\Gamma(j+\gamma+1)}{(-1)^j\Gamma(2j+\gamma+\delta+1)}\right)_{j=0}^{n}.$$

PROOF. The Jacobi polynomials $\bar{P}_n^{(\alpha,\beta)}$ satisfy the differential equation

$$(1-x^2)y''(x) - \big((\alpha+\beta+2)x+\alpha-\beta\big)y'(x) + n(n+\alpha+\beta+1)y(x) = 0,$$

with $y = \bar{P}_n^{(\alpha,\beta)}$, and similarly the Jacobi polynomials $\bar{P}_n^{(\gamma,\delta)}$ satisfy

$$(1-x^2)y''(x) - \big((\gamma+\delta+2)x+\gamma-\delta\big)y'(x) + n(n+\gamma+\delta+1)y(x) = 0,$$

with $y = \bar{P}_n^{(\gamma,\delta)}$. Thus, the corresponding differential operators are given by

$$\begin{aligned}\mathcal{D}^{(\alpha,\beta)} &= -(1-x^2)\frac{\mathrm{d}^2}{\mathrm{d}x^2} + ((\alpha+\beta+2)x+\alpha-\beta)\frac{\mathrm{d}}{\mathrm{d}x},\\ \mathcal{D}^{(\gamma,\delta)} &= -(1-x^2)\frac{\mathrm{d}^2}{\mathrm{d}x^2} + ((\gamma+\delta+2)x+\gamma-\delta)\frac{\mathrm{d}}{\mathrm{d}x},\end{aligned}$$

and we have

$$\mathcal{D}^{(\alpha,\beta)} - \mathcal{D}^{(\gamma,\delta)} = \big((\alpha+\beta-\gamma-\delta)x+\alpha-\beta-\gamma+\delta\big)\frac{\mathrm{d}}{\mathrm{d}x}.$$

This implies

$$\begin{aligned} g_{i,j} &= \frac{\langle \bar{P}_i^{(\gamma,\delta)}, \mathcal{D}^{(\alpha,\beta)}(\bar{P}_j^{(\gamma,\delta)})\rangle}{\langle \bar{P}_i^{(\gamma,\delta)}, \bar{P}_i^{(\gamma,\delta)}\rangle}\\ &= \frac{\langle \bar{P}_i^{(\gamma,\delta)}, \mathcal{D}^{(\gamma,\delta)}(\bar{P}_j^{(\gamma,\delta)})\rangle}{\langle \bar{P}_i^{(\gamma,\delta)}, \bar{P}_i^{(\gamma,\delta)}\rangle} + \frac{\langle \bar{P}_i^{(\gamma,\delta)}, (\mathcal{D}^{(\alpha,\beta)}-\mathcal{D}^{(\gamma,\delta)})(\bar{P}_j^{(\gamma,\delta)})\rangle}{\langle \bar{P}_i^{(\gamma,\delta)}, \bar{P}_i^{(\gamma,\delta)}\rangle},\\ &= \frac{\langle \bar{P}_i^{(\gamma,\delta)}, \mathcal{D}^{(\gamma,\delta)}(\bar{P}_j^{(\gamma,\delta)})\rangle}{\langle \bar{P}_i^{(\gamma,\delta)}, \bar{P}_i^{(\gamma,\delta)}\rangle} + \frac{\langle \bar{P}_i^{(\gamma,\delta)}, \big((\alpha+\beta-\gamma-\delta)x+\alpha-\beta-\gamma+\delta\big)\frac{\mathrm{d}}{\mathrm{d}x}\bar{P}_j^{(\gamma,\delta)}\rangle}{\langle \bar{P}_i^{(\gamma,\delta)}, \bar{P}_i^{(\gamma,\delta)}\rangle}.\end{aligned}$$

Since the polynomial $\bar{P}_j^{(\gamma,\delta)}$ is an eigenfunction of the operator $\mathcal{D}^{(\gamma,\delta)}$ to the eigenvalue $\lambda_j = j(j+\gamma+\delta+1)$, we verify for the first summand that

$$\frac{\langle \bar{P}_i^{(\gamma,\delta)}, \mathcal{D}^{(\gamma,\delta)}(\bar{P}_j^{(\gamma,\delta)})\rangle}{\langle \bar{P}_i^{(\gamma,\delta)}, \bar{P}_i^{(\gamma,\delta)}\rangle} = j(j+\gamma+\delta+1)\frac{\langle \bar{P}_i^{(\gamma,\delta)}, \bar{P}_j^{(\gamma,\delta)}\rangle}{\langle \bar{P}_i^{(\gamma,\delta)}, \bar{P}_i^{(\gamma,\delta)}\rangle} = j(j+\gamma+\delta+1)\delta_{i,j}.$$

For the second part, we note that with the help of Lemma 2.14 and the derivative identity from Theorem 1.75, it is clear that

$$\frac{\langle \bar{P}_i^{(\gamma,\delta)}, ((\alpha+\beta-\gamma-\delta)x+\alpha-\beta-\gamma+\delta)\frac{\mathrm{d}}{\mathrm{d}x}\bar{P}_j^{(\gamma,\delta)}\rangle}{\langle \bar{P}_i^{(\gamma,\delta)}, \bar{P}_i^{(\gamma,\delta)}\rangle}$$

$$= \begin{cases} j(\alpha+\beta-\gamma-\delta), & \text{if } i=j, \\ (\alpha-\gamma)\dfrac{2^j\Gamma(j+1)\Gamma(j+\delta+1)\Gamma(2i+\gamma+\delta+2)}{2^i\Gamma(i+1)\Gamma(i+\delta+1)\Gamma(2j+\gamma+\delta+1)}+ & \\ (-1)^{i+j}(\beta-\delta)\dfrac{2^j\Gamma(j+1)\Gamma(j+\gamma+1)\Gamma(2i+\gamma+\delta+2)}{2^i\Gamma(i+1)\Gamma(i+\gamma+1)\Gamma(2j+\gamma+\delta+1)}, & \text{if } i<j. \\ 0, & \text{else.} \end{cases}$$

This proves the desired representation of the matrix $\bar{\mathbf{G}}^{(\alpha,\beta)\to(\gamma,\delta)}$. □

Gegenbauer polynomials

Theorem 2.23 *Let $\{\bar{C}_n^{(\alpha)}\}_{n\in\mathbb{N}_0}$ and $\{\bar{C}_n^{(\beta)}\}_{n\in\mathbb{N}_0}$, with $\alpha,\beta>-1/2$, be two sequences of monic Gegenbauer polynomials. Then the corresponding matrix $\bar{\mathbf{G}}^{(\alpha)\to(\beta)}$ is checkerboard-like diagonal plus upper (1)-generator representable semiseparable,*

$$\bar{\mathbf{G}}^{(\alpha)\to(\beta)} = \operatorname{diag}(\mathbf{d}) + 4(\alpha-\beta)\operatorname{triuc}(\mathbf{u}\,\mathbf{v}^{\mathrm{T}},1),$$

with the vectors $\mathbf{d}$, $\mathbf{u}$, and $\mathbf{v}$ given by

$$\mathbf{d} = \big(j(j+2\alpha)\big)_{j=0}^n,$$

$$\mathbf{u} = \left(\frac{2^j\Gamma(j+\beta+1)}{\Gamma(j+1)}\right)_{j=0}^n, \qquad \mathbf{v} = \left(\frac{\Gamma(j+1)}{2^j\Gamma(j+\beta)}\right)_{j=0}^n.$$

PROOF. Theorem 2.22 for the Gegenbauer case yields

$$\mathbf{d} = \big(j(j+2\alpha)\big)_{j=0}^n,$$

$$\mathbf{u} = \left(\frac{\Gamma(2j+2\beta+1)}{2^j\Gamma(j+1)\Gamma(j+\beta+1/2)}\right)_{j=0}^n, \qquad \mathbf{v} = \left(\frac{2^j\Gamma(j+1)\Gamma(j+\beta+1/2)}{\Gamma(2j+2\beta)}\right)_{j=0}^n.$$

The well known identity $\Gamma(2z) = 2^{2z-1}\Gamma(z)\Gamma(z+1/2)/\sqrt{\pi}$ implies the desired representation. □

We omit construction the matrix $\mathbf{G}$ for the conversion between Chebyshev and Legendre polynomials since these can be obtained as special cases that can be obtained from the last result.

2.4.4 Scaling the eigenvectors. We have calculated the generator representation of the matrix $\mathbf{G}$. As mentioned before, the scaling of the target polynomials is already encoded in this representation, but we still need side conditions to ensure the desired scaling of the source polynomials $\{p_n\}_{n\in\mathbb{N}_0}$. Clearly, changing the scaling of the source polynomials is equivalent to rescaling the columns of the eigenvector matrix of $\mathbf{G}$, that is, the desired connection matrix $\mathbf{K}$. The following result provides the condition that allows for the correct scaling of these columns.

Lemma 2.24 *Let $\{p_n\}_{n\in\mathbb{N}_0}$ and $\{q_n\}_{n\in\mathbb{N}_0}$ be two families of orthogonal polynomials that have, respectively, k_n and $\hat{k}_n$ as their leading coefficients. Then the diagonal entries $\kappa_{i,i}$,* *for $i = 0, 1, \ldots,$ in the connection matrix $\mathbf{K} = (\kappa_{i,j})$ between the sequence $\{p_n\}_{n\in\mathbb{N}_0}$ and the sequence $\{q_n\}_{n\in\mathbb{N}_0}$ are given by*

$$\kappa_{i,i} = \frac{k_i}{\hat{k}_i}.$$

PROOF. By Definition, we have

$$\kappa_{i,i} = \frac{\langle q_i, p_i\rangle}{\langle q_i, q_i\rangle}, \qquad \text{with } i = 0, 1, \ldots,$$

where $\langle \cdot\,, \cdot\rangle$ is the inner product with respect to which the target polynomials $\{q_n\}_{n\in\mathbb{N}_0}$ are orthogonal. Since the polynomial q_i is orthogonal to every polynomial of strictly smaller degree, this can be rewritten as

$$\kappa_{i,i} = \frac{\langle q_i, p_i\rangle}{\langle q_i, q_i\rangle} = \frac{k_i}{\hat{k}_i}\frac{\langle q_i, q_i\rangle}{\langle q_i, q_i\rangle} = \frac{k_i}{\hat{k}_i}.$$

□

2.4.5 Other normalizations. The expressions provided in this section are for the matrix $\bar{\mathbf{G}}$ which corresponds to the monic variants of the respective orthogonal polynomials. To obtain the expressions for the standard and the normalized variants, we can use that

$$\bar{g}_{i,j} = \frac{\langle \bar{q}_i, \mathcal{D}(\bar{q}_j)\rangle}{\langle \bar{q}_i, \bar{q}_i\rangle} = \frac{k_i^2}{k_i k_j}\frac{\langle q_i, \mathcal{D}(q_j)\rangle}{\langle q_i, q_i\rangle} = \frac{k_i}{k_j} g_{i,j},$$

$$g_{i,j} = \frac{\langle q_i, \mathcal{D}(q_j)\rangle}{\langle q_i, q_i\rangle} = \frac{\sqrt{h_i h_j}}{h_i}\frac{\langle \tilde{q}_i, \mathcal{D}(\tilde{q}_j)\rangle}{\langle \tilde{q}_i, \tilde{q}_i\rangle} = \sqrt{\frac{h_j}{h_i}}\tilde{g}_{i,j}.$$

This gives

$$g_{i,j} = \frac{k_j}{k_i}\bar{g}_{i,j}, \qquad \tilde{g}_{i,j} = \sqrt{\frac{h_i}{h_j}} g_{i,j}.$$

Here, as before, k_i and h_i denote the leading coefficients and squared norms, respectively, of the polynomials $\{q_n\}_{n\in\mathbb{N}_0}$ in the standard normalization.

2.4.6 Numerical results. We are ready to test the divide-and-conquer method from Section 2.3 for the computation of the connection between classical orthogonal polynomials. To this end, we evaluate several test cases for which we take as input the coefficients x_j from an expansion in a sequence of classical orthogonal polynomials $\{p_n\}_{n\in\mathbb{N}_0}$ of the form

$$f = \sum_{j=0}^{n} x_j p_j.$$

Then we compute as output the coefficients y_j in the expansion

$$f = \sum_{j=0}^{n} y_j q_j,$$

with a different sequence of classical orthogonal polynomials $\{q_n\}_{n\in\mathbb{N}_0}$. This is done by applying the appropriate connection matrix $\mathbf{K} = (\kappa_{i,j})_{i,j=0}^{n}$ to the vector $\mathbf{x} = (x_j)_{j=0}^{n}$ to obtain the vector $\mathbf{y} = (y_j)_{j=0}^{n}$, i.e.,

$$\mathbf{y} = \mathbf{K}\,\mathbf{x}.$$

We have a choice of different methods, for example the divide-and-conquer method from Section 2.3 which uses the results on the matrix $\mathbf{G}$ from the last section.

All methods have been implemented in C and tested on an Intel Core 2 Duo 2.66 GHz MacBook Pro with 4GB RAM running Mac OS X 10.6.3 in double precision arithmetic. We have used Apple's `llvm-gcc-4.2` compiler with the optimization options `-O3 -fomit-frame-pointer -malign-double -ffast-math -mtune=core2 -march=core2`. To measure time spans the CPU cycle counters interface from the popular FFTW library was used; see [22]. The input coefficients x_i were drawn from a uniform quasi-random distribution in the interval $[-1, 1]$ using the standard C function `drand48()`.

To assess the accuracy of the numerical results, we used the component-wise error measure E_∞^{c} and the relative infinity norm E_∞ which are given by

$$E_\infty^{\mathrm{c}} := \max_{i=0,\ldots,n} \frac{|y_i^* - y_i|}{|y_i^{\mathrm{a}}|}, \qquad E_\infty := \frac{\|\mathbf{y}^* - \mathbf{y}\|_\infty}{\|\mathbf{y}^*\|_\infty}, \tag{2.13}$$

with the vectors

$$\mathbf{y}^* = (y_i^*)_{i=0}^n, \quad \mathbf{y} = (y_i)_{i=0}^n, \quad \text{and} \quad \mathbf{y}^{\mathrm{a}} = (y_i^{\mathrm{a}})_{i=0}^n.$$

The vector $\mathbf{y}^*$ stands for reference results which we computed on a PowerBook G4 which supports a quadruple precision datatype. Compared to double precision, this means roughly 16 additional significant decimal figures in the calculation which should be enough to provide correctly rounded results in double precision for the tests conducted. We computed the reference values by calculating the entries of each connection matrix $\mathbf{K}$ using the explicit expressions found in Chapter 1. We then applied the matrix $\mathbf{K}$ directly to the input coefficients $\mathbf{x}$. Therefore, this will be called the *direct method* (in quadruple precision). The vector $\mathbf{y}$ contains the results computed in double precision with any algorithm that was tested. For the error measure E_∞^{c}, we also need the values

$$y_i^{\mathrm{a}} := \sum_{j=0}^{i} |\kappa_{i,j}||x_j|$$

that are defined according to the results y_i given by

$$y_i = \sum_{j=0}^{i} \kappa_{i,j} x_j. \tag{2.14}$$

The error E_∞^{c} thus reflects the standard error bound that holds for the computation of the values y_i according to (2.14); see [34]. Ideally, we would have $E_\infty^{\mathrm{c}} \approx \epsilon$, where ϵ is the machine epsilon, typically $\epsilon \approx 2 \times 10^{-16}$ in double precision. In detail, the tested algorithms were the following:

The direct method: Each connection matrix $\mathbf{K} = (\kappa_{i,j})_{i,j=0}^n$ is calculated in full using the expressions for its entries $\kappa_{i,j}$ derived in Chapter 1. Then the matrix $\mathbf{K}$ is applied to the input vector $\mathbf{x}$, that is, we compute $\mathbf{y} = \mathbf{K}\mathbf{x}$ the usual way.

The usmv-direct method: Each connection matrix $\mathbf{K}$ is calculated in full as the properly scaled eigenvector matrix of the corresponding upper generator representable semiseparable matrix $\mathbf{G}$ as derived in Section 2.4.2. To compute this eigendecomposition, a C-translated version of the LAPACK routine `dgeev`, which uses the QR algorithm, is used. Then the matrix $\mathbf{K}$ is applied to the input vector $\mathbf{x}$, that is, we compute $\mathbf{y} = \mathbf{K}\mathbf{x}$ the usual way.

The usmv-fmm method: The matrix-vector product $\mathbf{y} = \mathbf{K}\mathbf{x}$ is computed using the FMM-accelerated divide-and-conquer method for the eigendecomposition of upper generator representable semiseparable matrices from Section 2.3, applied to the corresponding matrix $\mathbf{G}$ as derived in Section 2.4.2. That is, all data necessary for the hierarchical representation of the matrix $\mathbf{K}$, see, e.g., (2.8), is calculated and

Table 2.1: Summary of arithmetic cost and memory requirements of the tested algorithms.

	direct	usmv-direct	usmv-fmm
Cost (pre-computation)	$\mathcal{O}(n^2)$	$\mathcal{O}(n^3)$	$\mathcal{O}(n^2)$ or $\mathcal{O}(n \log n \log(1/\varepsilon))$
Cost (transformation)	$\mathcal{O}(n^2)$	$\mathcal{O}(n^2)$	$\mathcal{O}(n \log n \log(1/\varepsilon))$
Memory (both)	$\mathcal{O}(n^2)$	$\mathcal{O}(n^2)$	$\mathcal{O}(n \log n \log(1/\varepsilon))$

stored beforehand. Then the matrix $\mathbf{K}$ is applied using the FMM for all Cauchy-like matrices that appear. All other matrix-vector products are carried out the usual way. The parameters were chosen to maximize performance on the system used for testing, that is, for matrices with dimensions smaller than 256×256 the eigendecomposition was computed with the routine `dgeev` as in the usmv-direct method, and the desired accuracy ε of the FMM was chosen to be comparable to the machine epsilon ϵ.

Each of the algorithms has a pre-computation stage where data is stored that needs to be calculated only once. The actual transformation consists of the steps that depend on the input coefficients. In the following, we analyze the different arithmetic cost and memory requirements of the various methods. This is also summarized in Table 2.1.

direct method: For a transform of size n, the direct method needs $\mathcal{O}(n^2)$ arithmetic operations for the pre-computation stage and an amount of memory in the same order. The transformation is computed with $\mathcal{O}(n^2)$ arithmetic operations.

usmv-direct method: The usmv-direct method has a higher arithmetic cost than the direct method, as one needs $\mathcal{O}(n^3)$ operations to calculate the necessary eigendecomposition of the matrix $\mathbf{G}$. This led to unacceptable time requirements for the pre-computation stage whenever $n > 4096$. In other terms, this method is equivalent to the direct method.

usmv-fmm method: The usmv-fmm method needs $\mathcal{O}(n^2)$ arithmetic operations for pre-computation, but the amount of memory needed is in the order $\mathcal{O}(n \log n \log(1/\varepsilon))$. The arithmetic cost of the pre-computation stage could actually be lowered to $\mathcal{O}(n \log n \log(1/\varepsilon))$ with the help of the FMM. This acceleration, however, was not implemented since this possibly requires some care to guarantee the desired accuracy of the pre-computed values. Otherwise, the error observed in the actual calculation of the transformation could be polluted by errors in the pre-computation. The transformation needs only $\mathcal{O}(n \log n \log(1/\varepsilon))$ arithmetic operations. Again, ε denotes the desired accuracy which can be controlled by the user.

We should note that we did not attempt to optimize the pre-computation stage, so the times observed there should be taken with care. The time needed to compute the actual transformation, however, should give a good indication of the different performance among the tested algorithms.
Tests were conducted for Laguerre, Jacobi, and Gegenbauer polynomials, where we note that the latter are just a special case of Jacobi polynomials. For each type, we selected a number of representative test cases, that is, different combinations of the respective parameters carried by the polynomials. For Jacobi polynomials $\{P_n^{(\alpha,\beta)}\}_{n \in \mathbb{N}_0}$, we considered only changes to the parameter α. Changes to the second parameter β are entirely equivalent by Lemma (1.49) and would give similar results. Changes to both parameters could be computed by combining the two cases, or, by working with the corresponding

matrix $\mathbf{G}$ directly, that has its semiseparability rank $p = 2$. Both variants, however, have the same arithmetic cost.

Test results

Figure 2.5 reports time measurements for the pre-computation stage and the actual transformation for Laguerre, Jacobi, and Gegenbauer polynomials for a single test case. The results numerically confirm the expected asymptotic bounds, perhaps not entirely for the usmv-fmm method when applied to Gegenbauer polynomials (last row), which still seems to be influenced by other factors. The usmv-fmm method is attractively fast at any stage, with a visible advantage when n is larger than 512.
More detailed results for a number of representative test cases can be found in Tables 2.2 through 2.13. Shown there are also the error measures E_∞^{c} and E_∞. The general conclusion that can be drawn is that these are close to the theoretical optimum in double precision for all tested methods. The methods based on the eigendecomposition of upper generator representable semiseparable matrices show very competitive results.
Closer inspection, however, reveals that there are one or two test cases for each type of orthogonal polynomials where the error is substantially larger for the usmv-fmm method compared to the rest. Notably, this occurs when the "distance" between the parameters of source and target polynomials is relatively large. This motivates closer investigation of the relationship between this "distance" and the accuracy achieved by the usmv-fmm method. Of course, it is no problem to split a single transformation into a suitable number of other transformations that realize "smaller" steps to together compute the same result. For example, instead of transforming from, say, the polynomials $\{L_n^{(9.7)}\}_{n\in\mathbb{N}_0}$ to the polynomials $\{L_n^{(5.5)}\}_{n\in\mathbb{N}_0}$, we can equally first transform from $\{L_n^{(9.7)}\}_{n\in\mathbb{N}_0}$ to the intermediate polynomials $\{L_n^{(7.5)}\}_{n\in\mathbb{N}_0}$, and then from there to the target polynomials $\{L_n^{(5.5)}\}_{n\in\mathbb{N}_0}$. This is easily implemented, but changes the asymptotic behavior of the usmv-fmm method. For Laguerre polynomials, for example, the transformation from the polynomials $\{L_n^{(\alpha)}\}_{n\in\mathbb{N}_0}$ to the polynomials $\{L_n^{(\beta)}\}_{n\in\mathbb{N}_0}$ is then be computed with $\mathcal{O}(|\alpha-\beta|n\log n\log(1/\varepsilon))$ arithmetic operations instead of $\mathcal{O}(n\log n\log(1/\varepsilon))$, if the "distance", in terms of the involved parameters, between the source and target polynomials for a single transformation is kept below a fixed upper bound $s > 0$. The amount of memory needed rises analogously. But the splitting of a single larger step into many smaller steps indeed improves the achieved accuracy and makes it comparable again to the other methods. This is evident from the results in Tables 2.5, 2.9, and 2.13 compared to those in Tables 2.4, 2.8, and 2.12. We found a value of $s = 1$ to provide the optimal balance between numerical stability and performance in our tests. See Figure 2.6 for a comparison of the error measure E_∞^{c} and the time to compute the transformation versus the step size s for the mentioned scenario. The splitting provides an instant remedy for the infelicities observed for larger steps, but this comes at the expense of linearly more arithmetic operations and memory as the step size grows.

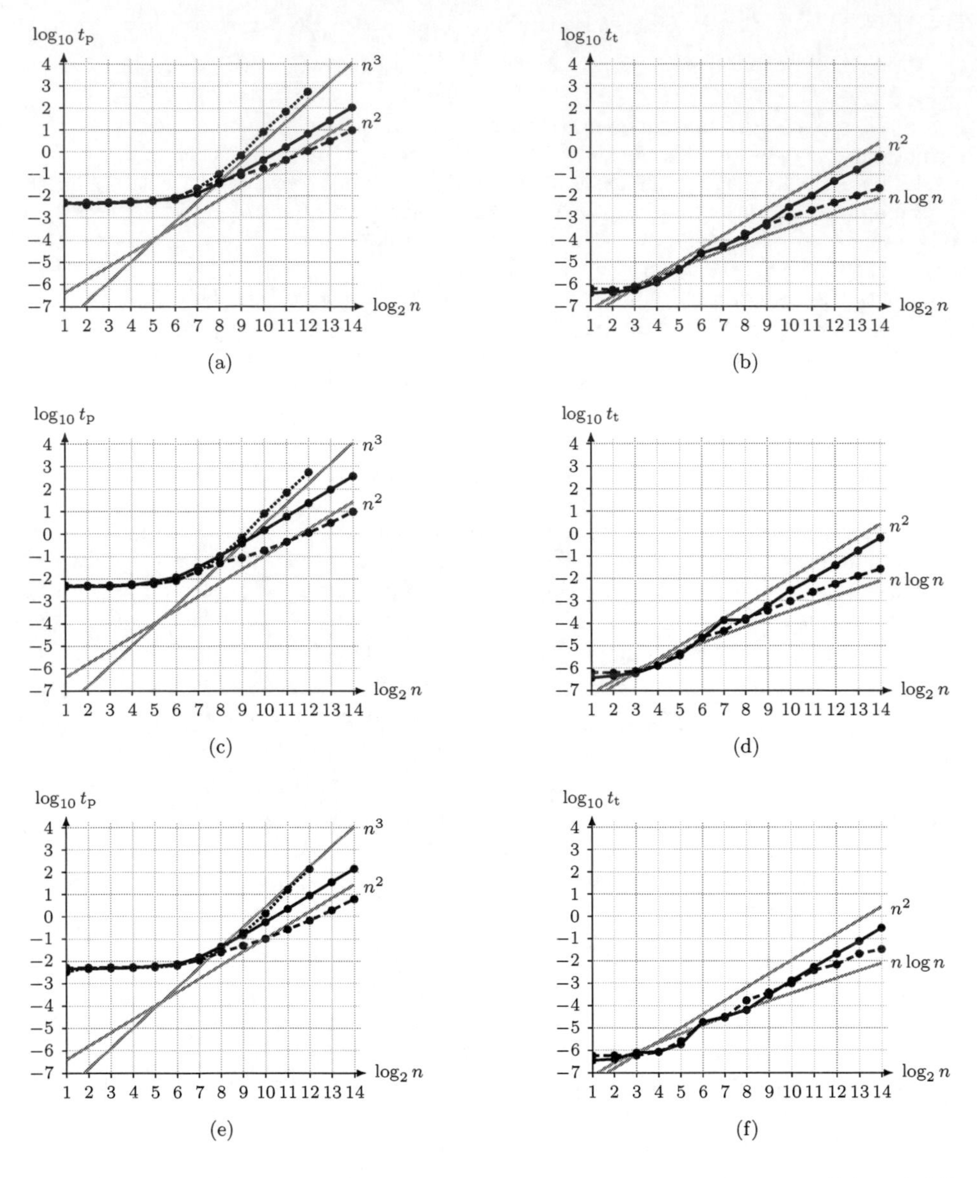

Figure 2.5: Shown from top to bottom are time measurements for Laguerre ((a) and (b)), Jacobi ((c) and (d)), and Gegenbauer polynomials ((e) and (f)), each of which correspond to the first test case reported in Tables 2.2 to 2.13, respectively. Left side: The times t_p for the pre-computation stage as a function of the transform length n. Shown are the direct method (solid), the usmv-direct method (dotted), and the usmv-fmm method (dashed) with accuracy controlling parameter $p = 18$ and arbitrarily large step size s. The gray lines are to facilitate recognition of the asymptotic behavior. Right side: Times t_t for the computation of the actual transformation as a function of the transform length n. Shown are the direct method (solid) and the usmv-fmm method (dashed).

α	β	n	t_{p}	t_{p}/n^2	t_{t}	t_{t}/n^2	E_∞^{c}	E_∞
-0.5	-0.7	256	3.7E-02	5.6E-07	1.4E-04	2.1E-09	7.9E-16	1.2E-15
-0.5	-0.7	512	1.2E-01	4.4E-07	5.9E-04	2.3E-09	1.3E-15	1.7E-15
-0.5	-0.7	1024	4.2E-01	4.0E-07	3.1E-03	3.0E-09	1.7E-15	3.2E-15
-0.5	-0.7	2048	1.7E+00	4.0E-07	1.0E-02	2.4E-09	2.3E-15	4.3E-15
-0.5	-0.7	4096	6.6E+00	3.9E-07	4.8E-02	2.9E-09	3.1E-15	6.6E-15
-0.5	-0.7	8192	2.6E+01	3.9E-07	1.6E-01	2.4E-09	4.4E-15	1.1E-14
-0.5	-0.7	16384	1.1E+02	3.9E-07	6.2E-01	2.3E-09	6.2E-15	1.5E-14
-0.5	0.2	256	4.6E-02	7.0E-07	1.4E-04	2.1E-09	1.6E-15	1.6E-15
-0.5	0.2	512	1.5E-01	5.8E-07	5.8E-04	2.2E-09	2.7E-15	1.8E-15
-0.5	0.2	1024	5.6E-01	5.3E-07	2.9E-03	2.8E-09	3.4E-15	2.7E-15
-0.5	0.2	2048	2.2E+00	5.2E-07	1.0E-02	2.5E-09	5.9E-15	3.7E-15
-0.5	0.2	4096	8.7E+00	5.2E-07	4.0E-02	2.4E-09	9.0E-15	6.3E-15
-0.5	0.2	8192	3.5E+01	5.2E-07	1.5E-01	2.3E-09	1.7E-14	1.0E-14
-0.5	0.2	16384	1.4E+02	5.2E-07	6.2E-01	2.3E-09	2.1E-14	1.5E-14
0.2	-0.5	256	3.6E-02	5.5E-07	1.5E-04	2.3E-09	3.4E-16	9.5E-16
0.2	-0.5	512	1.2E-01	4.4E-07	6.0E-04	2.3E-09	2.8E-16	1.2E-15
0.2	-0.5	1024	4.2E-01	4.0E-07	3.1E-03	3.0E-09	2.7E-16	2.0E-15
0.2	-0.5	2048	1.7E+00	4.0E-07	1.0E-02	2.5E-09	2.6E-16	3.9E-15
0.2	-0.5	4096	6.6E+00	3.9E-07	3.9E-02	2.3E-09	3.6E-16	4.3E-15
0.2	-0.5	8192	2.6E+01	3.9E-07	1.5E-01	2.3E-09	4.0E-16	9.7E-15
0.2	-0.5	16384	1.0E+02	3.8E-07	6.2E-01	2.3E-09	4.4E-16	1.5E-14
0.2	1.1	256	4.6E-02	7.1E-07	1.6E-04	2.4E-09	1.5E-15	8.2E-16
0.2	1.1	512	1.5E-01	5.5E-07	6.9E-04	2.6E-09	2.2E-15	2.0E-15
0.2	1.1	1024	5.5E-01	5.2E-07	3.2E-03	3.1E-09	4.2E-15	3.2E-15
0.2	1.1	2048	2.2E+00	5.1E-07	1.5E-02	3.5E-09	6.1E-15	4.0E-15
0.2	1.1	4096	8.6E+00	5.1E-07	3.9E-02	2.3E-09	1.1E-14	6.7E-15
0.2	1.1	8192	3.4E+01	5.1E-07	1.5E-01	2.3E-09	1.8E-14	1.0E-14
0.2	1.1	16384	1.4E+02	5.1E-07	6.1E-01	2.3E-09	2.4E-14	1.3E-14
5.6	7.8	256	4.3E-02	6.6E-07	1.3E-04	2.0E-09	5.7E-15	4.0E-15
5.6	7.8	512	1.4E-01	5.2E-07	5.6E-04	2.1E-09	6.2E-15	4.1E-15
5.6	7.8	1024	5.2E-01	5.0E-07	3.0E-03	2.9E-09	7.1E-15	4.8E-15
5.6	7.8	2048	2.1E+00	4.9E-07	1.3E-02	3.0E-09	8.4E-15	5.3E-15
5.6	7.8	4096	8.2E+00	4.9E-07	3.9E-02	2.3E-09	1.2E-14	8.8E-15
5.6	7.8	8192	3.3E+01	4.9E-07	1.5E-01	2.3E-09	1.4E-14	8.6E-15
5.6	7.8	16384	1.3E+02	4.9E-07	6.1E-01	2.3E-09	2.2E-14	1.3E-14
9.7	5.5	256	3.6E-02	5.5E-07	1.3E-04	1.9E-09	4.7E-16	3.3E-15
9.7	5.5	512	1.2E-01	4.4E-07	6.1E-04	2.3E-09	1.5E-15	3.7E-15
9.7	5.5	1024	4.4E-01	4.2E-07	3.3E-03	3.1E-09	6.5E-16	5.7E-15
9.7	5.5	2048	1.7E+00	4.0E-07	1.2E-02	2.7E-09	1.0E-15	6.3E-15
9.7	5.5	4096	6.7E+00	4.0E-07	3.9E-02	2.3E-09	1.0E-15	7.2E-15
9.7	5.5	8192	2.7E+01	4.0E-07	1.5E-01	2.3E-09	3.8E-16	1.3E-14
9.7	5.5	16384	1.1E+02	4.0E-07	6.1E-01	2.3E-09	3.8E-16	2.2E-14

Table 2.2: Test results for the connection between the Laguerre polynomials $\{L_n^{(\alpha)}\}_{n\in\mathbb{N}_0}$ and $\{L_n^{(\beta)}\}_{n\in\mathbb{N}_0}$ computed with the direct method for different transform sizes n. Shown are the times for pre-computation t_{p} and for the computation of the actual transform t_{t}. Both are also shown after division through the expected asymptotic expression in terms of the transform size n. Furthermore, the component-wise error E_∞^{c} and the relative infinity norm error E_∞ are reported.

α	β	n	t_p	t_p/n^3	t_t	t_t/n^2	E_∞^c	E_∞
-0.5	-0.7	256	9.9E-02	5.9E-09	1.6E-04	2.5E-09	7.9E-16	1.2E-15
-0.5	-0.7	512	6.9E-01	5.1E-09	1.0E-03	3.9E-09	1.1E-15	1.9E-15
-0.5	-0.7	1024	8.1E+00	7.5E-09	2.7E-03	2.6E-09	1.8E-15	3.4E-15
-0.5	-0.7	2048	6.8E+01	7.9E-09	1.0E-02	2.4E-09	2.1E-15	4.1E-15
-0.5	-0.7	4096	5.5E+02	7.9E-09	5.2E-02	3.1E-09	3.0E-15	6.7E-15
-0.5	-0.7	8192	-	-	-	-	-	-
-0.5	-0.7	16384	-	-	-	-	-	-
-0.5	0.2	256	1.1E-01	6.8E-09	1.9E-04	2.8E-09	1.6E-15	1.3E-15
-0.5	0.2	512	6.8E-01	5.1E-09	1.2E-03	4.4E-09	2.7E-15	1.8E-15
-0.5	0.2	1024	8.1E+00	7.6E-09	2.9E-03	2.8E-09	3.2E-15	2.5E-15
-0.5	0.2	2048	6.8E+01	7.9E-09	1.0E-02	2.5E-09	6.1E-15	3.7E-15
-0.5	0.2	4096	5.4E+02	7.9E-09	4.5E-02	2.7E-09	9.0E-15	6.4E-15
-0.5	0.2	8192	-	-	-	-	-	-
-0.5	0.2	16384	-	-	-	-	-	-
0.2	-0.5	256	9.9E-02	5.9E-09	1.6E-04	2.4E-09	2.1E-16	1.1E-15
0.2	-0.5	512	6.8E-01	5.1E-09	1.1E-03	4.1E-09	2.9E-16	1.3E-15
0.2	-0.5	1024	8.1E+00	7.5E-09	2.9E-03	2.8E-09	1.9E-16	2.6E-15
0.2	-0.5	2048	6.8E+01	7.9E-09	1.0E-02	2.5E-09	2.8E-16	4.2E-15
0.2	-0.5	4096	5.5E+02	8.0E-09	5.2E-02	3.1E-09	4.8E-16	3.5E-15
0.2	-0.5	8192	-	-	-	-	-	-
0.2	-0.5	16384	-	-	-	-	-	-
0.2	1.1	256	1.1E-01	6.6E-09	1.6E-04	2.4E-09	1.4E-15	8.2E-16
0.2	1.1	512	6.8E-01	5.1E-09	1.1E-03	4.2E-09	2.0E-15	1.8E-15
0.2	1.1	1024	8.1E+00	7.5E-09	2.9E-03	2.8E-09	4.0E-15	3.1E-15
0.2	1.1	2048	6.8E+01	7.9E-09	1.0E-02	2.4E-09	6.1E-15	4.0E-15
0.2	1.1	4096	5.4E+02	7.9E-09	5.2E-02	3.1E-09	1.1E-14	6.5E-15
0.2	1.1	8192	-	-	-	-	-	-
0.2	1.1	16384	-	-	-	-	-	-
5.6	7.8	256	1.0E-01	6.2E-09	1.7E-04	2.7E-09	1.7E-15	1.3E-15
5.6	7.8	512	6.8E-01	5.1E-09	1.1E-03	4.1E-09	2.4E-15	1.5E-15
5.6	7.8	1024	8.1E+00	7.5E-09	2.9E-03	2.8E-09	4.1E-15	2.2E-15
5.6	7.8	2048	6.8E+01	7.9E-09	1.0E-02	2.5E-09	6.1E-15	3.6E-15
5.6	7.8	4096	5.5E+02	8.0E-09	4.6E-02	2.7E-09	9.6E-15	6.5E-15
5.6	7.8	8192	-	-	-	-	-	-
5.6	7.8	16384	-	-	-	-	-	-
9.7	5.5	256	1.0E-01	6.1E-09	1.7E-04	2.6E-09	4.7E-16	3.0E-15
9.7	5.5	512	6.8E-01	5.1E-09	1.1E-03	4.1E-09	1.6E-15	4.3E-15
9.7	5.5	1024	8.1E+00	7.6E-09	2.9E-03	2.7E-09	3.9E-16	5.9E-15
9.7	5.5	2048	6.8E+01	7.9E-09	1.0E-02	2.4E-09	1.0E-15	8.5E-15
9.7	5.5	4096	5.5E+02	7.9E-09	4.6E-02	2.7E-09	9.7E-16	7.5E-15
9.7	5.5	8192	-	-	-	-	-	-
9.7	5.5	16384	-	-	-	-	-	-

Table 2.3: Test results for the connection between the Laguerre polynomials $\{L_n^{(\alpha)}\}_{n\in\mathbb{N}_0}$ and $\{L_n^{(\beta)}\}_{n\in\mathbb{N}_0}$ computed with the usmv-direct method for different transform sizes n. Shown are the times for pre-computation t_p and for the computation of the actual transform t_t. Both are also shown after division through the expected asymptotic expression in terms of the transform size n. Furthermore, the component-wise error E_∞^c and the relative infinity norm error E_∞ are reported. For sizes $n > 4096$ the transformation was not computed due to unacceptably large computation times in comparison to the direct method; see Table 2.2.

α	β	n	t_p	t_p/n^2	t_t	$t_\mathrm{t}/(n \log n)$	E_∞^c	E_∞
-0.5	-0.7	256	4.5E-02	6.8E-07	1.9E-04	1.4E-07	5.2E-16	6.8E-16
-0.5	-0.7	512	8.7E-02	3.3E-07	4.5E-04	1.4E-07	6.2E-16	6.8E-16
-0.5	-0.7	1024	1.8E-01	1.7E-07	1.1E-03	1.6E-07	2.0E-15	1.9E-15
-0.5	-0.7	2048	4.2E-01	1.0E-07	2.3E-03	1.5E-07	1.1E-15	1.7E-15
-0.5	-0.7	4096	1.1E+00	6.5E-08	5.1E-03	1.5E-07	2.7E-15	4.2E-15
-0.5	-0.7	8192	3.1E+00	4.6E-08	1.1E-02	1.4E-07	4.8E-15	1.0E-14
-0.5	-0.7	16384	9.6E+00	3.6E-08	2.3E-02	1.5E-07	5.1E-15	1.0E-14
-0.5	0.2	256	4.8E-02	7.3E-07	2.2E-04	1.6E-07	1.6E-14	3.3E-15
-0.5	0.2	512	8.7E-02	3.3E-07	4.3E-04	1.3E-07	2.1E-14	5.4E-15
-0.5	0.2	1024	1.7E-01	1.6E-07	9.6E-04	1.4E-07	1.8E-14	6.0E-15
-0.5	0.2	2048	4.3E-01	1.0E-07	2.3E-03	1.5E-07	2.6E-14	7.9E-15
-0.5	0.2	4096	1.1E+00	6.5E-08	5.1E-03	1.5E-07	3.9E-14	6.9E-15
-0.5	0.2	8192	3.1E+00	4.6E-08	1.1E-02	1.4E-07	4.1E-14	1.2E-14
-0.5	0.2	16384	9.5E+00	3.6E-08	2.3E-02	1.5E-07	6.0E-14	4.2E-14
0.2	-0.5	256	4.4E-02	6.7E-07	1.9E-04	1.3E-07	9.2E-16	4.1E-15
0.2	-0.5	512	9.0E-02	3.4E-07	4.0E-04	1.2E-07	1.3E-15	7.4E-15
0.2	-0.5	1024	1.8E-01	1.7E-07	9.9E-04	1.4E-07	1.8E-15	1.4E-14
0.2	-0.5	2048	4.2E-01	1.0E-07	2.3E-03	1.5E-07	1.5E-15	1.4E-14
0.2	-0.5	4096	1.1E+00	6.5E-08	5.1E-03	1.5E-07	1.6E-15	1.3E-14
0.2	-0.5	8192	3.1E+00	4.6E-08	1.1E-02	1.4E-07	2.3E-15	1.3E-14
0.2	-0.5	16384	9.6E+00	3.6E-08	2.3E-02	1.5E-07	1.5E-15	1.2E-14
0.2	1.1	256	4.6E-02	7.1E-07	1.9E-04	1.4E-07	4.9E-14	8.5E-15
0.2	1.1	512	8.3E-02	3.2E-07	4.0E-04	1.2E-07	5.0E-14	2.0E-14
0.2	1.1	1024	1.8E-01	1.8E-07	1.0E-03	1.4E-07	8.3E-14	1.6E-14
0.2	1.1	2048	4.3E-01	1.0E-07	2.3E-03	1.5E-07	1.5E-13	1.9E-14
0.2	1.1	4096	1.1E+00	6.5E-08	5.1E-03	1.5E-07	2.2E-13	2.5E-14
0.2	1.1	8192	3.1E+00	4.6E-08	1.2E-02	1.6E-07	3.8E-13	2.6E-14
0.2	1.1	16384	9.0E+00	3.6E-08	2.3E-02	1.5E-07	2.4E-13	5.1E-14
5.6	7.8	256	4.8E-02	7.3E-07	2.3E-04	1.6E-07	1.3E-11	5.4E-12
5.6	7.8	512	8.4E-02	3.2E-07	4.2E-04	1.3E-07	7.0E-10	1.4E-10
5.6	7.8	1024	1.9E-01	1.8E-07	1.1E-03	1.5E-07	2.9E-09	1.4E-09
5.6	7.8	2048	4.2E-01	1.0E-07	2.3E-03	1.5E-07	1.5E-08	6.0E-09
5.6	7.8	4096	1.1E+00	6.5E-08	5.0E-03	1.5E-07	3.2E-08	1.7E-08
5.6	7.8	8192	3.1E+00	4.6E-08	1.1E-02	1.4E-07	3.2E-07	1.4E-07
5.6	7.8	16384	9.6E+00	3.6E-08	2.3E-02	1.5E-07	3.7E-07	1.5E-07
9.7	5.5	256	4.6E-02	7.0E-07	1.9E-04	1.4E-07	4.3E-07	4.1E-06
9.7	5.5	512	8.4E-02	3.2E-07	4.2E-04	1.3E-07	6.5E-06	4.9E-05
9.7	5.5	1024	1.7E-01	1.6E-07	1.0E-03	1.4E-07	1.6E-05	2.8E-04
9.7	5.5	2048	4.3E-01	1.0E-07	2.3E-03	1.5E-07	3.0E-04	9.0E-03
9.7	5.5	4096	1.1E+00	6.5E-08	5.2E-03	1.5E-07	1.9E-01	1.3E+01
9.7	5.5	8192	3.1E+00	4.6E-08	1.1E-02	1.4E-07	1.9E+02	2.9E+04
9.7	5.5	16384	9.5E+00	3.6E-08	2.3E-02	1.5E-07	6.5E+09	2.3E+12

Table 2.4: Test results for the connection between the Laguerre polynomials $\{L_n^{(\alpha)}\}_{n\in\mathbb{N}_0}$ and $\{L_n^{(\beta)}\}_{n\in\mathbb{N}_0}$ computed with the usmv-fmm method with accuracy controlling parameter $p = 18$ and arbitrarily large step size s for different transform sizes n. Shown are the times for pre-computation t_p and for the computation of the actual transform t_t. Both are also shown after division through the expected asymptotic expression in terms of the transform size n. Furthermore, the component-wise error E_∞^c and the relative infinity norm error E_∞ are reported.

α	β	n	t_p	t_p/n^2	t_t	$t_t/(n \log n)$	E_∞^c	E_∞
-0.5	-0.7	256	4.5E-02	6.9E-07	1.9E-04	1.4E-07	5.2E-16	6.8E-16
-0.5	-0.7	512	8.9E-02	3.4E-07	4.1E-04	1.3E-07	6.2E-16	6.8E-16
-0.5	-0.7	1024	1.8E-01	1.7E-07	9.6E-04	1.4E-07	2.0E-15	1.9E-15
-0.5	-0.7	2048	4.3E-01	1.0E-07	2.3E-03	1.5E-07	1.1E-15	1.7E-15
-0.5	-0.7	4096	1.1E+00	6.5E-08	5.1E-03	1.5E-07	2.7E-15	4.2E-15
-0.5	-0.7	8192	3.1E+00	4.6E-08	1.0E-02	1.4E-07	4.8E-15	1.0E-14
-0.5	-0.7	16384	9.5E+00	3.6E-08	2.3E-02	1.5E-07	5.1E-15	1.0E-14
-0.5	0.2	256	4.4E-02	6.8E-07	2.0E-04	1.4E-07	1.6E-14	3.3E-15
-0.5	0.2	512	8.4E-02	3.2E-07	4.1E-04	1.3E-07	2.1E-14	5.4E-15
-0.5	0.2	1024	1.8E-01	1.7E-07	9.7E-04	1.4E-07	1.8E-14	6.0E-15
-0.5	0.2	2048	4.2E-01	1.0E-07	2.4E-03	1.5E-07	2.6E-14	7.9E-15
-0.5	0.2	4096	1.1E+00	6.5E-08	4.9E-03	1.4E-07	3.9E-14	6.9E-15
-0.5	0.2	8192	3.1E+00	4.6E-08	1.0E-02	1.4E-07	4.1E-14	1.2E-14
-0.5	0.2	16384	9.5E+00	3.6E-08	2.3E-02	1.5E-07	6.0E-14	4.2E-14
0.2	-0.5	256	4.8E-02	7.4E-07	2.3E-04	1.6E-07	9.2E-16	4.1E-15
0.2	-0.5	512	8.3E-02	3.2E-07	3.8E-04	1.2E-07	1.3E-15	7.4E-15
0.2	-0.5	1024	1.7E-01	1.6E-07	9.6E-04	1.4E-07	1.8E-15	1.4E-14
0.2	-0.5	2048	4.2E-01	1.0E-07	2.3E-03	1.5E-07	1.5E-15	1.4E-14
0.2	-0.5	4096	1.1E+00	6.6E-08	5.0E-03	1.5E-07	1.6E-15	1.3E-14
0.2	-0.5	8192	3.1E+00	4.6E-08	1.0E-02	1.4E-07	2.3E-15	1.3E-14
0.2	-0.5	16384	9.6E+00	3.6E-08	2.3E-02	1.5E-07	1.5E-15	1.2E-14
0.2	1.1	256	4.8E-02	7.3E-07	2.2E-04	1.6E-07	4.9E-14	8.7E-15
0.2	1.1	512	8.4E-02	3.2E-07	3.9E-04	1.2E-07	5.0E-14	2.0E-14
0.2	1.1	1024	1.7E-01	1.7E-07	1.0E-03	1.4E-07	8.3E-14	1.6E-14
0.2	1.1	2048	4.2E-01	1.0E-07	2.3E-03	1.5E-07	1.5E-13	1.9E-14
0.2	1.1	4096	1.1E+00	6.5E-08	5.0E-03	1.5E-07	2.2E-13	2.5E-14
0.2	1.1	8192	3.1E+00	4.6E-08	1.1E-02	1.5E-07	3.8E-13	2.6E-14
0.2	1.1	16384	9.5E+00	3.6E-08	2.3E-02	1.5E-07	2.4E-13	5.1E-14
5.6	7.8	256	7.7E-02	1.2E-06	5.6E-04	4.0E-07	2.7E-14	9.1E-15
5.6	7.8	512	1.5E-01	5.7E-07	1.4E-03	4.5E-07	1.4E-13	2.9E-14
5.6	7.8	1024	3.2E-01	3.0E-07	3.1E-03	4.4E-07	9.3E-14	2.2E-14
5.6	7.8	2048	8.7E-01	2.1E-07	6.6E-03	4.2E-07	4.1E-13	5.5E-14
5.6	7.8	4096	2.3E+00	1.4E-07	1.4E-02	4.1E-07	2.6E-13	6.3E-14
5.6	7.8	8192	6.9E+00	1.0E-07	3.1E-02	4.1E-07	2.3E-13	7.0E-14
5.6	7.8	16384	2.2E+01	8.3E-08	6.8E-02	4.3E-07	1.2E-12	7.8E-14
9.7	5.5	256	1.8E-01	2.7E-06	8.8E-04	6.2E-07	7.1E-16	6.0E-15
9.7	5.5	512	3.7E-01	1.4E-06	2.3E-03	7.3E-07	1.9E-15	1.4E-14
9.7	5.5	1024	7.9E-01	7.5E-07	4.8E-03	6.8E-07	2.2E-15	3.7E-14
9.7	5.5	2048	2.0E+00	4.7E-07	1.0E-02	6.6E-07	2.8E-15	8.3E-14
9.7	5.5	4096	4.9E+00	2.9E-07	2.4E-02	7.0E-07	2.8E-15	1.7E-13
9.7	5.5	8192	1.3E+01	2.0E-07	5.1E-02	6.9E-07	1.9E-15	2.3E-13
9.7	5.5	16384	4.0E+01	1.5E-07	1.1E-01	7.1E-07	1.9E-15	3.6E-13

Table 2.5: Test results for the connection between the Laguerre polynomials $\{L_n^{(\alpha)}\}_{n\in\mathbb{N}_0}$ and $\{L_n^{(\beta)}\}_{n\in\mathbb{N}_0}$ computed with the usmv-fmm method with accuracy controlling parameter $p = 18$ and maximum step size $s = 1$ for different transform sizes n. Shown are the times for pre-computation t_p and for the computation of the actual transform t_t. Both are also shown after division through the expected asymptotic expression in terms of the transform size n. Furthermore, the component-wise error E_∞^c and the relative infinity norm error E_∞ are reported.

α	β	γ	δ	n	t_{p}	t_{p}/n^2	t_{t}	$t_{\mathrm{t}}/(n\log n)$	E_∞^{c}	E_∞
-0.7	2.0	-0.9	2.0	256	1.0E-01	1.6E-06	1.4E-04	2.2E-09	1.0E-15	1.1E-15
-0.7	2.0	-0.9	2.0	512	3.8E-01	1.4E-06	6.1E-04	2.3E-09	7.4E-16	1.2E-15
-0.7	2.0	-0.9	2.0	1024	1.5E+00	1.4E-06	3.0E-03	2.8E-09	8.9E-16	1.1E-15
-0.7	2.0	-0.9	2.0	2048	5.9E+00	1.4E-06	1.0E-02	2.4E-09	7.2E-16	1.0E-15
-0.7	2.0	-0.9	2.0	4096	2.3E+01	1.4E-06	3.9E-02	2.3E-09	6.1E-16	1.2E-15
-0.7	2.0	-0.9	2.0	8192	9.3E+01	1.4E-06	1.7E-01	2.5E-09	6.2E-16	1.3E-15
-0.7	2.0	-0.9	2.0	16384	3.7E+02	1.4E-06	6.5E-01	2.4E-09	7.4E-16	1.4E-15
-0.7	2.0	0.0	2.0	256	1.0E-01	1.5E-06	1.4E-04	2.2E-09	1.3E-15	7.4E-16
-0.7	2.0	0.0	2.0	512	3.7E-01	1.4E-06	6.1E-04	2.3E-09	1.2E-15	9.8E-16
-0.7	2.0	0.0	2.0	1024	1.4E+00	1.4E-06	2.7E-03	2.6E-09	9.9E-16	7.5E-16
-0.7	2.0	0.0	2.0	2048	5.8E+00	1.4E-06	1.0E-02	2.5E-09	1.0E-15	7.6E-16
-0.7	2.0	0.0	2.0	4096	2.3E+01	1.4E-06	3.9E-02	2.3E-09	1.1E-15	9.3E-16
-0.7	2.0	0.0	2.0	8192	9.2E+01	1.4E-06	2.0E-01	3.0E-09	1.1E-15	8.6E-16
-0.7	2.0	0.0	2.0	16384	3.7E+02	1.4E-06	6.5E-01	2.4E-09	1.2E-15	8.1E-16
0.0	2.0	-0.7	2.0	256	9.3E-02	1.4E-06	1.3E-04	2.0E-09	5.0E-16	6.9E-16
0.0	2.0	-0.7	2.0	512	3.4E-01	1.3E-06	6.1E-04	2.3E-09	4.4E-16	1.2E-15
0.0	2.0	-0.7	2.0	1024	1.3E+00	1.3E-06	3.0E-03	2.9E-09	7.2E-16	1.0E-15
0.0	2.0	-0.7	2.0	2048	5.2E+00	1.3E-06	1.0E-02	2.4E-09	4.0E-16	2.1E-15
0.0	2.0	-0.7	2.0	4096	2.1E+01	1.2E-06	3.9E-02	2.3E-09	3.5E-16	3.1E-15
0.0	2.0	-0.7	2.0	8192	8.2E+01	1.2E-06	1.7E-01	2.5E-09	4.1E-16	3.0E-15
0.0	2.0	-0.7	2.0	16384	3.2E+02	1.2E-06	6.1E-01	2.3E-09	4.1E-16	3.2E-15
0.0	2.0	0.9	2.0	256	9.9E-02	1.5E-06	1.3E-04	2.0E-09	1.1E-15	8.5E-16
0.0	2.0	0.9	2.0	512	3.7E-01	1.4E-06	6.9E-04	2.6E-09	1.7E-15	1.1E-15
0.0	2.0	0.9	2.0	1024	1.4E+00	1.4E-06	3.0E-03	2.9E-09	1.3E-15	9.4E-16
0.0	2.0	0.9	2.0	2048	5.8E+00	1.4E-06	1.0E-02	2.4E-09	1.5E-15	1.0E-15
0.0	2.0	0.9	2.0	4096	2.3E+01	1.4E-06	3.9E-02	2.3E-09	1.6E-15	9.8E-16
0.0	2.0	0.9	2.0	8192	9.2E+01	1.4E-06	2.0E-01	3.0E-09	2.1E-15	1.2E-15
0.0	2.0	0.9	2.0	16384	3.7E+02	1.4E-06	6.4E-01	2.4E-09	1.7E-15	1.1E-15
5.4	2.0	7.6	2.0	256	1.1E-01	1.7E-06	1.4E-04	2.1E-09	1.7E-15	1.6E-15
5.4	2.0	7.6	2.0	512	4.1E-01	1.6E-06	6.6E-04	2.5E-09	2.0E-15	1.0E-15
5.4	2.0	7.6	2.0	1024	1.6E+00	1.5E-06	3.0E-03	2.8E-09	2.3E-15	1.3E-15
5.4	2.0	7.6	2.0	2048	6.4E+00	1.5E-06	1.0E-02	2.4E-09	2.0E-15	1.6E-15
5.4	2.0	7.6	2.0	4096	2.6E+01	1.5E-06	3.9E-02	2.3E-09	1.8E-15	1.5E-15
5.4	2.0	7.6	2.0	8192	1.0E+02	1.5E-06	2.0E-01	3.0E-09	2.2E-15	1.4E-15
5.4	2.0	7.6	2.0	16384	4.1E+02	1.5E-06	6.4E-01	2.4E-09	2.3E-15	1.8E-15
8.6	2.0	4.3	2.0	256	1.1E-01	1.6E-06	1.4E-04	2.1E-09	1.3E-15	1.4E-15
8.6	2.0	4.3	2.0	512	3.9E-01	1.5E-06	6.1E-04	2.3E-09	2.0E-15	8.3E-15
8.6	2.0	4.3	2.0	1024	1.5E+00	1.4E-06	3.0E-03	2.8E-09	2.0E-15	6.4E-15
8.6	2.0	4.3	2.0	2048	6.0E+00	1.4E-06	1.0E-02	2.5E-09	2.7E-15	5.0E-15
8.6	2.0	4.3	2.0	4096	2.4E+01	1.4E-06	4.2E-02	2.5E-09	1.8E-15	1.3E-14
8.6	2.0	4.3	2.0	8192	9.5E+01	1.4E-06	2.1E-01	3.1E-09	1.4E-15	6.5E-15
8.6	2.0	4.3	2.0	16384	3.8E+02	1.4E-06	6.4E-01	2.4E-09	1.2E-15	1.1E-14

Table 2.6: Test results for the connection between the Jacobi polynomials $\{P_n^{(\alpha,\beta)}\}_{n\in\mathbb{N}_0}$ and $\{P_n^{(\gamma,\beta)}\}_{n\in\mathbb{N}_0}$ computed with the direct method for different transform sizes n. Shown are the times for pre-computation t_{p} and for the computation of the actual transform t_{t}. Both are also shown after division through the expected asymptotic expression in terms of the transform size n. Furthermore, the component-wise error E_∞^{c} and the relative infinity norm error E_∞ are reported.

α	β	γ	δ	n	t_{p}	t_{p}/n^2	t_{t}	$t_{\mathrm{t}}/(n\log n)$	E_∞^{c}	E_∞
-0.7	2.0	-0.9	2.0	256	9.7E-02	5.8E-09	1.6E-04	2.4E-09	4.5E-15	1.8E-15
-0.7	2.0	-0.9	2.0	512	6.7E-01	5.0E-09	9.9E-04	3.8E-09	4.3E-15	3.8E-15
-0.7	2.0	-0.9	2.0	1024	8.0E+00	7.5E-09	2.8E-03	2.7E-09	1.4E-14	8.1E-15
-0.7	2.0	-0.9	2.0	2048	6.8E+01	7.9E-09	1.0E-02	2.4E-09	1.7E-14	1.8E-14
-0.7	2.0	-0.9	2.0	4096	5.5E+02	7.9E-09	3.9E-02	2.3E-09	5.3E-14	3.2E-14
-0.7	2.0	-0.9	2.0	8192	-	-	-	-	-	-
-0.7	2.0	-0.9	2.0	16384	-	-	-	-	-	-
-0.7	2.0	0.0	2.0	256	9.5E-02	5.7E-09	1.6E-04	2.5E-09	8.9E-15	4.8E-15
-0.7	2.0	0.0	2.0	512	6.7E-01	5.0E-09	1.1E-03	4.1E-09	1.7E-14	1.1E-14
-0.7	2.0	0.0	2.0	1024	8.0E+00	7.5E-09	2.9E-03	2.7E-09	3.9E-14	2.4E-14
-0.7	2.0	0.0	2.0	2048	6.8E+01	7.9E-09	1.0E-02	2.4E-09	8.7E-14	4.6E-14
-0.7	2.0	0.0	2.0	4096	5.4E+02	7.9E-09	3.9E-02	2.3E-09	1.9E-13	1.0E-13
-0.7	2.0	0.0	2.0	8192	-	-	-	-	-	-
-0.7	2.0	0.0	2.0	16384	-	-	-	-	-	-
0.0	2.0	-0.7	2.0	256	9.5E-02	5.7E-09	1.8E-04	2.8E-09	6.2E-16	1.4E-15
0.0	2.0	-0.7	2.0	512	6.7E-01	5.0E-09	1.0E-03	3.9E-09	5.8E-16	1.5E-15
0.0	2.0	-0.7	2.0	1024	8.0E+00	7.5E-09	2.8E-03	2.7E-09	7.9E-16	1.6E-15
0.0	2.0	-0.7	2.0	2048	6.7E+01	7.8E-09	1.0E-02	2.4E-09	6.4E-16	3.4E-15
0.0	2.0	-0.7	2.0	4096	5.5E+02	7.9E-09	4.2E-02	2.5E-09	8.0E-16	5.9E-15
0.0	2.0	-0.7	2.0	8192	-	-	-	-	-	-
0.0	2.0	-0.7	2.0	16384	-	-	-	-	-	-
0.0	2.0	0.9	2.0	256	1.0E-01	6.0E-09	1.6E-04	2.4E-09	2.6E-15	1.3E-15
0.0	2.0	0.9	2.0	512	6.7E-01	5.0E-09	1.1E-03	4.1E-09	2.8E-15	1.7E-15
0.0	2.0	0.9	2.0	1024	8.0E+00	7.4E-09	2.9E-03	2.8E-09	2.3E-15	1.7E-15
0.0	2.0	0.9	2.0	2048	6.7E+01	7.8E-09	1.0E-02	2.4E-09	2.8E-15	1.8E-15
0.0	2.0	0.9	2.0	4096	5.5E+02	7.9E-09	3.9E-02	2.3E-09	3.0E-15	1.5E-15
0.0	2.0	0.9	2.0	8192	-	-	-	-	-	-
0.0	2.0	0.9	2.0	16384	-	-	-	-	-	-
5.4	2.0	7.6	2.0	256	9.5E-02	5.7E-09	1.7E-04	2.6E-09	1.8E-14	1.1E-14
5.4	2.0	7.6	2.0	512	6.7E-01	5.0E-09	1.1E-03	4.0E-09	3.6E-14	2.0E-14
5.4	2.0	7.6	2.0	1024	8.0E+00	7.5E-09	2.8E-03	2.7E-09	8.9E-14	4.4E-14
5.4	2.0	7.6	2.0	2048	6.8E+01	7.9E-09	1.0E-02	2.4E-09	1.5E-13	9.1E-14
5.4	2.0	7.6	2.0	4096	5.5E+02	7.9E-09	3.9E-02	2.3E-09	4.5E-13	1.8E-13
5.4	2.0	7.6	2.0	8192	-	-	-	-	-	-
5.4	2.0	7.6	2.0	16384	-	-	-	-	-	-
8.6	2.0	4.3	2.0	256	1.0E-01	6.2E-09	1.6E-04	2.5E-09	8.4E-15	3.2E-15
8.6	2.0	4.3	2.0	512	6.8E-01	5.0E-09	1.0E-03	4.0E-09	1.6E-14	4.2E-14
8.6	2.0	4.3	2.0	1024	8.0E+00	7.5E-09	2.9E-03	2.7E-09	3.7E-14	1.1E-13
8.6	2.0	4.3	2.0	2048	6.8E+01	7.9E-09	1.0E-02	2.4E-09	1.3E-13	4.1E-14
8.6	2.0	4.3	2.0	4096	5.5E+02	7.9E-09	3.9E-02	2.3E-09	1.8E-13	5.7E-13
8.6	2.0	4.3	2.0	8192	-	-	-	-	-	-
8.6	2.0	4.3	2.0	16384	-	-	-	-	-	-

Table 2.7: Test results for the connection between the Jacobi polynomials $\{P_n^{(\alpha,\beta)}\}_{n\in\mathbb{N}_0}$ and $\{P_n^{(\gamma,\beta)}\}_{n\in\mathbb{N}_0}$ computed with the usmv-direct method for different transform sizes n. Shown are the times for pre-computation t_{p} and for the computation of the actual transform t_{t}. Both are also shown after division through the expected asymptotic expression in terms of the transform size n. Furthermore, the component-wise error E_∞^{c} and the relative infinity norm error E_∞ are reported. For sizes $n > 4096$ the transformation was not computed due to unacceptably large computation times in comparison to the direct method; see Table 2.6.

α	β	γ	δ	n	t_{p}	t_{p}/n^2	t_{t}	$t_{\mathrm{t}}/(n \log n)$	E_{∞}^{c}	E_{∞}
-0.7	2.0	-0.9	2.0	256	5.0E-02	7.6E-07	1.7E-04	1.2E-07	4.3E-15	1.8E-15
-0.7	2.0	-0.9	2.0	512	8.7E-02	3.3E-07	3.7E-04	1.2E-07	4.4E-15	3.5E-15
-0.7	2.0	-0.9	2.0	1024	1.8E-01	1.8E-07	9.7E-04	1.4E-07	1.4E-14	7.7E-15
-0.7	2.0	-0.9	2.0	2048	4.4E-01	1.1E-07	2.5E-03	1.6E-07	1.7E-14	1.9E-14
-0.7	2.0	-0.9	2.0	4096	1.1E+00	6.6E-08	5.7E-03	1.7E-07	5.3E-14	3.5E-14
-0.7	2.0	-0.9	2.0	8192	3.1E+00	4.7E-08	1.3E-02	1.8E-07	1.8E-13	6.4E-14
-0.7	2.0	-0.9	2.0	16384	9.7E+00	3.6E-08	2.7E-02	1.7E-07	1.7E-13	1.3E-13
-0.7	2.0	0.0	2.0	256	5.0E-02	7.6E-07	1.6E-04	1.1E-07	8.9E-15	4.8E-15
-0.7	2.0	0.0	2.0	512	8.8E-02	3.3E-07	3.7E-04	1.2E-07	1.9E-14	1.0E-14
-0.7	2.0	0.0	2.0	1024	1.8E-01	1.7E-07	9.9E-04	1.4E-07	3.8E-14	2.4E-14
-0.7	2.0	0.0	2.0	2048	4.4E-01	1.1E-07	2.5E-03	1.6E-07	8.8E-14	4.6E-14
-0.7	2.0	0.0	2.0	4096	1.1E+00	6.6E-08	5.6E-03	1.7E-07	1.9E-13	9.8E-14
-0.7	2.0	0.0	2.0	8192	3.1E+00	4.6E-08	1.3E-02	1.7E-07	3.8E-13	1.8E-13
-0.7	2.0	0.0	2.0	16384	9.7E+00	3.6E-08	2.6E-02	1.6E-07	7.3E-13	3.9E-13
0.0	2.0	-0.7	2.0	256	4.6E-02	7.1E-07	1.4E-04	9.8E-08	8.1E-16	3.1E-15
0.0	2.0	-0.7	2.0	512	8.7E-02	3.3E-07	3.6E-04	1.1E-07	8.8E-16	3.6E-15
0.0	2.0	-0.7	2.0	1024	1.8E-01	1.7E-07	1.0E-03	1.4E-07	7.9E-16	3.3E-15
0.0	2.0	-0.7	2.0	2048	4.4E-01	1.1E-07	2.5E-03	1.6E-07	8.9E-16	4.0E-15
0.0	2.0	-0.7	2.0	4096	1.1E+00	6.6E-08	5.6E-03	1.7E-07	8.7E-16	6.1E-15
0.0	2.0	-0.7	2.0	8192	3.1E+00	4.6E-08	1.3E-02	1.8E-07	1.1E-15	9.4E-15
0.0	2.0	-0.7	2.0	16384	9.7E+00	3.6E-08	2.6E-02	1.6E-07	2.0E-15	2.0E-14
0.0	2.0	0.9	2.0	256	4.5E-02	6.9E-07	1.4E-04	9.6E-08	6.8E-15	2.0E-15
0.0	2.0	0.9	2.0	512	8.8E-02	3.3E-07	3.9E-04	1.2E-07	1.6E-14	4.9E-15
0.0	2.0	0.9	2.0	1024	1.8E-01	1.7E-07	9.4E-04	1.3E-07	2.8E-14	7.5E-15
0.0	2.0	0.9	2.0	2048	4.4E-01	1.0E-07	2.4E-03	1.6E-07	5.2E-14	9.7E-15
0.0	2.0	0.9	2.0	4096	1.1E+00	6.6E-08	5.8E-03	1.7E-07	3.8E-14	2.1E-14
0.0	2.0	0.9	2.0	8192	3.1E+00	4.6E-08	1.3E-02	1.8E-07	7.4E-14	3.0E-14
0.0	2.0	0.9	2.0	16384	9.7E+00	3.6E-08	2.6E-02	1.6E-07	2.2E-13	4.5E-14
5.4	2.0	7.6	2.0	256	4.7E-02	7.2E-07	1.4E-04	9.6E-08	3.4E-13	3.0E-14
5.4	2.0	7.6	2.0	512	8.9E-02	3.4E-07	3.8E-04	1.2E-07	3.2E-12	1.4E-12
5.4	2.0	7.6	2.0	1024	1.8E-01	1.7E-07	9.9E-04	1.4E-07	1.8E-11	1.1E-11
5.4	2.0	7.6	2.0	2048	4.5E-01	1.1E-07	2.5E-03	1.6E-07	6.5E-10	3.0E-10
5.4	2.0	7.6	2.0	4096	1.1E+00	6.7E-08	5.6E-03	1.7E-07	1.2E-09	6.9E-10
5.4	2.0	7.6	2.0	8192	3.1E+00	4.7E-08	1.3E-02	1.8E-07	8.2E-09	2.7E-09
5.4	2.0	7.6	2.0	16384	9.7E+00	3.6E-08	2.6E-02	1.6E-07	1.2E-08	4.1E-09
8.6	2.0	4.3	2.0	256	4.9E-02	7.5E-07	1.5E-04	1.0E-07	3.1E-08	6.5E-09
8.6	2.0	4.3	2.0	512	9.0E-02	3.4E-07	3.9E-04	1.2E-07	1.9E-07	2.7E-06
8.6	2.0	4.3	2.0	1024	1.8E-01	1.7E-07	9.9E-04	1.4E-07	3.9E-06	3.5E-05
8.6	2.0	4.3	2.0	2048	4.5E-01	1.1E-07	2.5E-03	1.6E-07	2.1E-04	1.9E-03
8.6	2.0	4.3	2.0	4096	1.1E+00	6.7E-08	5.6E-03	1.6E-07	6.6E-03	4.8E-01
8.6	2.0	4.3	2.0	8192	3.1E+00	4.7E-08	1.2E-02	1.7E-07	4.6E+00	1.9E+01
8.6	2.0	4.3	2.0	16384	9.7E+00	3.6E-08	2.6E-02	1.6E-07	1.6E+10	7.5E+09

Table 2.8: Test results for the connection between the Jacobi polynomials $\{P_n^{(\alpha,\beta)}\}_{n\in\mathbb{N}_0}$ and $\{P_n^{(\gamma,\beta)}\}_{n\in\mathbb{N}_0}$ computed with the usmv-fmm method with accuracy controlling parameter $p = 18$ and arbitrarily large step size s for different transform sizes n. Shown are the times for pre-computation t_{p} and for the computation of the actual transform t_{t}. Both are also shown after division through the expected asymptotic expression in terms of the transform size n. Furthermore, the component-wise error E_{∞}^{c} and the relative infinity norm error E_{∞} are reported.

α	β	γ	δ	n	t_{p}	t_{p}/n^2	t_{t}	$t_{\text{t}}/(n \log n)$	E_{∞}^{c}	E_{∞}
-0.7	2.0	-0.9	2.0	256	4.5E-02	6.9E-07	1.4E-04	9.6E-08	4.3E-15	1.8E-15
-0.7	2.0	-0.9	2.0	512	9.2E-02	3.5E-07	3.6E-04	1.1E-07	4.4E-15	3.5E-15
-0.7	2.0	-0.9	2.0	1024	1.8E-01	1.7E-07	1.1E-03	1.6E-07	1.4E-14	7.7E-15
-0.7	2.0	-0.9	2.0	2048	4.4E-01	1.0E-07	2.5E-03	1.6E-07	1.7E-14	1.9E-14
-0.7	2.0	-0.9	2.0	4096	1.1E+00	6.7E-08	5.7E-03	1.7E-07	5.3E-14	3.5E-14
-0.7	2.0	-0.9	2.0	8192	3.1E+00	4.7E-08	1.3E-02	1.7E-07	1.8E-13	6.4E-14
-0.7	2.0	-0.9	2.0	16384	9.7E+00	3.6E-08	2.6E-02	1.7E-07	1.7E-13	1.3E-13
-0.7	2.0	0.0	2.0	256	4.7E-02	7.2E-07	1.3E-04	9.4E-08	8.9E-15	4.8E-15
-0.7	2.0	0.0	2.0	512	8.6E-02	3.3E-07	3.7E-04	1.2E-07	2.0E-14	1.0E-14
-0.7	2.0	0.0	2.0	1024	1.9E-01	1.8E-07	9.8E-04	1.4E-07	3.8E-14	2.4E-14
-0.7	2.0	0.0	2.0	2048	4.4E-01	1.0E-07	2.5E-03	1.6E-07	8.8E-14	4.6E-14
-0.7	2.0	0.0	2.0	4096	1.1E+00	6.6E-08	5.6E-03	1.6E-07	1.9E-13	9.8E-14
-0.7	2.0	0.0	2.0	8192	3.1E+00	4.6E-08	1.3E-02	1.8E-07	3.8E-13	1.8E-13
-0.7	2.0	0.0	2.0	16384	9.6E+00	3.6E-08	2.7E-02	1.7E-07	7.3E-13	3.9E-13
0.0	2.0	-0.7	2.0	256	4.5E-02	6.8E-07	1.4E-04	1.0E-07	8.1E-16	3.1E-15
0.0	2.0	-0.7	2.0	512	9.1E-02	3.5E-07	3.5E-04	1.1E-07	8.8E-16	3.6E-15
0.0	2.0	-0.7	2.0	1024	1.8E-01	1.7E-07	9.9E-04	1.4E-07	7.9E-16	3.3E-15
0.0	2.0	-0.7	2.0	2048	4.4E-01	1.1E-07	2.5E-03	1.6E-07	8.9E-16	4.0E-15
0.0	2.0	-0.7	2.0	4096	1.1E+00	6.7E-08	5.9E-03	1.7E-07	8.7E-16	6.1E-15
0.0	2.0	-0.7	2.0	8192	3.1E+00	4.6E-08	1.3E-02	1.8E-07	1.1E-15	9.5E-15
0.0	2.0	-0.7	2.0	16384	9.7E+00	3.6E-08	2.6E-02	1.6E-07	2.0E-15	2.0E-14
0.0	2.0	0.9	2.0	256	4.7E-02	7.2E-07	1.4E-04	9.6E-08	6.8E-15	2.0E-15
0.0	2.0	0.9	2.0	512	8.7E-02	3.3E-07	3.6E-04	1.1E-07	1.6E-14	4.9E-15
0.0	2.0	0.9	2.0	1024	1.8E-01	1.7E-07	9.8E-04	1.4E-07	2.8E-14	7.5E-15
0.0	2.0	0.9	2.0	2048	4.4E-01	1.1E-07	2.5E-03	1.6E-07	5.2E-14	9.7E-15
0.0	2.0	0.9	2.0	4096	1.1E+00	6.6E-08	5.8E-03	1.7E-07	3.7E-14	2.1E-14
0.0	2.0	0.9	2.0	8192	3.1E+00	4.7E-08	1.3E-02	1.8E-07	7.4E-14	3.0E-14
0.0	2.0	0.9	2.0	16384	9.6E+00	3.6E-08	2.6E-02	1.6E-07	2.2E-13	4.5E-14
5.4	2.0	7.6	2.0	256	1.1E-01	1.7E-06	4.5E-04	3.1E-07	1.7E-14	1.2E-14
5.4	2.0	7.6	2.0	512	2.3E-01	8.9E-07	1.4E-03	4.4E-07	3.0E-14	1.7E-14
5.4	2.0	7.6	2.0	1024	5.0E-01	4.8E-07	3.4E-03	4.7E-07	6.2E-14	4.7E-14
5.4	2.0	7.6	2.0	2048	1.3E+00	3.1E-07	7.4E-03	4.7E-07	1.3E-13	7.6E-14
5.4	2.0	7.6	2.0	4096	3.3E+00	2.0E-07	1.6E-02	4.8E-07	3.6E-13	1.5E-13
5.4	2.0	7.6	2.0	8192	9.3E+00	1.4E-07	3.6E-02	4.9E-07	1.2E-12	3.5E-13
5.4	2.0	7.6	2.0	16384	2.9E+01	1.1E-07	7.8E-02	4.9E-07	1.3E-12	6.4E-13
8.6	2.0	4.3	2.0	256	1.8E-01	2.7E-06	9.2E-04	6.5E-07	7.6E-15	5.1E-15
8.6	2.0	4.3	2.0	512	3.8E-01	1.4E-06	2.4E-03	7.6E-07	1.5E-14	2.8E-14
8.6	2.0	4.3	2.0	1024	8.3E-01	8.0E-07	5.3E-03	7.5E-07	3.9E-14	6.0E-14
8.6	2.0	4.3	2.0	2048	2.1E+00	5.1E-07	1.2E-02	7.4E-07	1.3E-13	3.6E-14
8.6	2.0	4.3	2.0	4096	5.5E+00	3.3E-07	2.6E-02	7.7E-07	1.2E-13	2.7E-13
8.6	2.0	4.3	2.0	8192	1.6E+01	2.3E-07	6.0E-02	8.1E-07	5.4E-13	2.0E-13
8.6	2.0	4.3	2.0	16384	4.8E+01	1.8E-07	1.3E-01	8.1E-07	6.3E-13	1.1E-12

Table 2.9: Test results for the connection between the Jacobi polynomials $\{P_n^{(\alpha,\beta)}\}_{n\in\mathbb{N}_0}$ and $\{P_n^{(\gamma,\beta)}\}_{n\in\mathbb{N}_0}$ computed with the usmv-fmm method with accuracy controlling parameter $p = 18$ and maximum step size $s = 1$ for different transform sizes n. Shown are the times for pre-computation t_{p} and for the computation of the actual transform t_{t}. Both are also shown after division through the expected asymptotic expression in terms of the transform size n. Furthermore, the component-wise error E_{∞}^{c} and the relative infinity norm error E_{∞} are reported.

α	β	n	t_{p}	t_{p}/n^2	t_{t}	t_{t}/n^2	E_∞^{c}	E_∞
-0.2	-0.4	256	4.5E-02	6.9E-07	6.3E-05	9.6E-10	5.7E-16	7.2E-16
-0.2	-0.4	512	1.5E-01	5.6E-07	2.9E-04	1.1E-09	6.5E-16	5.9E-16
-0.2	-0.4	1024	5.6E-01	5.4E-07	1.3E-03	1.2E-09	7.1E-16	7.2E-16
-0.2	-0.4	2048	2.2E+00	5.3E-07	5.3E-03	1.3E-09	6.6E-16	5.7E-16
-0.2	-0.4	4096	8.8E+00	5.2E-07	2.1E-02	1.2E-09	6.5E-16	6.9E-16
-0.2	-0.4	8192	3.5E+01	5.2E-07	7.6E-02	1.1E-09	6.1E-16	1.0E-15
-0.2	-0.4	16384	1.4E+02	5.2E-07	3.0E-01	1.1E-09	6.5E-16	9.7E-16
-0.2	0.5	256	5.0E-02	7.6E-07	6.5E-05	9.9E-10	9.2E-16	8.0E-17
-0.2	0.5	512	1.6E-01	6.2E-07	2.6E-04	9.8E-10	8.5E-16	2.2E-16
-0.2	0.5	1024	6.1E-01	5.8E-07	1.4E-03	1.3E-09	8.5E-16	1.2E-16
-0.2	0.5	2048	2.4E+00	5.7E-07	5.2E-03	1.2E-09	1.0E-15	8.0E-16
-0.2	0.5	4096	9.6E+00	5.7E-07	2.1E-02	1.2E-09	9.1E-16	1.7E-16
-0.2	0.5	8192	3.8E+01	5.7E-07	7.6E-02	1.1E-09	9.0E-16	1.2E-16
-0.2	0.5	16384	1.5E+02	5.7E-07	3.0E-01	1.1E-09	8.9E-16	5.2E-16
0.5	-0.2	256	4.4E-02	6.7E-07	6.2E-05	9.4E-10	3.8E-16	4.6E-16
0.5	-0.2	512	1.5E-01	5.7E-07	3.1E-04	1.2E-09	3.5E-16	8.0E-16
0.5	-0.2	1024	5.6E-01	5.4E-07	1.7E-03	1.7E-09	4.7E-16	9.2E-16
0.5	-0.2	2048	2.2E+00	5.3E-07	5.2E-03	1.2E-09	4.7E-16	1.1E-15
0.5	-0.2	4096	8.8E+00	5.2E-07	2.1E-02	1.2E-09	5.8E-16	1.0E-15
0.5	-0.2	8192	3.5E+01	5.2E-07	7.6E-02	1.1E-09	3.3E-16	1.2E-15
0.5	-0.2	16384	1.4E+02	5.1E-07	3.0E-01	1.1E-09	3.1E-16	1.6E-15
0.5	1.4	256	5.2E-02	8.0E-07	6.6E-05	1.0E-09	8.9E-16	2.3E-16
0.5	1.4	512	1.7E-01	6.3E-07	2.6E-04	9.8E-10	1.3E-15	8.0E-16
0.5	1.4	1024	6.3E-01	6.0E-07	1.4E-03	1.3E-09	1.1E-15	3.0E-16
0.5	1.4	2048	2.5E+00	6.0E-07	5.0E-03	1.2E-09	1.1E-15	2.1E-16
0.5	1.4	4096	1.0E+01	5.9E-07	1.9E-02	1.1E-09	1.0E-15	1.5E-16
0.5	1.4	8192	4.0E+01	5.9E-07	7.5E-02	1.1E-09	1.2E-15	3.0E-16
0.5	1.4	16384	1.6E+02	5.9E-07	3.0E-01	1.1E-09	1.0E-15	2.7E-16
5.9	8.1	256	5.0E-02	7.6E-07	6.1E-05	9.3E-10	8.5E-16	9.4E-16
5.9	8.1	512	1.7E-01	6.5E-07	2.6E-04	9.9E-10	9.7E-16	3.4E-16
5.9	8.1	1024	6.5E-01	6.2E-07	1.4E-03	1.3E-09	8.7E-16	4.5E-16
5.9	8.1	2048	2.6E+00	6.1E-07	5.3E-03	1.3E-09	1.1E-15	1.0E-15
5.9	8.1	4096	1.0E+01	6.1E-07	1.9E-02	1.1E-09	1.0E-15	2.2E-16
5.9	8.1	8192	4.1E+01	6.1E-07	9.3E-02	1.4E-09	1.1E-15	2.5E-16
5.9	8.1	16384	1.6E+02	6.0E-07	3.0E-01	1.1E-09	1.3E-15	4.9E-16
9.0	4.8	256	5.0E-02	7.6E-07	6.4E-05	9.8E-10	1.1E-15	1.0E-15
9.0	4.8	512	1.6E-01	6.0E-07	2.5E-04	9.4E-10	1.1E-15	7.1E-15
9.0	4.8	1024	6.0E-01	5.7E-07	1.5E-03	1.4E-09	9.7E-16	2.3E-15
9.0	4.8	2048	2.4E+00	5.6E-07	5.3E-03	1.3E-09	9.5E-16	1.7E-15
9.0	4.8	4096	9.4E+00	5.6E-07	2.0E-02	1.2E-09	1.2E-15	3.6E-15
9.0	4.8	8192	3.7E+01	5.6E-07	7.7E-02	1.1E-09	1.1E-15	2.2E-15
9.0	4.8	16384	1.5E+02	5.6E-07	3.0E-01	1.1E-09	1.0E-15	4.5E-15

Table 2.10: Test results for the connection between the Gegenbauer polynomials $\{C_n^{(\alpha)}\}_{n\in\mathbb{N}_0}$ and $\{C_n^{(\beta)}\}_{n\in\mathbb{N}_0}$ computed with the direct method for different transform sizes n. Shown are the times for pre-computation t_{p} and for the computation of the actual transform t_{t}. Both are also shown after division through the expected asymptotic expression in terms of the transform size n. Furthermore, the component-wise error E_∞^{c} and the relative infinity norm error E_∞ are reported.

α	β	n	t_{p}	t_{p}/n^3	t_{t}	t_{t}/n^2	E_∞^{c}	E_∞
-0.2	-0.4	256	3.6E-02	2.1E-09	6.7E-05	1.0E-09	2.4E-15	1.3E-15
-0.2	-0.4	512	1.8E-01	1.4E-09	4.3E-04	1.6E-09	4.5E-15	2.4E-15
-0.2	-0.4	1024	1.4E+00	1.3E-09	1.6E-03	1.5E-09	4.5E-15	4.3E-15
-0.2	-0.4	2048	1.6E+01	1.9E-09	5.2E-03	1.2E-09	1.1E-14	7.7E-15
-0.2	-0.4	4096	1.4E+02	2.0E-09	1.9E-02	1.1E-09	3.1E-14	1.6E-14
-0.2	-0.4	8192	-	-	-	-	-	-
-0.2	-0.4	16384	-	-	-	-	-	-
-0.2	0.5	256	3.8E-02	2.3E-09	6.7E-05	1.0E-09	4.8E-15	2.1E-16
-0.2	0.5	512	1.9E-01	1.4E-09	4.2E-04	1.6E-09	9.5E-15	2.2E-16
-0.2	0.5	1024	1.4E+00	1.3E-09	1.6E-03	1.5E-09	2.6E-14	2.0E-16
-0.2	0.5	2048	1.6E+01	1.9E-09	5.1E-03	1.2E-09	4.6E-14	7.8E-16
-0.2	0.5	4096	1.4E+02	2.0E-09	1.9E-02	1.1E-09	9.1E-14	5.3E-16
-0.2	0.5	8192	-	-	-	-	-	-
-0.2	0.5	16384	-	-	-	-	-	-
0.5	-0.2	256	3.6E-02	2.2E-09	6.9E-05	1.0E-09	7.8E-16	8.7E-16
0.5	-0.2	512	1.8E-01	1.3E-09	4.2E-04	1.6E-09	5.8E-16	1.4E-15
0.5	-0.2	1024	1.4E+00	1.3E-09	1.6E-03	1.5E-09	6.2E-16	9.9E-16
0.5	-0.2	2048	1.6E+01	1.9E-09	5.2E-03	1.2E-09	8.2E-16	2.4E-15
0.5	-0.2	4096	1.4E+02	2.0E-09	1.9E-02	1.2E-09	9.9E-16	2.2E-15
0.5	-0.2	8192	-	-	-	-	-	-
0.5	-0.2	16384	-	-	-	-	-	-
0.5	1.4	256	3.5E-02	2.1E-09	7.5E-05	1.1E-09	1.6E-15	3.2E-16
0.5	1.4	512	1.8E-01	1.4E-09	4.2E-04	1.6E-09	2.4E-15	1.2E-15
0.5	1.4	1024	1.4E+00	1.3E-09	1.6E-03	1.5E-09	2.2E-15	4.0E-16
0.5	1.4	2048	1.6E+01	1.9E-09	5.1E-03	1.2E-09	2.0E-15	4.2E-16
0.5	1.4	4096	1.4E+02	2.0E-09	1.9E-02	1.1E-09	2.5E-15	4.0E-16
0.5	1.4	8192	-	-	-	-	-	-
0.5	1.4	16384	-	-	-	-	-	-
5.9	8.1	256	3.5E-02	2.1E-09	6.2E-05	9.5E-10	1.1E-14	1.0E-14
5.9	8.1	512	1.8E-01	1.4E-09	4.3E-04	1.6E-09	2.2E-14	5.9E-15
5.9	8.1	1024	1.4E+00	1.3E-09	1.5E-03	1.5E-09	4.9E-14	6.4E-15
5.9	8.1	2048	1.6E+01	1.9E-09	5.0E-03	1.2E-09	9.1E-14	1.2E-14
5.9	8.1	4096	1.4E+02	2.0E-09	1.9E-02	1.1E-09	2.3E-13	4.6E-15
5.9	8.1	8192	-	-	-	-	-	-
5.9	8.1	16384	-	-	-	-	-	-
9.0	4.8	256	3.8E-02	2.2E-09	7.8E-05	1.2E-09	1.6E-15	2.7E-15
9.0	4.8	512	1.8E-01	1.4E-09	4.2E-04	1.6E-09	1.3E-15	1.6E-14
9.0	4.8	1024	1.4E+00	1.3E-09	1.6E-03	1.5E-09	1.4E-15	4.0E-15
9.0	4.8	2048	1.6E+01	1.9E-09	5.2E-03	1.2E-09	1.8E-15	5.3E-15
9.0	4.8	4096	1.4E+02	2.0E-09	1.9E-02	1.1E-09	1.8E-15	6.7E-15
9.0	4.8	8192	-	-	-	-	-	-
9.0	4.8	16384	-	-	-	-	-	-

Table 2.11: Test results for the connection between the Gegenbauer polynomials $\{C_n^{(\alpha)}\}_{n\in\mathbb{N}_0}$ and $\{C_n^{(\beta)}\}_{n\in\mathbb{N}_0}$ computed with the usmv-direct method for different transform sizes n. Shown are the times for pre-computation t_{p} and for the computation of the actual transform t_{t}. Both are also shown after division through the expected asymptotic expression in terms of the transform size n. Furthermore, the component-wise error E_∞^{c} and the relative infinity norm error E_∞ are reported. For sizes $n > 4096$ the transformation was not computed due to unacceptably large computation times in comparison to the direct method; see Table 2.10.

α	β	n	t_{p}	t_{p}/n^2	t_{t}	$t_{\mathrm{t}}/(n \log n)$	E_∞^{c}	E_∞
-0.2	-0.4	256	2.5E-02	3.8E-07	1.7E-04	1.2E-07	2.4E-15	1.4E-15
-0.2	-0.4	512	5.0E-02	1.9E-07	3.8E-04	1.2E-07	4.0E-15	2.3E-15
-0.2	-0.4	1024	1.0E-01	9.7E-08	9.9E-04	1.4E-07	4.4E-15	4.2E-15
-0.2	-0.4	2048	2.7E-01	6.3E-08	3.8E-03	2.4E-07	1.1E-14	7.5E-15
-0.2	-0.4	4096	6.6E-01	4.0E-08	6.9E-03	2.0E-07	3.1E-14	1.6E-14
-0.2	-0.4	8192	1.9E+00	2.9E-08	2.1E-02	2.8E-07	7.4E-14	3.5E-14
-0.2	-0.4	16384	6.1E+00	2.3E-08	3.3E-02	2.1E-07	1.4E-13	6.1E-14
-0.2	0.5	256	2.4E-02	3.7E-07	1.5E-04	1.1E-07	1.1E-14	4.3E-16
-0.2	0.5	512	4.7E-02	1.8E-07	3.7E-04	1.2E-07	1.1E-14	2.0E-15
-0.2	0.5	1024	1.0E-01	9.6E-08	1.0E-03	1.4E-07	2.8E-14	3.7E-16
-0.2	0.5	2048	2.5E-01	6.0E-08	3.8E-03	2.5E-07	4.9E-14	2.0E-15
-0.2	0.5	4096	6.6E-01	4.0E-08	9.1E-03	2.7E-07	9.2E-14	2.8E-15
-0.2	0.5	8192	1.9E+00	2.9E-08	2.1E-02	2.8E-07	2.2E-13	1.4E-15
-0.2	0.5	16384	6.1E+00	2.3E-08	3.3E-02	2.1E-07	4.1E-13	5.5E-15
0.5	-0.2	256	1.9E-02	2.9E-07	1.4E-04	9.8E-08	2.1E-15	4.0E-15
0.5	-0.2	512	4.8E-02	1.8E-07	3.8E-04	1.2E-07	7.6E-16	2.1E-15
0.5	-0.2	1024	1.0E-01	9.9E-08	1.0E-03	1.4E-07	1.3E-15	5.1E-15
0.5	-0.2	2048	2.5E-01	6.1E-08	3.7E-03	2.4E-07	1.3E-15	3.2E-15
0.5	-0.2	4096	6.7E-01	4.0E-08	8.9E-03	2.6E-07	1.4E-15	1.2E-14
0.5	-0.2	8192	1.9E+00	2.9E-08	2.0E-02	2.8E-07	1.9E-15	8.7E-15
0.5	-0.2	16384	6.1E+00	2.3E-08	3.3E-02	2.0E-07	2.4E-15	1.0E-14
0.5	1.4	256	2.4E-02	3.7E-07	1.6E-04	1.2E-07	1.2E-14	9.1E-16
0.5	1.4	512	4.9E-02	1.9E-07	3.6E-04	1.1E-07	8.2E-14	2.3E-15
0.5	1.4	1024	1.0E-01	9.7E-08	9.9E-04	1.4E-07	9.8E-14	6.1E-16
0.5	1.4	2048	2.5E-01	6.0E-08	3.8E-03	2.4E-07	9.8E-14	7.7E-16
0.5	1.4	4096	6.6E-01	3.9E-08	9.1E-03	2.7E-07	9.2E-14	1.9E-15
0.5	1.4	8192	1.9E+00	2.9E-08	2.1E-02	2.8E-07	1.6E-13	1.5E-15
0.5	1.4	16384	6.1E+00	2.3E-08	3.3E-02	2.1E-07	1.5E-13	5.7E-15
5.9	8.1	256	2.4E-02	3.7E-07	1.7E-04	1.2E-07	4.3E-13	3.7E-13
5.9	8.1	512	5.0E-02	1.9E-07	3.9E-04	1.2E-07	7.4E-12	1.5E-13
5.9	8.1	1024	1.0E-01	1.0E-07	1.1E-03	1.6E-07	1.3E-10	1.3E-12
5.9	8.1	2048	2.6E-01	6.1E-08	3.9E-03	2.5E-07	4.8E-10	1.7E-12
5.9	8.1	4096	6.6E-01	3.9E-08	6.7E-03	2.0E-07	1.5E-09	2.6E-13
5.9	8.1	8192	1.9E+00	2.9E-08	1.6E-02	2.1E-07	6.9E-09	6.3E-13
5.9	8.1	16384	6.1E+00	2.3E-08	3.3E-02	2.1E-07	1.0E-07	1.5E-11
9.0	4.8	256	2.5E-02	3.8E-07	1.8E-04	1.3E-07	5.6E-09	6.4E-09
9.0	4.8	512	4.8E-02	1.8E-07	3.8E-04	1.2E-07	4.9E-09	1.6E-07
9.0	4.8	1024	1.1E-01	1.1E-07	1.1E-03	1.6E-07	2.1E-07	2.5E-06
9.0	4.8	2048	2.6E-01	6.3E-08	3.8E-03	2.5E-07	1.2E-04	3.3E-04
9.0	4.8	4096	6.6E-01	4.0E-08	6.8E-03	2.0E-07	4.5E-03	2.7E-02
9.0	4.8	8192	1.9E+00	2.9E-08	2.0E-02	2.8E-07	4.4E-01	3.1E+00
9.0	4.8	16384	6.1E+00	2.3E-08	3.3E-02	2.1E-07	1.5E+02	3.3E+03

Table 2.12: Test results for the connection between the Gegenbauer polynomials $\{C_n^{(\alpha)}\}_{n\in\mathbb{N}_0}$ and $\{C_n^{(\beta)}\}_{n\in\mathbb{N}_0}$ computed with the usmv-fmm method with accuracy controlling parameter $p = 18$ and arbitrarily large step size s for different transform sizes n. Shown are the times for pre-computation t_{p} and for the computation of the actual transform t_{t}. Both are also shown after division through the expected asymptotic expression in terms of the transform size n. Furthermore, the component-wise error E_∞^{c} and the relative infinity norm error E_∞ are reported.

α	β	n	t_p	t_p/n^2	t_t	$t_\mathrm{t}/(n\log n)$	E_∞^c	E_∞
-0.2	-0.4	256	2.4E-02	3.7E-07	1.7E-04	1.2E-07	2.4E-15	1.4E-15
-0.2	-0.4	512	5.2E-02	2.0E-07	4.6E-04	1.4E-07	4.0E-15	2.3E-15
-0.2	-0.4	1024	1.1E-01	1.0E-07	1.0E-03	1.5E-07	4.4E-15	4.2E-15
-0.2	-0.4	2048	2.5E-01	6.0E-08	3.8E-03	2.4E-07	1.1E-14	7.5E-15
-0.2	-0.4	4096	6.7E-01	4.0E-08	8.9E-03	2.6E-07	3.1E-14	1.6E-14
-0.2	-0.4	8192	1.9E+00	2.9E-08	2.1E-02	2.8E-07	7.4E-14	3.5E-14
-0.2	-0.4	16384	6.1E+00	2.3E-08	4.5E-02	2.8E-07	1.4E-13	6.1E-14
-0.2	0.5	256	2.5E-02	3.7E-07	1.7E-04	1.2E-07	1.1E-14	4.3E-16
-0.2	0.5	512	4.8E-02	1.8E-07	5.3E-04	1.7E-07	1.1E-14	2.0E-15
-0.2	0.5	1024	1.0E-01	9.8E-08	1.0E-03	1.4E-07	2.8E-14	3.7E-16
-0.2	0.5	2048	2.5E-01	6.0E-08	2.9E-03	1.9E-07	4.9E-14	2.0E-15
-0.2	0.5	4096	6.7E-01	4.0E-08	9.0E-03	2.6E-07	9.2E-14	2.8E-15
-0.2	0.5	8192	1.9E+00	2.9E-08	1.6E-02	2.1E-07	2.2E-13	1.4E-15
-0.2	0.5	16384	6.1E+00	2.3E-08	4.5E-02	2.8E-07	4.1E-13	5.5E-15
0.5	-0.2	256	2.4E-02	3.7E-07	1.6E-04	1.2E-07	2.1E-15	4.0E-15
0.5	-0.2	512	4.7E-02	1.8E-07	5.4E-04	1.7E-07	7.6E-16	2.2E-15
0.5	-0.2	1024	1.0E-01	9.7E-08	1.0E-03	1.5E-07	1.3E-15	5.1E-15
0.5	-0.2	2048	2.6E-01	6.2E-08	3.7E-03	2.4E-07	1.3E-15	3.2E-15
0.5	-0.2	4096	6.6E-01	4.0E-08	6.7E-03	2.0E-07	1.4E-15	1.2E-14
0.5	-0.2	8192	1.9E+00	2.9E-08	2.3E-02	3.1E-07	1.9E-15	8.7E-15
0.5	-0.2	16384	6.1E+00	2.3E-08	4.5E-02	2.8E-07	2.4E-15	9.9E-15
0.5	1.4	256	2.5E-02	3.8E-07	1.6E-04	1.1E-07	1.2E-14	9.1E-16
0.5	1.4	512	4.8E-02	1.8E-07	5.4E-04	1.7E-07	8.3E-14	2.3E-15
0.5	1.4	1024	1.0E-01	9.9E-08	1.0E-03	1.4E-07	9.8E-14	6.1E-16
0.5	1.4	2048	2.5E-01	6.0E-08	3.8E-03	2.4E-07	9.8E-14	7.7E-16
0.5	1.4	4096	6.6E-01	4.0E-08	9.0E-03	2.6E-07	9.2E-14	1.9E-15
0.5	1.4	8192	1.9E+00	2.9E-08	2.1E-02	2.8E-07	1.6E-13	1.5E-15
0.5	1.4	16384	6.1E+00	2.3E-08	4.5E-02	2.8E-07	1.5E-13	5.7E-15
5.9	8.1	256	5.5E-02	8.4E-07	3.4E-04	2.4E-07	1.1E-14	1.1E-14
5.9	8.1	512	1.2E-01	4.4E-07	1.2E-03	3.7E-07	3.5E-14	5.9E-15
5.9	8.1	1024	2.8E-01	2.6E-07	3.5E-03	4.9E-07	8.0E-14	6.4E-15
5.9	8.1	2048	7.2E-01	1.7E-07	1.1E-02	7.1E-07	6.3E-14	1.1E-14
5.9	8.1	4096	1.9E+00	1.2E-07	1.9E-02	5.5E-07	1.7E-13	2.5E-15
5.9	8.1	8192	5.7E+00	8.5E-08	5.9E-02	7.9E-07	4.9E-13	6.6E-15
5.9	8.1	16384	1.8E+01	6.8E-08	1.3E-01	8.2E-07	7.8E-13	4.3E-15
9.0	4.8	256	7.8E-02	1.2E-06	6.0E-04	4.2E-07	1.1E-15	1.3E-15
9.0	4.8	512	1.9E-01	7.1E-07	2.1E-03	6.6E-07	1.6E-15	1.5E-14
9.0	4.8	1024	4.5E-01	4.3E-07	5.5E-03	7.8E-07	1.3E-15	5.8E-15
9.0	4.8	2048	1.2E+00	2.8E-07	1.8E-02	1.1E-06	1.1E-15	6.5E-15
9.0	4.8	4096	3.2E+00	1.9E-07	4.2E-02	1.2E-06	1.8E-15	2.1E-14
9.0	4.8	8192	9.6E+00	1.4E-07	9.7E-02	1.3E-06	1.9E-15	3.3E-14
9.0	4.8	16384	3.0E+01	1.1E-07	2.2E-01	1.4E-06	2.1E-15	6.5E-14

Table 2.13: Test results for the connection between the Gegenbauer polynomials $\{C_n^{(\alpha)}\}_{n\in\mathbb{N}_0}$ and $\{C_n^{(\beta)}\}_{n\in\mathbb{N}_0}$ computed with the usmv-fmm method with accuracy controlling parameter $p = 18$ and maximum step size $s = 1$ for different transform sizes n. Shown are the times for pre-computation t_p and for the computation of the actual transform t_t. Both are also shown after division through the expected asymptotic expression in terms of the transform size n. Furthermore, the component-wise error E_∞^c and the relative infinity norm error E_∞ are reported.

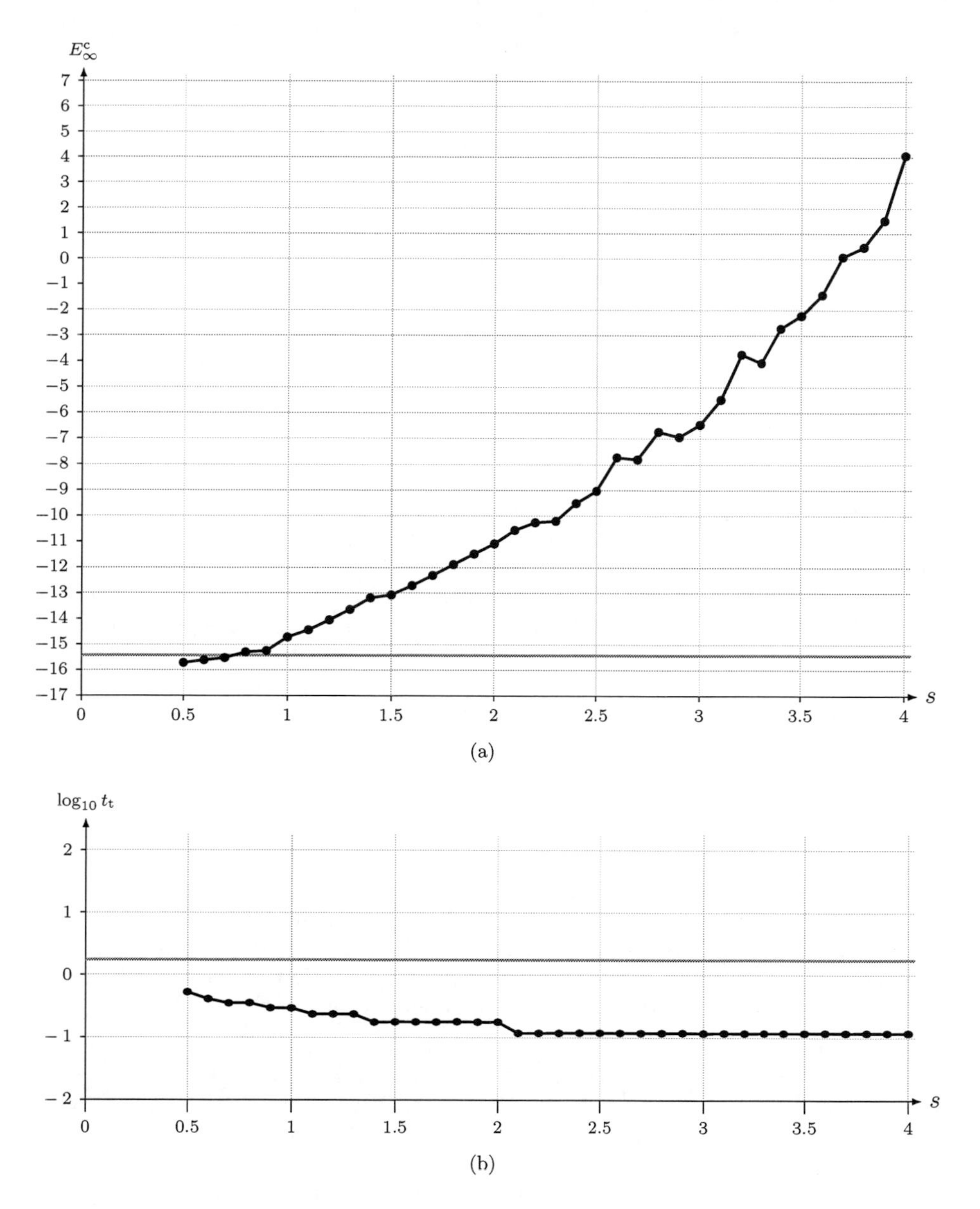

Figure 2.6: Error measure E^{c}_{∞} versus the step size s (a) and computation time t_t for a single transformation versus the step size s (b) for the last test case for Laguerre polynomials reported in Tables 2.2 to 2.5, that is, for the conversion from the polynomials $\{L_n^{(9.7)}\}_{n\in\mathbb{N}_0}$ to the polynomials $\{L_n^{(5.5)}\}_{n\in\mathbb{N}_0}$, at transform size $n = 16384$. Shown are the usmv-fmm method with accuracy controlling parameter $p = 18$ (solid, black) and the corresponding errors and times for the direct method (solid, dark gray).

2.5 The Eigendecomposition of symmetric semiseparable matrices

In this section a divide-and-conquer method to compute the eigendecomposition of extended symmetric diagonal plus generator representable semiseparable matrices is reviewed. It is similar to the method for diagonal plus upper or lower generator representable semiseparable matrices from Section 2.3. Historically though, the method described in this section was developed earlier by Chandrasekaran and Gu [9] for symmetric block-diagonal plus $(1, 1)$-generator representable semiseparable matrices. Similar techniques also appeared in [60]. In addition to describing the original algorithm, we also give a simple extension to higher semiseparability ranks.
As before, the divide-and-conquer method will not only allow us to compute eigendecompositions explicitly, but also to efficiently pre-compute enough data that allows to apply an $n \times n$ eigenvector matrix to any vector with $\mathcal{O}(n \log n \log(1/\varepsilon))$ arithmetic operations, instead of the usual $\mathcal{O}(n^2)$. The reason for this is the same as before, that is, usage of the fast multipole method to accelerate the calculation of certain matrix-vector products.
Among the differences to the triangular case is that the divide-and-conquer method for extended symmetric diagonal plus generator representable semiseparable matrices requires a few extra considerations. To guarantee numerical stability, some modifications to otherwise straightforward procedures are necessary. We will mention these below as we walk through the procedure.
Any symmetric matrix, and in particular, any extended symmetric diagonal plus generator representable semiseparable matrix, has real eigenvalues. The spectral theorem [26, Theorem 8.1.1, p. 393] asserts that a symmetric eigendecomposition of the form

$$\mathbf{A} = \mathbf{Q}\,\mathbf{\Lambda}\,\mathbf{Q}^{\mathrm{T}} \tag{2.15}$$

exists, with an orthogonal eigenvector matrix $\mathbf{Q}$ and a real diagonal eigenvalue matrix $\mathbf{\Lambda}$. In the following, we describe the divide-and-conquer method from [9] to compute the desired eigendecomposition.

2.5.1 Divide-and-conquer method. We first describe the divide-and-conquer method for symmetric diagonal plus $(1, 1)$-generator representable matrices $\mathbf{A}$. As before, in the *divide phase*, the matrix $\mathbf{A}$ is recursively divided into smaller matrices until these are sufficiently small. In the *conquer phase*, the eigendecompositions of these small matrices are successively recombined to obtain the eigendecomposition of the full matrix $\mathbf{A}$.

Divide phase

Given a symmetric diagonal plus $(1, 1)$-generator representable semiseparable matrix $\mathbf{A} = \operatorname{diag}\mathbf{d} + \operatorname{triu}(\mathbf{u}\mathbf{v}^{\mathrm{T}}, 1) + \operatorname{tril}(\mathbf{u}\mathbf{v}^{\mathrm{T}}, -1)$, we would like to write this in the form of several smaller matrices of the same type. This can be done in the following way. Take $\delta = \pm 1$ to be a freely chosen scalar. Split each of the vectors $\mathbf{d}$, $\mathbf{u}$, $\mathbf{v}$ into two vectors with the first $\lfloor n/2 \rfloor$ components in the first vector, and the remaining components in the second vector. That is, define $\mathbf{d}_1$, $\mathbf{d}_2$, $\mathbf{u}_1$, $\mathbf{u}_2$, $\mathbf{v}_1$, $\mathbf{v}_2$ and $\mathbf{w}$ such that

$$\mathbf{d} = \begin{pmatrix} \mathbf{d}_1 \\ \mathbf{d}_2 \end{pmatrix}, \qquad \mathbf{u} = \begin{pmatrix} \mathbf{u}_1 \\ \mathbf{u}_2 \end{pmatrix}, \qquad \mathbf{v} = \begin{pmatrix} \mathbf{v}_1 \\ \mathbf{v}_2 \end{pmatrix}, \qquad \mathbf{w} = \begin{pmatrix} \delta\,\mathbf{u}_1 \\ \mathbf{v}_2 \end{pmatrix}.$$

We can now write $\mathbf{A}$ in the form

$$\mathbf{A} = \begin{pmatrix} \hat{\mathbf{A}}_1 & \mathbf{0} \\ \mathbf{0} & \hat{\mathbf{A}}_2 \end{pmatrix} + \delta\,\mathbf{w}\mathbf{w}^{\mathrm{T}},$$

where $\hat{\mathbf{A}}_1$ and $\hat{\mathbf{A}}_2$ are defined to the symmetric diagonal plus $(1,1)$-generator representable semiseparable matrices

$$\hat{\mathbf{A}}_1 = \operatorname{diag}\left(\mathbf{d_1} - \delta \operatorname{diag}(\mathbf{u}_1\mathbf{u}_1^{\mathrm{T}})\right) + \operatorname{triu}\left(\mathbf{u}_1(\mathbf{v}_1 - \delta\mathbf{u}_1)^{\mathrm{T}}, 1\right) + \operatorname{tril}\left((\mathbf{v}_1 - \delta\mathbf{u}_1)\mathbf{u}_1^{\mathrm{T}}, -1\right),$$
$$\hat{\mathbf{A}}_2 = \operatorname{diag}\left(\mathbf{d_2} - \delta \operatorname{diag}(\mathbf{v}_2\mathbf{v}_2^{\mathrm{T}})\right) + \operatorname{triu}\left((\mathbf{u}_2 - \delta\mathbf{v}_2)\mathbf{v}_2^{\mathrm{T}}, 1\right) + \operatorname{tril}\left(\mathbf{v}_2(\mathbf{u}_2 - \delta\mathbf{v}_2)^{\mathrm{T}}, -1\right).$$

This result can be easily verified. Each of the matrices $\hat{\mathbf{A}}_1$ and $\hat{\mathbf{A}}_2$ may now be decomposed into a similar pattern.

Conquer phase

In the conquer phase we regroup the subproblems into larger problems. Suppose that two symmetric diagonal plus $(1,1)$-generator representable semiseparable matrices $\hat{\mathbf{A}}_1$ and $\hat{\mathbf{A}}_2$, obtained from the divide phase, have the eigendecomposition

$$\hat{\mathbf{A}}_1 = \mathbf{Q}_1\mathbf{\Lambda}_1\mathbf{Q}_1^{\mathrm{T}} \quad \text{and} \quad \hat{\mathbf{A}}_2 = \mathbf{Q}_2\mathbf{\Lambda}_2\mathbf{Q}_2^{\mathrm{T}},$$

with the real, diagonal eigenvalue matrices $\mathbf{\Lambda}_1$ and $\mathbf{\Lambda}_2$ and the orthogonal eigenvector matrices $\mathbf{Q}_1$ and $\mathbf{Q}_2$, then the matrix $\mathbf{A}$ has the representation

$$\mathbf{A} = \begin{pmatrix} \mathbf{Q}_1 & \mathbf{0} \\ \mathbf{0} & \mathbf{Q}_2 \end{pmatrix} (\mathbf{\Lambda} + \delta\,\mathbf{z}\,\mathbf{z}^{\mathrm{T}}) \begin{pmatrix} \mathbf{Q}_1 & \mathbf{0} \\ \mathbf{0} & \mathbf{Q}_2 \end{pmatrix}^{\mathrm{T}},$$

where δ is defined as it was for the divide phase and $\mathbf{\Lambda}$ is the diagonal matrix and $\mathbf{z}$ a vector defined by

$$\mathbf{\Lambda} = \begin{pmatrix} \mathbf{\Lambda}_1 & \mathbf{0} \\ \mathbf{0} & \mathbf{\Lambda}_2 \end{pmatrix} \quad \text{and} \quad \mathbf{z} = \begin{pmatrix} \mathbf{Q}_1 & \mathbf{0} \\ \mathbf{0} & \mathbf{Q}_2 \end{pmatrix}^{\mathrm{T}} \mathbf{w}.$$

Suppose that we can efficiently compute the eigendecomposition of the symmetric rank-one modified diagonal matrix $\mathbf{\Lambda} + \delta\,\mathbf{z}\mathbf{z}^{\mathrm{T}}$ which we write as

$$\mathbf{\Lambda} + \delta\,\mathbf{z}\,\mathbf{z}^{\mathrm{T}} = \mathbf{P}\,\mathbf{\Omega}\,\mathbf{P}^{\mathrm{T}}.$$

Then we can write the eigendecomposition of the full matrix $\mathbf{A}$ as

$$\mathbf{A} = \begin{pmatrix} \mathbf{Q}_1 & \mathbf{0} \\ \mathbf{0} & \mathbf{Q}_2 \end{pmatrix} \mathbf{P}\,\mathbf{\Omega}\,\mathbf{P}^{\mathrm{T}} \begin{pmatrix} \mathbf{Q}_1 & \mathbf{0} \\ \mathbf{0} & \mathbf{Q}_2 \end{pmatrix}^{\mathrm{T}} = \mathbf{Q}\,\mathbf{\Omega}\,\mathbf{Q}^{\mathrm{T}}, \quad \text{with } \mathbf{Q} = \begin{pmatrix} \mathbf{Q}_1 & \mathbf{0} \\ \mathbf{0} & \mathbf{Q}_2 \end{pmatrix} \mathbf{P}.$$

Similar to the triangular case, it remains to explain how the eigendecomposition of the symmetric rank-one modified diagonal matrix $\mathbf{\Lambda} + \delta\,\mathbf{z}\,\mathbf{z}^{\mathrm{T}}$ can be computed, and how the eigenvector matrix $\mathbf{P}$ can be applied efficiently. This is done in the following.

2.5.2 Symmetric rank-one modified eigenproblem. The problem of determining the eigendecomposition of a symmetric rank-one modified diagonal matrix $\mathbf{\Lambda} + \delta\,\mathbf{z}\,\mathbf{z}^{\mathrm{T}}$ was formulated by Golub [25] and subsequently investigated in [8, 14, 15, 29, 37, 71]. It is valid to assume that all diagonal entries of the diagonal matrix $\mathbf{\Lambda}$ are numerically distinct and that all entries in the vector $\mathbf{z}$ are bounded away from zero. If not, one can use the deflation procedure described in [15], e.g., with the criterion used in [29], to arrange this. This is an important detail for a successful implementation, as deflation is often necessary in practical situations. Moreover, we require that, by permutations, we have ordered the diagonal entries of $\mathbf{\Lambda}$ increasingly. The following theorem, restating results found in [25] and [8], characterizes the structure of the desired eigendecomposition.

Theorem 2.25 *Let $\mathbf{\Lambda}$ be a diagonal matrix with entries $\lambda_1 < \lambda_2 < \cdots < \lambda_n$, $\mathbf{z} = (z_1, z_2, \ldots, z_n)^{\mathrm{T}}$ a vector with non-zero entries, and $\delta \neq 0$. Then for the symmetric rank-one modified diagonal matrix $\mathbf{B} = \mathbf{\Lambda} + \delta\,\mathbf{z}\,\mathbf{z}^{\mathrm{T}}$ the following results hold:*

(i) *The eigenvalues* $\omega_1, \omega_2, \ldots, \omega_n$ *of the matrix* $\mathbf{B}$ *satisfy the interlacing property*

$$\begin{cases} \lambda_1 < \omega_1 < \lambda_2 < \omega_2 < \cdots < \lambda_n < \omega_n < \lambda_n + \delta\, \mathbf{z}^{\mathrm{T}}\mathbf{z}, & \text{if } \delta > 0, \\ \lambda_1 + \delta\, \mathbf{z}^{\mathrm{T}}\mathbf{z} < \omega_1 < \lambda_1 < \omega_2 < \lambda_2 < \cdots < \omega_n < \lambda_n, & \text{if } \delta < 0. \end{cases}$$

(ii) *The eigenvalues* $\omega_1, \omega_2, \ldots, \omega_n$ *of the matrix* $\mathbf{B}$ *are solutions to the secular equation*

$$1 + \delta \sum_{j=1}^{n} \frac{z_j^2}{\lambda_j - \omega} = 0. \tag{2.16}$$

(iii) *For each eigenvalue* ω_j*, the two corresponding unit-length eigenvectors* $\mathbf{p}_j$ *are given by*

$$\mathbf{p}_j = \pm \left(\sum_{i=1}^{n} \frac{z_i^2}{(\lambda_i - \omega_j)^2} \right)^{-1/2} \times \left(\frac{z_1}{\lambda_1 - \omega_j}, \frac{z_2}{\lambda_2 - \omega_j}, \ldots, \frac{z_n}{\lambda_n - \omega_j} \right)^{\mathrm{T}}.$$

The eigenvalues $\omega_1, \omega_2, \ldots, \omega_n$ of the matrix $\mathbf{B}$ can be efficiently obtained as the zeros of the rational equation (2.16) by using iterative methods; these are described, e.g., in [8, 57]. The popular subroutine library LAPACK implements the method described in [57]. A modified C translation was adopted by our own implementation. The eigenvectors $\mathbf{p}_1, \mathbf{p}_2, \ldots, \mathbf{p}_n$, have explicit expressions in terms of the entries of the vector $\mathbf{z}$, the diagonal entries λ_j and the eigenvalues ω_j for $j = 1, 2, \ldots, n$.

2.5.3 Efficient application of the eigenvector matrix. Again, an efficient method is needed to apply the eigenvector matrices $\mathbf{P}$ and $\mathbf{P}^{-1} = \mathbf{P}^{\mathrm{T}}$, obtained from the symmetric rank-one modified diagonal matrix $\mathbf{D} + \mathbf{z}\mathbf{z}^{\mathrm{T}}$, to an arbitrary vector. As before in Section 2.3.3, we can observe that the matrices $\mathbf{P}$ and $\mathbf{P}^{\mathrm{T}}$ are Cauchy-like matrices. This allows us to use the FMM to accelerate the calculation of matrix-vector products to $\mathcal{O}(n \log(1/\varepsilon))$ arithmetic operations, where ε is the desired accuracy. The last statement should be taken with care as the usual implementation of the FMM can poorly fail to provide acceptable accuracy. To see this, notice that the diagonal entries $\lambda_1, \lambda_2, \ldots, \lambda_n$ and the eigenvalues $\omega_1, \omega_2, \ldots, \omega_n$ can be very close to each other, and that all these quantities can contain rounding errors. This issue is usually not present in the triangular case. It implies that the calculation of differences of the form

$$\lambda_j - \omega_j \quad \text{or} \quad \lambda_{j+1} - \omega_j$$

can be subject to catastrophic cancellations. Therefore, a modification to the FMM was proposed in [30]. The iterative methods used to find the eigenvalues ω_j can be reformulated to return a representation of the form $\omega_j = \lambda_j + \mu_j$ or $\omega_j = \lambda_{j+1} - \mu_j$, respectively. That is, each eigenvalue ω_j is written as one of the enclosing diagonal entries, λ_j or λ_{j+1}, plus a possibly small perturbation μ_j. For the FMM, this representation is then used to compute all required differences of the form $c - \omega_j$ with any number c to high relative accuracy. For example, if $c = \omega_{j+1} = \lambda_{j+1} + \mu_{j+1}$ and $\omega_j = \lambda_{j+1} - \mu_j$, then the difference $\omega_{j+1} - \omega j$ can be obtained as

$$\omega_{j+1} - \omega j = \mu_{j+1} + \mu_j$$

to high relative accuracy. For the details, we refer the reader to [30].

2.5.4 Complexity of the divide-and-conquer algorithm. The cost analysis for the whole divide-and-conquer method is identical to that for triangular diagonal plus generator representable semiseparable matrices; see Section 2.3.4. A factorization of the eigenvector matrix $\mathbf{Q}$ into a product of block-diagonal matrices is obtained, for example,

$$\mathbf{Q} = \begin{pmatrix} \mathbf{Q}_{000} & & & \\ & \mathbf{Q}_{001} & & \\ & & \ddots & \\ & & & \mathbf{Q}_{111} \end{pmatrix} \begin{pmatrix} \mathbf{P}_{00} & & & \\ & \mathbf{P}_{01} & & \\ & & \mathbf{P}_{10} & \\ & & & \mathbf{P}_{11} \end{pmatrix} \begin{pmatrix} \mathbf{P}_{0} & \\ & \mathbf{P}_{1} \end{pmatrix} \mathbf{P}.$$

The divide-and-conquer method needs $O(n^2)$ arithmetic operations and an amount of memory in the order $O(n^2)$ to explicitly compute the full eigendecomposition $\mathbf{A} = \mathbf{Q}\,\boldsymbol{\Lambda}\,\mathbf{Q}^{\mathrm{T}}$ of the $n \times n$ matrix $\mathbf{A}$ with plain matrix-vector computations. To apply the eigenvector matrix $\mathbf{Q}$ to a vector takes $O(n^2)$ operations. With FMM-accelerated matrix-vector products, the matrix $\mathbf{Q}$ can be multiplied with any vector using only $\mathcal{O}\big(n \log n \log(1/\varepsilon)\big)$ arithmetic operations and an amount of memory in the same order. Recall also that we never need to setup the matrix $\mathbf{Q}$ explicitly.

2.5.5 Generalization to higher semiseparability ranks. We can generalize the discussed divide-and-conquer method for symmetric diagonal plus $(1,1)$-generator representable semiseparable matrices to higher semiseparability ranks. The modifications are similar to the triangular case. Any extended symmetric diagonal plus (p,p)-generator representable semiseparable matrix $\mathbf{A}$ can be written as

$$\mathbf{A} = \operatorname{diag}(\mathbf{d}) + \operatorname{triu}(\mathbf{U}\mathbf{V}^{\mathrm{T}}, 1) + \operatorname{tril}(\mathbf{V}\mathbf{U}^{\mathrm{T}}, -1),$$

with $n \times p$ matrices $\mathbf{U}$ and $\mathbf{V}$. The right-hand side may be decomposed into smaller matrices of the same type. To this end, take a diagonal matrix $\boldsymbol{\Delta} = \operatorname{diag}(\boldsymbol{\delta})$, built from a vector $\boldsymbol{\delta} = (\delta_1, \delta_2, \ldots, \delta_p)^{\mathrm{T}}$ with freely chosen entries $\delta_i = \pm 1$, $i = 1, 2, \ldots, p$, and define vectors $\mathbf{d}_1$, $\mathbf{d}_2$, and matrices $\mathbf{U}_1$, $\mathbf{U}_2$, $\mathbf{V}_1$, $\mathbf{V}_2$, and $\mathbf{W}$ similarly such that

$$\mathbf{d} = \begin{pmatrix} \mathbf{d}_1 \\ \mathbf{d}_2 \end{pmatrix}, \qquad \mathbf{U} = \begin{pmatrix} \mathbf{U}_1 \\ \mathbf{U}_2 \end{pmatrix}, \qquad \mathbf{V} = \begin{pmatrix} \mathbf{V}_1 \\ \mathbf{V}_2 \end{pmatrix}, \qquad \mathbf{W} = \begin{pmatrix} \boldsymbol{\Delta}\,\mathbf{U}_1 \\ \mathbf{V}_2 \end{pmatrix}.$$

Then the matrix $\mathbf{A}$ may be written as

$$\mathbf{A} = \begin{pmatrix} \hat{\mathbf{A}}_1 & \mathbf{0} \\ \mathbf{0} & \hat{\mathbf{A}}_2 \end{pmatrix} + \mathbf{W}\,\boldsymbol{\Delta}\,\mathbf{W}^{\mathrm{T}},$$

where $\hat{\mathbf{A}}_1$ and $\hat{\mathbf{A}}_2$ are defined to the extended symmetric diagonal plus (p,p)-generator representable semiseparable matrices

$$\begin{aligned} \hat{\mathbf{A}}_1 = {} & \operatorname{diag}\big(\mathbf{d}_1 - \operatorname{diag}(\mathbf{U}_1\,\boldsymbol{\Delta}\,\mathbf{U}_1^{\mathrm{T}})\big) + \operatorname{triu}\big(\mathbf{U}_1(\mathbf{V}_1 - \boldsymbol{\Delta}\,\mathbf{U}_1)^{\mathrm{T}}, 1\big) \\ & + \operatorname{tril}\big((\mathbf{V}_1 - \boldsymbol{\Delta}\,\mathbf{U}_1)\mathbf{U}_1^{\mathrm{T}}, -1\big), \\ \hat{\mathbf{A}}_2 = {} & \operatorname{diag}\big(\mathbf{d}_2 - \operatorname{diag}(\mathbf{V}_2\,\boldsymbol{\Delta}\,\mathbf{V}_2^{\mathrm{T}})\big) + \operatorname{triu}\big((\mathbf{U}_2 - \boldsymbol{\Delta}\,\mathbf{V}_2)\mathbf{V}_2^{\mathrm{T}}, 1\big) \\ & + \operatorname{tril}\big(\mathbf{V}_2(\mathbf{U}_2 - \boldsymbol{\Delta}\,\mathbf{V}_2)^{\mathrm{T}}, -1\big). \end{aligned}$$

In the conquer phase, we assume that the eigendecompositions of $\hat{\mathbf{A}}_1$ and $\hat{\mathbf{A}}_2$ are

$$\hat{\mathbf{A}}_1 = \mathbf{Q}_1\boldsymbol{\Lambda}_1\mathbf{Q}_1^{\mathrm{T}} \quad \text{and} \quad \hat{\mathbf{A}}_2 = \mathbf{Q}_2\boldsymbol{\Lambda}_2\mathbf{Q}_2^{\mathrm{T}}.$$

This implies the representation

$$\mathbf{A} = \begin{pmatrix} \mathbf{Q}_1 & \mathbf{0} \\ \mathbf{0} & \mathbf{Q}_2 \end{pmatrix} (\boldsymbol{\Lambda} + \mathbf{Z}\,\boldsymbol{\Delta}\,\mathbf{Z}^{\mathrm{T}}) \begin{pmatrix} \mathbf{Q}_1 & \mathbf{0} \\ \mathbf{0} & \mathbf{Q}_2 \end{pmatrix}^{\mathrm{T}},$$

where $\mathbf{\Delta}$ is defined as it was for the divide phase and $\mathbf{\Lambda}$ is a diagonal matrix and $\mathbf{Z}$ is a matrix defined by

$$\mathbf{\Lambda} = \begin{pmatrix} \mathbf{\Lambda}_1 & \mathbf{0} \\ \mathbf{0} & \mathbf{\Lambda}_2 \end{pmatrix} \quad \text{and} \quad \mathbf{Z} = \begin{pmatrix} \mathbf{Q}_1 & \mathbf{0} \\ \mathbf{0} & \mathbf{Q}_2 \end{pmatrix}^{\mathrm{T}} \mathbf{W}.$$

Let the eigendecomposition of the symmetric rank-p modified diagonal matrix $\mathbf{\Lambda}+\mathbf{Z}\,\mathbf{\Delta}\,\mathbf{Z}^{\mathrm{T}}$ be written as

$$\mathbf{\Lambda} + \mathbf{Z}\,\mathbf{\Delta}\,\mathbf{Z}^{\mathrm{T}} = \mathbf{P}\,\mathbf{\Omega}\,\mathbf{P}^{\mathrm{T}}.$$

Then we can write the eigendecomposition of $\mathbf{A}$ as

$$\mathbf{A} = \begin{pmatrix} \mathbf{Q}_1 & \mathbf{0} \\ \mathbf{0} & \mathbf{Q}_2 \end{pmatrix} \mathbf{P}\,\mathbf{\Omega}\,\mathbf{P}^{\mathrm{T}} \begin{pmatrix} \mathbf{Q}_1 & \mathbf{0} \\ \mathbf{0} & \mathbf{Q}_2 \end{pmatrix}^{\mathrm{T}} = \mathbf{Q}\,\mathbf{\Omega}\,\mathbf{Q}^{\mathrm{T}}, \qquad \text{with } \mathbf{Q} = \begin{pmatrix} \mathbf{Q}_1 & \mathbf{0} \\ \mathbf{0} & \mathbf{Q}_2 \end{pmatrix} \mathbf{P}. \tag{2.17}$$

It remains to efficiently compute the eigendecomposition of the symmetric rank-p modified diagonal matrix $\mathbf{\Lambda} + \mathbf{Z}\,\mathbf{\Delta}\,\mathbf{Z}^{\mathrm{T}}$. We write this in the equivalent form

$$\mathbf{\Lambda} + \mathbf{Z}\,\mathbf{\Delta}\,\mathbf{Z}^{\mathrm{T}} = \mathbf{\Lambda} + \delta_1 \mathbf{z}_1\,\mathbf{z}_1^{\mathrm{T}} + \delta_2 \mathbf{z}_2\,\mathbf{z}_2^{\mathrm{T}} + \cdots + \delta_r \mathbf{z}_p\,\mathbf{z}_p^{\mathrm{T}}, \tag{2.18}$$

and use Theorem 2.25 to cheaply obtain the eigendecomposition of the symmetric rank-one modified diagonal matrix $\mathbf{\Lambda} + \delta_1 \mathbf{z}_1\,\mathbf{z}_1^{\mathrm{T}}$ as

$$\mathbf{\Lambda} + \delta_1 \mathbf{z}_1\,\mathbf{z}_1^{\mathrm{T}} = \mathbf{P}_1\,\mathbf{\Omega}_1\,\mathbf{P}_1.$$

Then, (2.18) can be recast into

$$\mathbf{\Lambda} + \delta_1 \mathbf{z}_1\,\mathbf{z}_1^{\mathrm{T}} + \delta_2 \mathbf{z}_2\,\mathbf{z}_2^{\mathrm{T}} + \cdots + \delta_r \mathbf{z}_p\,\mathbf{z}_p^{\mathrm{T}} = \mathbf{P}_1(\mathbf{\Omega}_1 + \delta_2 \tilde{\mathbf{z}}_2\,\tilde{\mathbf{z}}_2^{\mathrm{T}} + \delta_3 \tilde{\mathbf{z}}_3\,\tilde{\mathbf{z}}_3^{\mathrm{T}} + \cdots + \delta_r \tilde{\mathbf{z}}_p\,\tilde{\mathbf{z}}_p^{\mathrm{T}})\mathbf{P}_1^{\mathrm{T}},$$

where the vectors $\tilde{\mathbf{z}}_2, \tilde{\mathbf{z}}_3, \ldots, \tilde{\mathbf{z}}_p$ are defined by

$$\tilde{\mathbf{z}}_i = \mathbf{P}_1^{\mathrm{T}}\,\mathbf{z}_i, \qquad \text{with } i = 2, 3, \ldots, p.$$

Now it remains to compute the eigendecomposition of the symmetric rank-$(p-1)$ modified diagonal matrix $\mathbf{\Omega}_1 + \delta_2 \tilde{\mathbf{z}}_2\,\tilde{\mathbf{z}}_2^{\mathrm{T}} + \delta_3 \tilde{\mathbf{z}}_3\,\tilde{\mathbf{z}}_3^{\mathrm{T}} + \cdots + \delta_r \tilde{\mathbf{z}}_p\,\tilde{\mathbf{z}}_p^{\mathrm{T}}$ which can be dealt with recursively. Finally, the eigendecomposition

$$\mathbf{\Lambda} + \mathbf{Z}\,\mathbf{\Delta}\,\mathbf{Z}^{\mathrm{T}} = \mathbf{P}\,\mathbf{\Omega}\,\mathbf{P}^{\mathrm{T}} = \mathbf{P}_1\,\mathbf{P}_2\,\cdots\,\mathbf{P}_p\,\mathbf{\Omega}_p\,\mathbf{P}_p^{\mathrm{T}}\,\mathbf{P}_{p-1}^{\mathrm{T}}\,\cdots\,\mathbf{P}_1^{\mathrm{T}}$$

is obtained, where all the matrices $\mathbf{P}_1, \mathbf{P}_2, \ldots, \mathbf{P}_p$ stem from the eigendecomposition of successively obtained symmetric rank-one modified diagonal matrices. Each of these matrices can be applied efficiently with the FMM.

2.6 Classical associated functions and semiseparable matrices

The divide-and-conquer method for the eigendecomposition of extended symmetric diagonal plus (p, p)-generator representable semiseparable matrices from the last section can be used for the connection problem between classical associated functions of different orders. Basic results about this problem have already been observed in Section 1.6.4. More precisely, we can obtain an efficient algorithm to apply the connection matrices between sequences of (generalized) associated functions of different orders derived from the same sequence of classical orthogonal polynomials. To this end, the corresponding connection matrices will be shown to contain eigenvectors of certain explicitly known matrices with a semiseparable structure.

2.6.1 Connection matrices and semiseparable matrices. Recall that our goal is to find an efficient algorithm to convert between (generalized) associated functions of different orders. For example, assume that a function f has a finite expansion in associated Laguerre functions $\{L_n^{(\alpha),m}\}_{n\in\mathbb{N}_0,\mu\leq n}$ of a certain order m, i.e.,

$$f = \sum_{j=\mu}^{n} x_j L_j^{(\alpha),m},$$

with known coefficients x_j. We want to compute the coefficients y_j in the expansion

$$f = \sum_{j=\hat{\mu}}^{n-(\mu-\hat{\mu})/2} y_j L_j^{(\alpha),\hat{m}},$$

where $\{L_n^{(\alpha),\hat{m}}\}_{n\in\mathbb{N}_0,\hat{\mu}\leq n}$ is another sequence of associated Laguerre functions that satisfies $m+\hat{m}$ even and $|\hat{\mu}| < |\mu|$; cf. Lemma 1.92. The situation for generalized associated Jacobi functions is analogous. There, one seeks to compute from an expansion of the form

$$f = \sum_{j=n^*}^{n} x_j P_j^{(\alpha,\beta),m,m'},$$

with known coefficients x_j, the coefficients y_j in the expansion

$$f = \sum_{j=\hat{n}^*}^{n} y_j P_j^{(\alpha,\beta),\hat{m},\hat{m}'},$$

where $\{P_n^{(\alpha,\beta),m,m'}\}_{n\in\mathbb{N}_0,n^*\leq n}$ and $\{P_n^{(\alpha,\beta),\hat{m},\hat{m}'}\}_{n\in\mathbb{N}_0,\hat{n}^*\leq n}$ are two sequences of generalized associated Jacobi functions such that $m+\hat{m}+m'+\hat{m}'$ is even, $|\hat{\mu}| < |\mu|$, and $|\hat{\nu}| < |\nu|$; see Lemma 1.94.

Note that associated Laguerre functions and generalized associated Jacobi functions cover all cases of associated functions treated in this work. Therefore, it is enough to state results only for these two types of associated functions. All other cases can be obtained from there. Again, we collect the coefficients x_j and y_j in two vectors $\mathbf{x}$ and $\mathbf{y}$,

$$\mathbf{x} = (x_\mu, x_{\mu+1}, \ldots, x_n)^{\mathrm{T}}, \qquad \mathbf{y} = (y_{\hat{\mu}}, y_{\hat{\mu}+1}, \ldots, y_{n-(\mu-\hat{\mu})/2})^{\mathrm{T}},$$

or

$$\mathbf{x} = (x_{n^*}, x_{n^*+1}, \ldots, x_n)^{\mathrm{T}}, \quad \mathbf{y} = (y_{\hat{n}^*}, y_{\hat{n}^*+1}, \ldots, y_n)^{\mathrm{T}},$$

respectively. Then the vector $\mathbf{y}$ can be obtained from the vector $\mathbf{x}$ by calculating the matrix-vector product

$$\mathbf{y} = \mathbf{K}\,\mathbf{x},$$

where $\mathbf{K}$ is the rectangular connection matrix between the two respective sequences of associated functions. For the efficient handling of this conversion, we rely on an approach similar to that used in Section 2.4 for the connection problem between classical orthogonal polynomials. A matrix $\mathbf{G}$ will be defined that has its properly scaled eigenvector matrix $\mathbf{Q}$ contain the desired connection matrix $\mathbf{K}$ as a submatrix. It will shown that the matrix $\mathbf{G}$ has a semiseparable structure which in turn allows us to apply the divide-and-conquer method from Section 2.5. This implies a fast algorithm to apply the $(n-\frac{\mu+\hat{\mu}}{2}+1)\times(n-\mu+1)$ or $(n-\hat{n}^*)\times(n-n^*+1)$ matrix $\mathbf{K}$, respectively, to any vector with $\mathcal{O}(n\log n\log(1/\varepsilon))$ arithmetic operations instead of $\mathcal{O}(n^2)$. The matrix $\mathbf{G}$ is defined in the following for associated Laguerre as well as for generalized associated Jacobi functions.

Definition 2.26 *Let* $\{L_n^{(\alpha),m}\}_{n\in\mathbb{N}_0,\mu\leq n}$ *be a sequence of associated Laguerre functions with* $\alpha > -1$ *which satisfy the hypergeometric-like differential equation*

$$xy''(x) + (1+\alpha-x)y'(x) + \left(n - \frac{\mu(2(x+\alpha)+\mu)}{4x}\right)y(x) = 0, \quad \text{with } y = L_n^{(\alpha),m}$$

and which have the corresponding differential operator $\mathcal{D}^{(\alpha),m}$ *given by*

$$\mathcal{D}^{(\alpha),m} = -x\frac{\mathrm{d}^2}{\mathrm{d}x^2} - (1+\alpha-x)\frac{\mathrm{d}}{\mathrm{d}x} + \frac{\mu(2(x+\alpha)+\mu)}{4x}.$$

Let $\{L_n^{(\alpha),\hat{m}}\}_{n\in\mathbb{N}_0,\hat{\mu}\leq n}$ *be a different sequence of associated Laguerre functions. Furthermore, denote by* $\langle\cdot,\cdot\rangle$ *the inner product with respect to which both sequences of associated functions are orthogonal, and by* $h_n^{(\alpha)} = h_n^{(\alpha),m} = h_n^{(\alpha),\hat{m}}$ *the respective squared norms. Then the matrix*

$$\mathbf{G}^{(\alpha),m\to\hat{m}} = \left(g_{i,j}^{(\alpha),m\to\hat{m}}\right)_{i,j=\hat{\mu}}^{n-(\mu-\hat{\mu})/2}$$

is defined by

$$g_{i,j}^{(\alpha),m\to\hat{m}} = \frac{\langle L_i^{(\alpha),\hat{m}}, \mathcal{D}^{(\alpha),m}(L_j^{(\alpha),\hat{m}})\rangle}{\sqrt{h_i^{(\alpha)}h_j^{(\alpha)}}}. \tag{2.19}$$

Definition 2.27 *Let* $\{P_n^{(\alpha,\beta),m,m'}\}_{n\in\mathbb{N}_0,n^*\leq n}$ *with* $\alpha,\beta > -1$ *be a sequence of generalized associated Jacobi functions which satisfy the hypergeometric-like differential equation*

$$\sigma(x)y''(x) + \tau(x)y'(x) + \big(\lambda_n + f_{\mu,\nu}(x)\big)y(x) = 0, \quad \text{with } y = P_n^{(\alpha,\beta),m,m'},$$

and which have the corresponding differential operator $\mathcal{D}^{(\alpha,\beta),m,m'}$ *defined by*

$$\mathcal{D}^{(\alpha,\beta),m,m'} = -\sigma(x)\frac{\mathrm{d}^2}{\mathrm{d}x^2} - \tau(x)\frac{\mathrm{d}}{\mathrm{d}x} - f_{\mu,\nu}(x).$$

Let $\{P_n^{(\alpha,\beta),\hat{m},\hat{m}'}\}_{n\in\mathbb{N}_0,\hat{n}^*\leq n}$ *be a different sequence of generalized associated Jacobi functions. Furthermore, denote by* $\langle\cdot,\cdot\rangle$ *the inner product with respect to which both sequences of generalized associated functions are orthogonal, and by* $h_n = h_n^{(\alpha,\beta),m,m'} = h_n^{(\alpha,\beta),\hat{m},\hat{m}'}$ *the respective squared norms. Then the matrix*

$$\mathbf{G}^{(\alpha,\beta),m,m'\to\hat{m},\hat{m}'} = \left(g_{i,j}^{(\alpha,\beta),m,m'\to\hat{m},\hat{m}'}\right)_{i,j=\hat{n}^*}^{n}$$

is defined by

$$g_{i,j}^{(\alpha,\beta),m,m'\to\hat{m},\hat{m}'} = \frac{\langle P_i^{(\alpha,\beta),\hat{m},\hat{m}'}, \mathcal{D}^{(\alpha,\beta),m,m'}(P_j^{(\alpha,\beta),\hat{m},\hat{m}'})\rangle}{\sqrt{h_i^{(\alpha,\beta)}h_j^{(\alpha,\beta)}}}. \tag{2.20}$$

As before, we drop superscripts whenever they are clear from the context. The matrix $\mathbf{G}$ can be shown to have eigenvectors that are the columns of the connection matrix $\mathbf{K}$ between the respective two sequences of associated functions. The proof of the following Lemma is omitted since it is almost identical to that of Lemma 2.19.

Lemma 2.28 *Assume that the preconditions of Definition 2.26 or Definition 2.27 hold. Then the columns of the connection matrix* $\mathbf{K} = (\boldsymbol{\kappa}_\mu, \boldsymbol{\kappa}_{\mu+1}, \ldots, \boldsymbol{\kappa}_n)$ *or, respectively,* $\mathbf{K} = (\boldsymbol{\kappa}_{n^*}, \boldsymbol{\kappa}_{n^*+1}, \ldots, \boldsymbol{\kappa}_n)$ *between the two sequences of (generalized) associated functions are eigenvectors of the respective matrix* $\mathbf{G}$, *i.e.,*

$$\mathbf{G}\boldsymbol{\kappa}_j = \lambda_j\boldsymbol{\kappa}_j,$$

where λ_j are the eigenvalues from the corresponding hypergeometric-like differential equation of the target polynomials.

Remark 2.29 Note that the columns of the connection matrix $\mathbf{K}$ cannot span the complete set of eigenspaces of the matrix $\mathbf{G}$. This is true since the matrix $\mathbf{G}$ is quadratic of size $\big(n-(\mu+\hat{\mu})/2+1\big)\times\big(n-(\mu+\hat{\mu})/2+1\big)$ or $(n-\hat{n}^*+1)\times(n-\hat{n}^*+1)$, respectively, while the matrix $\mathbf{K}$ is generally rectangular of size $(n-\hat{\mu}+1)\times(n-\mu+1)$ or $(n-\hat{n}^*+1)\times(n-n^*+1)$, respectively. Therefore, the matrix $\mathbf{K}$ has fewer columns than the matrix $\mathbf{G}$. Missing from the matrix $\mathbf{K}$ are the eigenvectors to the eigenvalues $\lambda_{\hat{\mu}},\lambda_{\hat{\mu}+1},\ldots,\lambda_{\mu-1}$ or $\lambda_{\hat{n}^*},\lambda_{\hat{n}^*+1},\ldots,\lambda_{n^*-1}$, respectively.

Remark 2.30 Lemma 2.28 implies that the matrix $\mathbf{G}$ has simple eigenvalues, since for all associated functions, the values λ_n, $n=0,1,\ldots$, from the hypergeometric-like differential equation are simple. This is, however, not a requirement for the divide-and-conquer method from Section 2.5.1.

It remains to derive the explicit expressions for the entries of the matrix $\mathbf{G}$ in each case. Also, the correct scaling of its eigenvectors needs to be determined. The scaling of the target sequence of associated functions is already encoded in the matrix $\mathbf{G}$, while the scaling of the source sequence is reflected in the scaling of the columns of the eigenvectors of $\mathbf{G}$.

2.6.2 Examples. In the following, explicit expressions for the entries of the matrix $\mathbf{G}$ are given for all cases of (generalized) associated functions.

Associated Laguerre functions

Theorem 2.31 *Let $\{L_n^{(\alpha),m}\}_{n\in\mathbb{N}_0,\mu\le n}$ and $\{L_n^{(\alpha),\hat{m}}\}_{n\in\mathbb{N}_0,\hat{\mu}\le n}$ with $\alpha>-1$ be two sequences of associated Laguerre functions such that $m+\hat{m}$ is an even number and $\hat{\mu}\le\mu$. Then the corresponding matrix $\mathbf{G}=(g_{i,j})_{i,j=\hat{\mu}}^{n-(\mu-\hat{\mu})/2}$ is symmetric diagonal plus $(1,1)$-generator representable semiseparable,*

$$\mathbf{G}=\operatorname{diag}(\mathbf{d})+\frac{\mu-\hat{\mu}}{4}\left(2+\frac{\mu-\hat{\mu}}{\alpha+\mu}\right)\left(\operatorname{triu}(\mathbf{u}\,\mathbf{v}^{\mathrm{T}},1)+\operatorname{tril}(\mathbf{v}\,\mathbf{u}^{\mathrm{T}},-1)\right),$$

with the vectors $\mathbf{d}$, $\mathbf{u}$, and $\mathbf{v}$ given by

$$\mathbf{d}=\left(j+\frac{\mu-\hat{\mu}}{4}\left(4+\frac{\mu-\hat{\mu}}{\alpha+\hat{\mu}}\right)\right)_{j=\hat{\mu}}^{n-(\mu-\hat{\mu})/2},$$

$$\mathbf{u}=\left(\sqrt{\frac{\Gamma(j+\alpha+1)}{\Gamma(j-\hat{\mu}+1)}}\right)_{j=\hat{\mu}}^{n-(\mu-\hat{\mu})/2},$$

$$\mathbf{v}=\left(\sqrt{\frac{\Gamma(j-\hat{\mu}+1)}{\Gamma(j+\alpha+1)}}\right)_{j=\hat{\mu}}^{n-(\mu-\hat{\mu})/2}.$$

PROOF. The associated Laguerre functions $L_n^{(\alpha),m}$ satisfy the differential equation

$$xy''(x)+(1+\alpha-x)y'(x)+\left(n-\frac{\mu\big(2(x+\alpha)+\mu\big)}{4x}\right)y(x)=0,\quad\text{with } y=L_n^{(\alpha),m},$$

and similarly the associated Laguerre functions $L_n^{(\alpha),\hat{m}}$ satisfy

$$xy''(x)+(1+\alpha-x)y'(x)+\left(n-\frac{\hat{\mu}\big(2(x+\alpha)+\hat{\mu}\big)}{4x}\right)y(x)=0,\quad\text{with } y=L_n^{(\alpha),\hat{m}}.$$

Thus, the corresponding differential operators are given by

$$\mathcal{D}^{(\alpha),m} = -x\frac{\mathrm{d}^2}{\mathrm{d}x^2} - (1+\alpha-x)\frac{\mathrm{d}}{\mathrm{d}x} + \frac{\mu(2(x+\alpha)+\mu)}{4x},$$
$$\mathcal{D}^{(\alpha),\hat{m}} = -x\frac{\mathrm{d}^2}{\mathrm{d}x^2} - (1+\alpha-x)\frac{\mathrm{d}}{\mathrm{d}x} + \frac{\hat{\mu}(2(x+\alpha)+\hat{\mu})}{4x},$$

and we have

$$\mathcal{D}^{(\alpha),m} - \mathcal{D}^{(\alpha),\hat{m}} = \frac{(\mu-\hat{\mu})(2(x+\alpha)+\mu+\hat{\mu})}{4x}.$$

This implies

$$\begin{aligned} g_{i,j} &= \frac{\langle L_i^{(\alpha),\hat{m}}, \mathcal{D}^{(\alpha),m}(L_j^{(\alpha),\hat{m}})\rangle}{\left(h_i^{(\alpha)}h_j^{(\alpha)}\right)^{1/2}} \\ &= \frac{\langle L_i^{(\alpha),\hat{m}}, \mathcal{D}^{(\alpha),\hat{m}}(L_j^{(\alpha),\hat{m}})\rangle}{\left(h_i^{(\alpha)}h_j^{(\alpha)}\right)^{1/2}} + \frac{\langle L_i^{(\alpha),\hat{m}}, (\mathcal{D}^{(\alpha),m} - \mathcal{D}^{(\alpha),\hat{m}})(L_j^{(\alpha),\hat{m}})\rangle}{\left(h_i^{(\alpha)}h_j^{(\alpha)}\right)^{1/2}} \\ &= \frac{\langle L_i^{(\alpha),\hat{m}}, \mathcal{D}^{(\alpha),\hat{m}}(L_j^{(\alpha),\hat{m}})\rangle}{\left(h_i^{(\alpha)}h_j^{(\alpha)}\right)^{1/2}} + \frac{\langle L_i^{(\alpha),\hat{m}}, \frac{(\mu-\hat{\mu})(2(x+\alpha)+\mu+\hat{\mu})}{4x} L_j^{(\alpha),\hat{m}}\rangle}{\left(h_i^{(\alpha)}h_j^{(\alpha)}\right)^{1/2}}. \end{aligned}$$

Since the associated Laguerre function $L_j^{(\alpha),\hat{m}}$ is an eigenfunction of the operator $\mathcal{D}^{(\alpha),\hat{m}}$ to the eigenvalue $\lambda_j = j$, we verify for the first summand that

$$\frac{\langle L_i^{(\alpha),\hat{m}}, \mathcal{D}^{(\alpha),\hat{m}}(L_j^{(\alpha),\hat{m}})\rangle}{\left(h_i^{(\alpha)}h_j^{(\alpha)}\right)^{1/2}} = j\frac{\langle L_i^{(\alpha),\hat{m}}, L_j^{(\alpha),\hat{m}}\rangle}{\left(h_i^{(\alpha)}h_j^{(\alpha)}\right)^{1/2}} = j\delta_{i,j}.$$

For the second summand, recall the definition of the associated Laguerre functions (1.6.3), i.e.,

$$L_i^{(\alpha),\hat{m}} = (-1)^{\hat{\mu}}\left(\frac{\Gamma(i-\hat{\mu}+1)}{\Gamma(i+1)}\right)^{1/2} x^{\hat{\mu}/2} L_{i-\hat{\mu}}^{(\alpha+\hat{\mu})}(x)$$

which gives

$$\begin{aligned} &\langle L_i^{(\alpha),\hat{m}}, \frac{(\mu-\hat{\mu})(2(x+\alpha)+\mu+\hat{\mu})}{4x} L_j^{(\alpha),\hat{m}}\rangle \\ &= \frac{1}{4}\left(\frac{\Gamma(i-\hat{\mu}+1)}{\Gamma(i+1)}\frac{\Gamma(j-\hat{\mu}+1)}{\Gamma(j+1)}\right)^{1/2} \\ &\quad\times \int_0^\infty L_{i-\hat{\mu}}^{(\alpha+\hat{\mu})}(x)\big((\mu-\hat{\mu})(2(x+\alpha)+\mu+\hat{\mu})\big)L_{j-\hat{\mu}}^{(\alpha+\hat{\mu})}(x)\, x^{\alpha+\hat{\mu}-1}\mathrm{e}^{-x}\,\mathrm{d}x. \end{aligned}$$

This shows, that the second summand is symmetric and therefore the matrix $\mathbf{G}$ must be symmetric. To calculate the last integral, we split it into two pieces

$$\begin{aligned}
&\int_0^\infty L_{i-\hat{\mu}}^{(\alpha+\hat{\mu})}(x)\big((\mu-\hat{\mu})(2(x+\alpha)+\mu+\hat{\mu})\big)L_{j-\hat{\mu}}^{(\alpha+\hat{\mu})}(x)\,x^{\alpha+\hat{\mu}-1}\mathrm{e}^{-x}\,\mathrm{d}x \\
&= 2(\mu-\hat{\mu})\int_0^\infty L_{i-\hat{\mu}}^{(\alpha+\hat{\mu})}(x)L_{j-\hat{\mu}}^{(\alpha+\hat{\mu})}(x)\,x^{\alpha+\hat{\mu}}\mathrm{e}^{-x}\,\mathrm{d}x \\
&\quad + (\mu-\hat{\mu})(2\alpha+\mu+\hat{\mu})\int_0^\infty L_{i-\hat{\mu}}^{(\alpha+\hat{\mu})}(x)L_{j-\hat{\mu}}^{(\alpha+\hat{\mu})}(x)\,x^{\alpha+\hat{\mu}-1}\mathrm{e}^{-x}\,\mathrm{d}x \\
&= 2(\mu-\hat{\mu})\frac{\Gamma(i+\alpha+1)}{\Gamma(i-\hat{\mu}+1)}\delta_{i,j} \\
&\quad + (\mu-\hat{\mu})(2\alpha+\mu+\hat{\mu})\int_0^\infty L_{i-\hat{\mu}}^{(\alpha+\hat{\mu})}(x)L_{j-\hat{\mu}}^{(\alpha+\hat{\mu})}(x)\,x^{\alpha+\hat{\mu}-1}\mathrm{e}^{-x}\,\mathrm{d}x.
\end{aligned}$$

For the remaining integral, we assume $i \leq j$ without loss of generality. We can use the connection coefficients $\kappa_{i,j}^{\mathrm{L},(\alpha+\hat{\mu})\to(\alpha+\hat{\mu}-1)} = 1$ to expand the polynomials $L_{i-\hat{\mu}}^{(\alpha+\hat{\mu})}(x)$ and $L_{j-\hat{\mu}}^{(\alpha+\hat{\mu})}(x)$ in the Laguerre polynomials $\{L_n^{(\alpha+\hat{\mu}-1)}\}_{n\in\mathbb{N}_0}$. By orthogonality, this gives

$$\begin{aligned}
&\int_0^\infty L_{i-\hat{\mu}}^{(\alpha+\hat{\mu})}(x)L_{j-\hat{\mu}}^{(\alpha+\hat{\mu})}(x)\,x^{\alpha+\hat{\mu}-1}\mathrm{e}^{-x}\,\mathrm{d}x \\
&= \sum_{k=0}^{i-\hat{\mu}}\int_0^\infty \left(L_k^{(\alpha+\hat{\mu}-1)}(x)\right)^2 x^{\alpha+\hat{\mu}-1}\mathrm{e}^{-x}\,\mathrm{d}x \\
&= \sum_{k=0}^{i-\hat{\mu}}\frac{\Gamma(k+\alpha+\hat{\mu})}{\Gamma(k+1)} \\
&= \frac{\Gamma(i+\alpha+1)}{(\alpha+\hat{\mu})\Gamma(i-\hat{\mu}+1)}.
\end{aligned}$$

Finally, we use

$$h_i^{(\alpha)} = \frac{\Gamma(i+\alpha+1)}{\Gamma(i+1)},$$

and obtain

$$\begin{aligned}
&\frac{\langle L_i^{(\alpha),\hat{m}}, \frac{(\mu-\hat{\mu})(2(x+\alpha)+\mu+\hat{\mu})}{4x}L_j^{(\alpha),\hat{m}}\rangle}{\sqrt{h_i^{(\alpha)}h_j^{(\alpha)}}} \\
&= \frac{1}{4}\left(\frac{\Gamma(i+\alpha+1)\Gamma(j-\hat{\mu}+1)}{\Gamma(j+\alpha+1)\Gamma(i-\hat{\mu}+1)}\right)^{1/2}\left(2+\frac{\mu-\hat{\mu}}{\alpha+\hat{\mu}}+2\delta_{i,j}\right).
\end{aligned}$$

This proves the desired representation of the matrix $\mathbf{G}$. □

Generalized associated Jacobi functions

Theorem 2.32 *Let $\{P_n^{(\alpha,\beta),m,m'}\}_{n\in\mathbb{N}_0,n^*\leq n}$ and $\{P_n^{(\alpha,\beta),\hat{m},\hat{m}'}\}_{n\in\mathbb{N}_0,\hat{n}^*\leq n}$ be two sequences of generalized associated Jacobi functions, with $\alpha,\beta > -1$, such that $m+\hat{m}+m'+\hat{m}'$ is an even number, $\hat{\mu} \leq \mu$ and $\hat{\nu} \leq \nu$. Then the corresponding matrix $\mathbf{G} = (g_{i,j})_{i,j=\hat{n}^*}^n$ is* *extended symmetric diagonal plus* $(2,2)$*-generator representable semiseparable,*

$$\begin{aligned}\mathbf{G} = \operatorname{diag}(\mathbf{d}) &+ \frac{(\mu-\hat{\mu})(2\alpha+\mu+\hat{\mu})}{4(\alpha+\hat{\mu})}\left(\operatorname{triu}(\mathbf{u}_1\,\mathbf{v}_1^{\mathrm{T}},1)+\operatorname{tril}(\mathbf{v}_1\,\mathbf{u}_1^{\mathrm{T}},-1)\right)\\ &+ \frac{(\nu-\hat{\nu})(2\beta+\nu+\hat{\nu})}{4(\beta+\hat{\nu})}\left(\operatorname{triu}(\mathbf{u}_2\,\mathbf{v}_2^{\mathrm{T}},1)+\operatorname{tril}(\mathbf{v}_2\,\mathbf{u}_2^{\mathrm{T}},-1)\right),\end{aligned}$$

with the vectors $\mathbf{d}$*,* $\mathbf{u}_1$*,* $\mathbf{v}_1$*,* $\mathbf{u}_2$*, and* $\mathbf{v}_2$ *given by*

$$\begin{aligned}\mathbf{d} &= \Bigg(j(j+\alpha+\beta+1)\\ &\qquad + (2j+\alpha+\beta+1)\left(\frac{(\mu-\hat{\mu})(2\alpha+\mu+\hat{\mu})}{4(\alpha+\hat{\mu})}+\frac{(\nu-\hat{\nu})(2\beta+\nu+\hat{\nu})}{4(\beta+\hat{\nu})}\right)\Bigg)_{j=\hat{n}^*}^{n},\\ \mathbf{u}_1 &= \left(\left(\frac{(2j+\alpha+\beta+1)\Gamma(j-\hat{n}^*+\alpha+\hat{\mu}+1)\Gamma(j+\hat{n}^*+\alpha+\beta+1)}{\Gamma(j-\hat{n}^*+1)\Gamma(j-\hat{n}^*+\beta+\hat{\nu}+1)}\right)^{1/2}\right)_{j=\hat{n}^*}^{n},\\ \mathbf{v}_1 &= \left(\left(\frac{(2j+\alpha+\beta+1)\Gamma(j-\hat{n}^*+1)\Gamma(j-\hat{n}^*+\beta+\hat{\nu}+1)}{\Gamma(j-\hat{n}^*+\alpha+\hat{\mu}+1)\Gamma(j+\hat{n}^*+\alpha+\beta+1)}\right)^{1/2}\right)_{j=\hat{n}^*}^{n},\\ \mathbf{u}_2 &= \left(\left(\frac{(2j+\alpha+\beta+1)\Gamma(j-\hat{n}^*+\beta+\hat{\nu}+1)\Gamma(j+\hat{n}^*+\alpha+\beta+1)}{(-1)^j\Gamma(j-\hat{n}^*+1)\Gamma(j-\hat{n}^*+\alpha+\hat{\mu}+1)}\right)^{1/2}\right)_{j=\hat{n}^*}^{n},\\ \mathbf{v}_2 &= \left(\left((-1)^j\frac{(2j+\alpha+\beta+1)\Gamma(j-\hat{n}^*+1)\Gamma(j-\hat{n}^*+\alpha+\hat{\mu}+1)}{\Gamma(j-\hat{n}^*+\beta+\hat{\nu}+1)\Gamma(j+\hat{n}^*+\alpha+\beta+1)}\right)^{1/2}\right)_{j=\hat{n}^*}^{n}.\end{aligned}$$

PROOF. The generalized associated Jacobi functions $P_n^{(\alpha,\beta),m,m'}$ satisfy the differential equation

$$\sigma(x)y''(x)+\tau(x)y'(x)+\big(\lambda_n+f_{\mu,\nu}(x)\big)y(x)=0,\quad \text{with } y=P_n^{(\alpha,\beta),m,m'},$$

where

$$\sigma(x)=1-x^2,\qquad \tau(x)=-(\alpha+\beta+2)x+\beta-\alpha,\qquad \lambda_n=n(n+\alpha+\beta+1),$$

and

$$f_{\mu,\nu}(x)=-\frac{\mu(2\alpha+\mu)}{2(1-x)}-\frac{\nu(2\beta+\nu)}{2(1+x)}.$$

Similarly the generalized associated Jacobi functions $P_n^{(\alpha,\beta),\hat{m},\hat{m}'}$ satisfy the differential equation

$$\sigma(x)y''(x)+\tau(x)y'(x)+\big(\lambda_n+f_{\hat{\mu},\hat{\nu}}(x)\big)y(x)=0,\quad \text{with } y=P_n^{(\alpha,\beta),m,m'}.$$

Thus, the corresponding differential operators are given by

$$\begin{aligned}\mathcal{D}^{(\alpha,\beta),m,m'} &= -\sigma(x)\frac{\mathrm{d}^2}{\mathrm{d}x^2}-\tau(x)\frac{\mathrm{d}}{\mathrm{d}x}-f_{\mu,\nu}(x),\\ \mathcal{D}^{(\alpha,\beta),\hat{m},\hat{m}'} &= -\sigma(x)\frac{\mathrm{d}^2}{\mathrm{d}x^2}-\tau(x)\frac{\mathrm{d}}{\mathrm{d}x}-f_{\hat{\mu},\hat{\nu}}(x),\end{aligned}$$

and we have

$$\begin{aligned}
&\mathcal{D}^{(\alpha,\beta),m,m'} - \mathcal{D}^{(\alpha,\beta),\hat{m},\hat{m}'} \\
&= f_{\hat{\mu},\hat{\nu}}(x) - f_{\mu,\nu}(x) \\
&= \frac{\mu(2\alpha+\mu)}{2(1-x)} + \frac{\nu(2\beta+\nu)}{2(1+x)} - \frac{\hat{\mu}(2\alpha+\hat{\mu})}{2(1-x)} - \frac{\hat{\nu}(2\beta+\hat{\nu})}{2(1+x)} \\
&= \frac{(\mu-\hat{\mu})(2\alpha+\mu+\hat{\mu})}{2(1-x)} + \frac{(\nu-\hat{\nu})(2\beta+\nu+\hat{\nu})}{2(1+x)}.
\end{aligned}$$

With $\hat{n}^* \leq i, j$, this implies

$$\begin{aligned}
g_{i,j} &= \frac{\langle P_i^{(\alpha,\beta),\hat{m},\hat{m}'}, \mathcal{D}^{(\alpha,\beta),m,m'}(P_j^{(\alpha,\beta),\hat{m},\hat{m}'})\rangle}{\sqrt{h_i^{(\alpha,\beta)} h_j^{(\alpha,\beta)}}} \\
&= \frac{\langle P_i^{(\alpha,\beta),\hat{m},\hat{m}'}, \mathcal{D}^{(\alpha,\beta),\hat{m},\hat{m}'}(P_j^{(\alpha,\beta),\hat{m},\hat{m}'})\rangle}{\sqrt{h_i^{(\alpha,\beta)} h_j^{(\alpha,\beta)}}} \\
&\quad + \frac{\langle P_i^{(\alpha,\beta),\hat{m},\hat{m}'}, (\mathcal{D}^{(\alpha,\beta),m,m'} - \mathcal{D}^{(\alpha,\beta),\hat{m},\hat{m}'})(P_j^{(\alpha,\beta),\hat{m},\hat{m}'})\rangle}{\sqrt{h_i^{(\alpha,\beta)} h_j^{(\alpha,\beta)}}} \\
&= \frac{\langle P_i^{(\alpha,\beta),\hat{m},\hat{m}'}, \mathcal{D}^{(\alpha,\beta),\hat{m},\hat{m}'}(P_j^{(\alpha,\beta),\hat{m},\hat{m}'})\rangle}{\sqrt{h_i^{(\alpha,\beta)} h_j^{(\alpha,\beta)}}} \\
&\quad + \frac{\langle P_i^{(\alpha,\beta),\hat{m},\hat{m}'}, \left(\frac{(\mu-\hat{\mu})(2\alpha+\mu+\hat{\mu})}{2(1-x)} + \frac{(\nu-\hat{\nu})(2\beta+\nu+\hat{\nu})}{2(1+x)}\right) P_j^{(\alpha,\beta),\hat{m},\hat{m}'}\rangle}{\sqrt{h_i^{(\alpha,\beta)} h_j^{(\alpha,\beta)}}}.
\end{aligned}$$

Since the generalized associated Jacobi function $P_j^{(\alpha,\beta),\hat{m},\hat{m}'}$ is an eigenfunction of the operator $\mathcal{D}^{(\alpha,\beta),\hat{m},\hat{m}'}$ to the eigenvalue $\lambda_j = j(j+\alpha+\beta+1)$, we verify for the first summand that

$$\frac{\langle P_i^{(\alpha,\beta),\hat{m},\hat{m}'}, \mathcal{D}^{\hat{m},\hat{m}'}(P_j^{(\alpha,\beta),\hat{m},\hat{m}'})\rangle}{\sqrt{h_i^{(\alpha,\beta)} h_j^{(\alpha,\beta)}}} = j(j+\alpha+\beta+1)\delta_{i,j}.$$

For the second part, we recall the definition of the generalized associated Jacobi functions (1.44),

$$P_n^{(\alpha,\beta),\hat{m},\hat{m}'}(x) = C_{n,\hat{\mu},\hat{\nu}}(1-x)^{\hat{\mu}/2}(1+x)^{\hat{\nu}/2} P_{n-\hat{n}^*}^{(\alpha+\hat{\mu},\beta+\hat{\nu})}(x),$$

with

$$\begin{aligned}
C_{n,\hat{\mu},\hat{\nu}} =& 2^{-\hat{n}^*}\left(\frac{\Gamma(n+\alpha+1)\Gamma(n+\beta+1)}{\Gamma(n+1)\Gamma(n+\alpha+\beta+1)}\right. \\
&\left.\times \frac{\Gamma(n-\hat{n}^*+1)\Gamma(n+\hat{n}^*+\alpha+\beta+1)}{\Gamma(n-\hat{n}^*+\alpha+\hat{\mu}+1)\Gamma(n-\hat{n}^*+\beta+\hat{\nu}+1)}\right)^{1/2}.
\end{aligned}$$

This gives

$$\langle P_i^{(\alpha,\beta),\hat{m},\hat{m}'}, \left(\frac{(\mu-\hat{\mu})(2\alpha+\mu+\hat{\mu})}{2(1-x)} + \frac{(\nu-\hat{\nu})(2\beta+\nu+\hat{\nu})}{2(1+x)} \right) P_j^{(\alpha,\beta),\hat{m},\hat{m}'} \rangle$$
$$= C_{i,\hat{\mu},\hat{\nu}} C_{j,\hat{\mu},\hat{\nu}} \int_{-1}^{1} P_{i-\hat{n}^*}^{(\alpha+\hat{\mu},\beta+\hat{\nu})}(x) \left(\frac{(\mu-\hat{\mu})(2\alpha+\mu+\hat{\mu})}{2(1-x)} + \frac{(\nu-\hat{\nu})(2\beta+\nu+\hat{\nu})}{2(1+x)} \right)$$
$$\times P_{j-\hat{n}^*}^{(\alpha+\hat{\mu},\beta+\hat{\nu})}(x)\,(1-x)^{\alpha+\hat{\mu}}(1+x)^{\beta+\hat{\nu}}\,\mathrm{d}x,$$

and shows that this is symmetric. Therefore the matrix $\mathbf{G}$ must be symmetric. Without loss of generality we assume that $i \leq j$ and start by calculating the integral,

$$I_{i,j} := \int_{-1}^{1} P_{i-\hat{n}^*}^{(\alpha+\hat{\mu},\beta+\hat{\nu})}(x) \frac{1}{1-x} P_{j-\hat{n}^*}^{(\alpha+\hat{\mu},\beta+\hat{\nu})}(x)\,(1-x)^{\alpha+\hat{\mu}}(1+x)^{\beta+\hat{\nu}}\,\mathrm{d}x$$
$$= \int_{-1}^{1} P_{i-\hat{n}^*}^{(\alpha+\hat{\mu},\beta+\hat{\nu})}(x) P_{j-\hat{n}^*}^{(\alpha+\hat{\mu},\beta+\hat{\nu})}(x)\,(1-x)^{\alpha+\hat{\mu}-1}(1+x)^{\beta+\hat{\nu}}\,\mathrm{d}x.$$

If $i = \hat{n}^*$, then the formula 7.391 4. in [27, p. 228] gives

$$I_{0,j} = \int_{-1}^{1} P_0^{(\alpha+\hat{\mu},\beta+\hat{\nu})}(x) P_{j-\hat{n}^*}^{(\alpha+\hat{\mu},\beta+\hat{\nu})}(x)\,(1-x)^{\alpha+\hat{\mu}-1}(1+x)^{\beta+\hat{\nu}}\,\mathrm{d}x$$
$$= 2^{\alpha+\hat{\mu}+\beta+\hat{\nu}}\Gamma(\alpha+\hat{\mu}) \frac{\Gamma(j-\hat{n}^*+\beta+\hat{\nu}+1)}{\Gamma(j+\hat{n}^*+\alpha+\beta+1)},$$

where we have used that $\hat{\mu}+\hat{\nu} = 2\hat{n}^*$. This will be used repeatedly throughout the rest of this proof. Similarly, if $i = j$ then formula 7.391 5. in [27, p. 228] gives

$$I_{i,i} = \int_{-1}^{1} \left(P_{i-\hat{n}^*}^{(\alpha+\hat{\mu},\beta+\hat{\nu})}(x) \right)^2 (1-x)^{\alpha+\hat{\mu}-1}(1+x)^{\beta+\hat{\nu}}\,\mathrm{d}x$$
$$= \frac{2^{\alpha+\hat{\mu}+\beta+\hat{\nu}}}{\alpha+\hat{\mu}} \frac{\Gamma(i-\hat{n}+\alpha+\hat{\mu}^*+1)}{\Gamma(i-\hat{n}^*+1)} \frac{\Gamma(i-\hat{n}^*+\beta+\hat{\nu}+1)}{\Gamma(i+\hat{n}^*+\alpha+\beta+1)}.$$

This motivates the hypothesis that $I_{i,j}$ may be written as

$$I_{i,j} = \int_{-1}^{1} P_{i-\hat{n}^*}^{(\alpha+\hat{\mu},\beta+\hat{\nu})}(x) P_{j-\hat{n}^*}^{(\alpha+\hat{\mu},\beta+\hat{\nu})}(x)\,(1-x)^{\alpha+\hat{\mu}-1}(1+x)^{\beta+\hat{\nu}}\,\mathrm{d}x$$
$$= \frac{2^{\alpha+\hat{\mu}+\beta+\hat{\nu}}}{\alpha+\hat{\mu}} \frac{\Gamma(i-\hat{n}^*+\alpha+\hat{\mu}+1)}{\Gamma(i-\hat{n}^*+1)} \frac{\Gamma(j-\hat{n}^*+\beta+\hat{\nu}+1)}{\Gamma(j+\hat{n}^*+\alpha+\beta+1)}, \tag{2.21}$$

which we prove by double induction. For an arbitrary but fixed value of i, assume that (2.21) is true for all integrals $I_{i',j'}$ with $i' < i$ and $j' \geq i'$. The value $I_{i,i}$ is already known, so we calculate $I_{i,i+1}$ by using the three-term recurrence

$$P_{i-\hat{n}^*+1}^{(\alpha+\hat{\mu},\beta+\hat{\nu})}(x) = (a_{i-\hat{n}^*}x - b_{i-\hat{n}^*}) P_{i-\hat{n}^*}^{(\alpha+\hat{\mu},\beta+\hat{\nu})}(x) - c_{i-\hat{n}^*} P_{i-\hat{n}^*-1}^{(\alpha+\hat{\mu},\beta+\hat{\nu})}(x),$$

with the coefficients

$$a_{i-\hat{n}^*} = \frac{(2i+\alpha+\beta+1)(2i+\alpha+\beta+2)}{2(i-\hat{n}^*+1)(i+\hat{n}^*+\alpha+\beta+1)}$$

$$b_{i-\hat{n}^*} = -\frac{(2i+\alpha+\beta+1)(\alpha+\hat{\mu}+\beta+\hat{\nu})(\alpha+\hat{\mu}-\beta-\hat{\nu})}{2(i-\hat{n}^*+1)(i+\hat{n}^*+\alpha+\beta+1)(2i+\alpha+\beta)}$$

$$c_{i-\hat{n}^*} = \frac{(i-\hat{n}^*+\alpha+\hat{\mu})(i-\hat{n}^*+\beta+\hat{\nu})(2i+\alpha+\beta+2)}{(i-\hat{n}^*+1)(i+\hat{n}^*+\alpha+\beta+1)(2i+\alpha+\beta)},$$

satisfied by the Jacobi polynomials $\{P_n^{(\alpha+\hat{\mu},\beta+\hat{\nu})}(x)\}_{n\in\mathbb{N}_0}$; cf. (1.22). Thus,

$$\begin{aligned}
I_{i,i+1} &= \int_{-1}^{1} P_{i-\hat{n}^*}^{(\alpha+\hat{\mu},\beta+\hat{\nu})}(x) P_{i+1-\hat{n}^*}^{(\alpha+\hat{\mu},\beta+\hat{\nu})}(x)\,(1-x)^{\alpha+\hat{\mu}-1}(1+x)^{\beta+\hat{\nu}}\,\mathrm{d}x \\
&= \int_{-1}^{1} P_{i-\hat{n}^*}^{(\alpha+\hat{\mu},\beta+\hat{\nu})}(x) \\
&\quad \times \left((a_{i-\hat{n}^*}x - b_{i-\hat{n}^*})P_{i-\hat{n}^*}^{(\alpha+\hat{\mu},\beta+\hat{\nu})}(x) - c_{i-\hat{n}^*}P_{i-\hat{n}^*-1}^{(\alpha+\hat{\mu},\beta+\hat{\nu})}(x)\right) \\
&\quad \times (1-x)^{\alpha+\hat{\mu}-1}(1+x)^{\beta+\hat{\nu}}\,\mathrm{d}x \\
&= -\,a_{i-\hat{n}^*}\int_{-1}^{1}\left(P_{i-\hat{n}^*}^{(\alpha+\hat{\mu},\beta+\hat{\nu})}(x)\right)^2(1-x)^{\alpha+\hat{\mu}}(1+x)^{\beta+\hat{\nu}}\,\mathrm{d}x \\
&\quad + (a_{i-\hat{n}^*} - b_{i-\hat{n}^*})\int_{-1}^{1}\left(P_{i-\hat{n}^*}^{(\alpha+\hat{\mu},\beta+\hat{\nu})}(x)\right)^2(1-x)^{\alpha+\hat{\mu}-1}(1+x)^{\beta+\hat{\nu}}\,\mathrm{d}x \\
&\quad - c_{i-\hat{n}^*}\int_{-1}^{1} P_{i-\hat{n}^*}^{(\alpha+\hat{\mu},\beta+\hat{\nu})}(x)P_{i-\hat{n}^*-1}^{(\alpha+\hat{\mu},\beta+\hat{\nu})}(x)(1-x)^{\alpha+\hat{\mu}-1}(1+x)^{\beta+\hat{\nu}}\,\mathrm{d}x \\
&= -\,a_{i-\hat{n}^*}\frac{2^{\alpha+\hat{\mu}+\beta+\hat{\nu}+1}\Gamma(i-\hat{n}^*+\alpha+\hat{\mu}+1)\Gamma(i-\hat{n}^*+\beta+\hat{\nu}+1)}{(2i+\alpha+\beta+1)\Gamma(i-\hat{n}^*+1)\Gamma(i+\hat{n}^*+\alpha+\beta+1)} \\
&\quad + (a_{i-\hat{n}^*} - b_{i-\hat{n}^*})I_{i,i} - c_{i-\hat{n}^*}I_{i-1,i}.
\end{aligned}$$

For the last identity, we have used that

$$\begin{aligned}
h_{i-\hat{n}^*}^{(\alpha+\hat{\mu},\beta+\hat{\mu})} &= \int_{-1}^{1}\left(P_{i-\hat{n}^*}^{(\alpha+\hat{\mu},\beta+\hat{\nu})}(x)\right)^2(1-x)^{\alpha+\hat{\mu}}(1+x)^{\beta+\hat{\nu}}\,\mathrm{d}x \\
&= \frac{2^{\alpha+\hat{\mu}+\beta+\hat{\nu}+1}\Gamma(i-\hat{n}^*+\alpha+\hat{\mu}+1)\Gamma(i-\hat{n}^*+\beta+\hat{\nu}+1)}{(2i+\alpha+\beta+1)\Gamma(i-\hat{n}^*+1)\Gamma(i+\hat{n}^*+\alpha+\beta+1)}.
\end{aligned}$$

With in the expressions for the coefficients $a_{i-\hat{n}^*}$, $a_{i-\hat{n}^*} - b_{i-\hat{n}^*}$, and the integrals $I_{i,i}$, and $I_{i-1,i}$, this can be simplified to

$$I_{i,i+1} = \frac{2^{\alpha+\hat{\mu}+\beta+\hat{\nu}}}{\alpha+\hat{\mu}}\frac{\Gamma(i-\hat{n}^*+\alpha+\hat{\mu}+1)}{\Gamma(i-\hat{n}^*+1)}\frac{\Gamma(i-\hat{n}^*+\beta+\hat{\nu}+2)}{\Gamma(i+\hat{n}^*+\alpha+\beta+2)}.$$

The general case for a fixed i but arbitrary $j \geq i$ can be proved by induction. To this end, fix an arbitrary $j > i+1$. Then

$$\begin{aligned}
I_{i,j} &= \int_{-1}^{1} P_{i-\hat{n}^*}^{(\alpha+\hat{\mu},\beta+\hat{\nu})}(x)P_{j-\hat{n}^*}^{(\alpha+\hat{\mu},\beta+\hat{\nu})}(x)\,(1-x)^{\alpha+\hat{\mu}-1}(1+x)^{\beta+\hat{\nu}}\,\mathrm{d}x \\
&= -\,a_{j-\hat{n}^*-1}\int_{-1}^{1} P_{i-\hat{n}^*}^{(\alpha+\hat{\mu},\beta+\hat{\nu})}(x)P_{j-\hat{n}^*-1}^{(\alpha+\hat{\mu},\beta+\hat{\nu})}(x)(1-x)^{\alpha+\hat{\mu}}(1+x)^{\beta+\hat{\nu}}\,\mathrm{d}x \\
&\quad + (a_{j-\hat{n}^*-1} - b_{j-\hat{n}^*-1}) \\
&\qquad \times \int_{-1}^{1} P_{i-\hat{n}^*}^{(\alpha+\hat{\mu},\beta+\hat{\nu})}(x)P_{j-\hat{n}^*-1}^{(\alpha+\hat{\mu},\beta+\hat{\nu})}(x)(1-x)^{\alpha+\hat{\mu}-1}(1+x)^{\beta+\hat{\nu}}\,\mathrm{d}x \\
&\quad - c_{j-\hat{n}^*-1}\int_{-1}^{1} P_{i-\hat{n}^*}^{(\alpha+\hat{\mu},\beta+\hat{\nu})}(x)P_{j-\hat{n}^*-2}^{(\alpha+\hat{\mu},\beta+\hat{\nu})}(x)(1-x)^{\alpha+\hat{\mu}-1}(1+x)^{\beta+\hat{\nu}}\,\mathrm{d}x \\
&= (a_{j-\hat{n}^*-1} - b_{j-\hat{n}^*-1})I_{i,j-1} - c_{j-\hat{n}^*-1}I_{i,j-2}.
\end{aligned}$$

Using the induction hypothesis for the integrals $I_{i,j-1}$ and $I_{i,j-2}$, this can be simplified to the desired expression

$$I_{i,j} = \frac{2^{\alpha+\hat{\mu}+\beta+\hat{\nu}}}{\alpha+\hat{\mu}} \frac{\Gamma(i-\hat{n}^*+\alpha+\hat{\mu}+1)}{\Gamma(i-\hat{n}^*+1)} \frac{\Gamma(j-\hat{n}^*+\beta+\hat{\nu}+1)}{\Gamma(j+\hat{n}^*+\alpha+\beta+1)}.$$

The proof of the formula for the integral $I_{i,j}$ is completed by the outer induction over i. We obtain the formula

$$\begin{aligned} C_{i,\hat{\mu},\hat{\nu}} C_{j,\hat{\mu},\hat{\nu}} & \int_{-1}^{1} P_{i-\hat{n}^*}^{(\alpha+\hat{\mu},\beta+\hat{\nu})}(x) \frac{(\mu-\hat{\mu})(2\alpha+\mu+\hat{\mu})}{2(1-x)} \\ & \qquad \times P_{j-\hat{n}^*}^{(\alpha+\hat{\mu},\beta+\hat{\nu})}(x)\,(1-x)^{\alpha+\hat{\mu}}(1+x)^{\beta+\hat{\nu}}\,\mathrm{d}x \\ = 2^{\alpha+\beta-1} & \frac{(\mu-\hat{\mu})(2\alpha+\mu+\hat{\mu})}{\alpha+\hat{\mu}} \\ & \times \left(\frac{\Gamma(i+\alpha+1)\Gamma(i+\beta+1)\Gamma(i-\hat{n}^*+\alpha+\hat{\mu}+1)\Gamma(i+\hat{n}^*+\alpha+\beta+1)}{\Gamma(i+1)\Gamma(i+\alpha+\beta+1)\Gamma(i-\hat{n}^*+1)\Gamma(i-\hat{n}^*+\beta+\hat{\nu}+1)} \right)^{1/2} \\ & \times \left(\frac{\Gamma(j+\alpha+1)\Gamma(j+\beta+1)\Gamma(j-\hat{n}^*+1)\Gamma(j-\hat{n}^*+\beta+\hat{\nu}+1)}{\Gamma(j+1)\Gamma(j+\alpha+\beta+1)\Gamma(j-\hat{n}^*+\alpha+\hat{\mu}+1)\Gamma(j+\hat{n}^*+\alpha+\beta+1)} \right)^{1/2}. \end{aligned} \tag{2.22}$$

It remains to calculate the integral

$$\begin{aligned} C_{i,\hat{\mu},\hat{\nu}} C_{j,\hat{\mu},\hat{\nu}} & \int_{-1}^{1} P_{i-\hat{n}^*}^{(\alpha+\hat{\mu},\beta+\hat{\nu})}(x) \frac{(\nu-\hat{\nu})(2\beta+\nu+\hat{\nu})}{2(1+x)} \\ & \qquad \times P_{j-\hat{n}^*}^{(\alpha+\hat{\mu},\beta+\hat{\nu})}(x)\,(1-x)^{\alpha+\hat{\mu}}(1+x)^{\beta+\hat{\nu}}\,\mathrm{d}x \\ = (-1)^{i-j} C_{i,\hat{\mu},\hat{\nu}} C_{j,\hat{\mu},\hat{\nu}} & \int_{-1}^{1} P_{i-\hat{n}^*}^{(\beta+\hat{\nu},\alpha+\hat{\mu})}(x) \frac{(\nu-\hat{\nu})(2\beta+\nu+\hat{\nu})}{2(1-x)} \\ & \qquad \times P_{j-\hat{n}^*}^{(\beta+\hat{\nu},\alpha+\hat{\mu})}(x)\,(1-x)^{\alpha+\hat{\mu}}(1+x)^{\beta+\hat{\nu}}\,\mathrm{d}x, \end{aligned}$$

where we have used the identity $P_n^{(\alpha,\beta)}(x) = (-1)^n P_n^{(\beta,\alpha)}(-x)$; cf. (1.21). This is obviously analogous to (2.22), so that we obtain

$$\begin{aligned} (-1)^{i-j} C_{i,\hat{\mu},\hat{\nu}} C_{j,\hat{\mu},\hat{\nu}} & \int_{-1}^{1} P_{i-\hat{n}^*}^{(\beta+\hat{\nu},\alpha+\hat{\mu})}(x) \frac{(\nu-\hat{\nu})(2\beta+\nu+\hat{\nu})}{2(1-x)} \\ & \qquad \times P_{j-\hat{n}^*}^{(\beta+\hat{\nu},\alpha+\hat{\mu})}(x)\,(1-x)^{\alpha+\hat{\mu}}(1+x)^{\beta+\hat{\nu}}\,\mathrm{d}x \\ = (-1)^{i-j} 2^{\alpha+\beta-1} & \frac{(\nu-\hat{\nu})(2\beta+\nu+\hat{\nu})}{\beta+\hat{\nu}} \\ & \times \left(\frac{\Gamma(i+\alpha+1)\Gamma(i+\beta+1)\Gamma(i-\hat{n}^*+\beta+\hat{\nu}+1)\Gamma(i+\hat{n}^*+\alpha+\beta+1)}{\Gamma(i+1)\Gamma(i+\alpha+\beta+1)\Gamma(i-\hat{n}^*+1)\Gamma(i-\hat{n}^*+\alpha+\hat{\mu}+1)} \right)^{1/2} \\ & \times \left(\frac{\Gamma(j+\alpha+1)\Gamma(j+\beta+1)\Gamma(j-\hat{n}^*+1)\Gamma(j-\hat{n}^*+\alpha+\hat{\mu}+1)}{\Gamma(j+1)\Gamma(j+\alpha+\beta+1)\Gamma(j-\hat{n}^*+\beta+\hat{\nu}+1)\Gamma(j+\hat{n}^*+\alpha+\beta+1)} \right)^{1/2}. \end{aligned}$$

Finally, we have

$$\frac{\langle P_i^{(\alpha,\beta),\hat{m},\hat{m}'}, \left(\frac{(\mu-\hat{\mu})(2\alpha+\mu+\hat{\mu})}{2(1-x)} + \frac{(\nu-\hat{\nu})(2\beta+\nu+\hat{\nu})}{2(1+x)}\right) P_j^{(\alpha,\beta),\hat{m},\hat{m}'}\rangle}{\sqrt{h_i^{(\alpha,\beta)} h_j^{(\alpha,\beta)}}}$$

$$\begin{aligned}
&= \frac{(\mu-\hat{\mu})(2\alpha+\mu+\hat{\mu})}{4(\alpha+\hat{\mu})} \\
&\quad \times \left(\frac{(2i+\alpha+\beta+1)\Gamma(i-\hat{n}^*+\alpha+\hat{\mu}+1)\Gamma(i+\hat{n}^*+\alpha+\beta+1)}{\Gamma(i-\hat{n}^*+1)\Gamma(i-\hat{n}^*+\beta+\hat{\nu}+1)}\right)^{1/2} \\
&\quad \times \left(\frac{(2j+\alpha+\beta+1)\Gamma(j-\hat{n}^*+1)\Gamma(j-\hat{n}^*+\beta+\hat{\nu}+1)}{\Gamma(j-\hat{n}^*+\alpha+\hat{\mu}+1)\Gamma(j+\hat{n}^*+\alpha+\beta+1)}\right)^{1/2} \\
&\quad + (-1)^{i-j}\frac{(\nu-\hat{\nu})(2\beta+\nu+\hat{\nu})}{4(\beta+\hat{\nu})} \\
&\quad \times \left(\frac{(2i+\alpha+\beta+1)\Gamma(i-\hat{n}^*+\beta+\hat{\nu}+1)\Gamma(i+\hat{n}^*+\alpha+\beta+1)}{\Gamma(i-\hat{n}^*+1)\Gamma(i-\hat{n}^*+\alpha+\hat{\mu}+1)}\right)^{1/2} \\
&\quad \times \left(\frac{(2j+\alpha+\beta+1)\Gamma(j-\hat{n}^*+1)\Gamma(j-\hat{n}^*+\alpha+\hat{\mu}+1)}{\Gamma(j-\hat{n}^*+\beta+\hat{\nu}+1)\Gamma(j+\hat{n}^*+\alpha+\beta+1)}\right)^{1/2},
\end{aligned}$$

which completes the proof. □

In the following, we give analogous results for the remaining classes of (generalized) associated functions. The proofs are omitted since these can all be derived as special cases from the results for generalized associated Jacobi functions.

Generalized associated Gegenbauer functions

Theorem 2.33 *Let $\{C_n^{(\alpha),m,m'}\}_{n\in\mathbb{N}_0,n^*\le n}$ and $\{C_n^{(\alpha),\hat{m},\hat{m}'}\}_{n\in\mathbb{N}_0,\hat{n}^*\le n}$ with $\alpha > -1/2$ and $\alpha \neq 0$ be two sequences of generalized associated Gegenbauer functions such that $m + \hat{m} + m' + \hat{m}'$ is an even number, $\hat{\mu} \le \mu$ and $\hat{\nu} \le \nu$. Then the corresponding matrix $\mathbf{G} = (g_{i,j})_{i,j=\hat{n}^*}^n$ is extended symmetric diagonal plus $(2,2)$-generator representable semiseparable,*

$$\begin{aligned}
\mathbf{G} = \operatorname{diag}(\mathbf{d}) &+ \frac{(\mu-\hat{\mu})(2\alpha+\mu+\hat{\mu}-1)}{2\alpha+2\hat{\mu}-1}\left(\operatorname{triu}(\mathbf{u}_1\mathbf{v}_1^{\mathrm{T}},1)+\operatorname{tril}(\mathbf{v}_1\mathbf{u}_1^{\mathrm{T}},-1)\right) \\
&+ \frac{(\nu-\hat{\nu})(2\alpha+\nu+\hat{\nu}-1)}{2\alpha+2\hat{\nu}-1}\left(\operatorname{triu}(\mathbf{u}_2\mathbf{v}_2^{\mathrm{T}},1)+\operatorname{tril}(\mathbf{v}_2\mathbf{u}_2^{\mathrm{T}},-1)\right),
\end{aligned}$$

with the vectors $\mathbf{d}$, $\mathbf{u}_1$, $\mathbf{v}_1$, $\mathbf{u}_2$, and $\mathbf{v}_2$ given by

$$\mathbf{d} = \left(j(j+2\alpha)+(j+\alpha)\left(\frac{(\mu-\hat{\mu})(2\alpha+\mu+\hat{\mu}-1)}{2\alpha+2\hat{\mu}-1}+\frac{(\nu-\hat{\nu})(2\alpha+\nu+\hat{\nu}-1)}{2\alpha+2\hat{\nu}-1}\right)\right)_{j=\hat{n}^*}^n,$$

$$\mathbf{u}_1 = \left(\left(\frac{(j+\alpha)\Gamma(j-\hat{n}^*+\alpha+\hat{\mu}+1/2)\Gamma(j+\hat{n}^*+2\alpha)}{\Gamma(j-\hat{n}^*+1)\Gamma(j-\hat{n}^*+\alpha+\hat{\nu}+1/2)}\right)^{1/2}\right)_{j=\hat{n}^*}^n,$$

$$\mathbf{v}_1 = \left(\left(\frac{(j+\alpha)\Gamma(j-\hat{n}^*+1)\Gamma(j-\hat{n}^*+\alpha+\hat{\nu}+1/2)}{\Gamma(j-\hat{n}^*+\alpha+\hat{\mu}+1/2)\Gamma(j+\hat{n}^*+2\alpha)}\right)^{1/2}\right)_{j=\hat{n}^*}^n,$$

$$\mathbf{u}_2 = \left(\left((-1)^j \frac{(j+\alpha)\Gamma(j-\hat{n}^*+\alpha+\hat{\nu}+1/2)\Gamma(j+\hat{n}^*+2\alpha)}{\Gamma(j-\hat{n}^*+1)\Gamma(j-\hat{n}^*+\alpha+\hat{\mu}+1/2)} \right)^{1/2} \right)_{j=\hat{n}^*}^{n},$$

$$\mathbf{v}_2 = \left(\left((-1)^j \frac{(j+\alpha)\Gamma(j-\hat{n}^*+1)\Gamma(j-\hat{n}^*+\alpha+\hat{\mu}+1/2)}{\Gamma(j-\hat{n}^*+\alpha+\hat{\nu}+1/2)\Gamma(j+\hat{n}^*+2\alpha)} \right)^{1/2} \right)_{j=\hat{n}^*}^{n}.$$

Generalized associated Legendre functions

Theorem 2.34 *Let $\{P_n^{m,m'}\}_{n\in\mathbb{N}_0, n^*\le n}$ and $\{P_n^{\hat{m},\hat{m}'}\}_{n\in\mathbb{N}_0,\hat{n}^*\le n}$ be two sequences of generalized associated Legendre functions such that $m+\hat{m}+m'+\hat{m}'$ is an even number, $\hat{\mu}\le\mu$ and $\hat{\nu}\le\nu$. Then the corresponding matrix $\mathbf{G}=(g_{i,j})_{i,j=\hat{n}^*}^{n}$ is extended symmetric diagonal plus $(2,2)$-generator representable semiseparable,*

$$\begin{aligned}\mathbf{G} = \operatorname{diag}(\mathbf{d}) &+ \frac{\mu^2-\hat{\mu}^2}{4\hat{\mu}}\left(\operatorname{triu}(\mathbf{u}_1\mathbf{v}_1^{\mathrm{T}},1)+\operatorname{tril}(\mathbf{v}_1\mathbf{u}_1^{\mathrm{T}},-1)\right)\\ &+\frac{\nu^2-\hat{\nu}^2}{4\hat{\nu}}\left(\operatorname{triu}(\mathbf{u}_2\mathbf{v}_2^{\mathrm{T}},1)+\operatorname{tril}(\mathbf{v}_2\mathbf{u}_2^{\mathrm{T}},-1)\right),\end{aligned}$$

with the vectors $\mathbf{d}$, $\mathbf{u}_1$, $\mathbf{v}_1$, $\mathbf{u}_2$, and $\mathbf{v}_2$ given by

$$\mathbf{d} = \left(j(j+1) + (2j+1)\left(\frac{\mu^2-\hat{\mu}^2}{4\hat{\mu}}+\frac{\nu^2-\hat{\nu}^2}{4\hat{\nu}}\right)\right)_{j=\hat{n}^*}^{n},$$

$$\mathbf{u}_1 = \left(\left(\frac{(2j+1)\Gamma(j-\hat{n}^*+\hat{\mu}+1)\Gamma(j+\hat{n}^*+1)}{\Gamma(j-\hat{n}^*+1)\Gamma(j-\hat{n}^*+\hat{\nu}+1)} \right)^{1/2} \right)_{j=\hat{n}^*}^{n},$$

$$\mathbf{v}_1 = \left(\left(\frac{(2j+1)\Gamma(j-\hat{n}^*+1)\Gamma(j-\hat{n}^*+\hat{\nu}+1)}{\Gamma(j-\hat{n}^*+\hat{\mu}+1)\Gamma(j+\hat{n}^*+1)} \right)^{1/2} \right)_{j=\hat{n}^*}^{n},$$

$$\mathbf{u}_2 = \left(\left((-1)^j \frac{(2j+1)\Gamma(j-\hat{n}^*+\hat{\nu}+1)\Gamma(j+\hat{n}^*+1)}{\Gamma(j-\hat{n}^*+1)\Gamma(j-\hat{n}^*+\hat{\mu}+1)} \right)^{1/2} \right)_{j=\hat{n}^*}^{n},$$

$$\mathbf{v}_2 = \left(\left((-1)^j \frac{(2j+1)\Gamma(j-\hat{n}^*+1)\Gamma(j-\hat{n}^*+\hat{\mu}+1)}{\Gamma(j-\hat{n}^*+\hat{\nu}+1)\Gamma(j+\hat{n}^*+1)} \right)^{1/2} \right)_{j=\hat{n}^*}^{n}.$$

Generalized associated Chebyshev functions of first kind

Theorem 2.35 *Let $\{T_n^{m,m'}\}_{n\in\mathbb{N}_0, n^*\le n}$ and $\{T_n^{\hat{m},\hat{m}'}\}_{n\in\mathbb{N}_0,\hat{n}^*\le n}$ be two sequences of generalized associated Chebyshev functions of first kind such that $m+\hat{m}+m'+\hat{m}'$ is an even number, $\hat{\mu}\le\mu$ and $\hat{\nu}\le\nu$. Then the corresponding matrix $\mathbf{G}=(g_{i,j})_{i,j=\hat{n}^*}^{n}$ is extended symmetric diagonal plus $(2,2)$-generator representable semiseparable,*

$$\begin{aligned}\mathbf{G} = \operatorname{diag}(\mathbf{d}) &+ \frac{(\mu-\hat{\mu})(\mu+\hat{\mu}-1)}{4\hat{\mu}-2}\left(\operatorname{triu}(\mathbf{u}_1\mathbf{v}_1^{\mathrm{T}},1)+\operatorname{tril}(\mathbf{v}_1\mathbf{u}_1^{\mathrm{T}},-1)\right)\\ &+\frac{(\nu-\hat{\nu})(\nu+\hat{\nu}-1)}{4\hat{\nu}-2}\left(\operatorname{triu}(\mathbf{u}_2\mathbf{v}_2^{\mathrm{T}},1)+\operatorname{tril}(\mathbf{v}_2\mathbf{u}_2^{\mathrm{T}},-1)\right),\end{aligned}$$

with the vectors $\mathbf{d}$, $\mathbf{u}_1$, $\mathbf{v}_1$, $\mathbf{u}_2$, *and* $\mathbf{v}_2$ *given by*

$$\mathbf{d} = \left(j^2 + 2j\left(\frac{(\mu-\hat{\mu})(\mu+\hat{\mu}-1)}{4\hat{\mu}-2} + \frac{(\nu-\hat{\nu})(\nu+\hat{\nu}-1)}{4\hat{\nu}-2} \right) \right)_{j=\hat{n}^*}^{n},$$

$$\mathbf{u}_1 = \left(\left(\frac{2j\Gamma(j-\hat{n}^*+\hat{\mu}+1/2)\Gamma(j+\hat{n}^*)}{\Gamma(j-\hat{n}^*+1)\Gamma(j-\hat{n}^*+\hat{\nu}+1/2)} \right)^{1/2} \right)_{j=\hat{n}^*}^{n},$$

$$\mathbf{v}_1 = \left(\left(\frac{2j\Gamma(j-\hat{n}^*+1)\Gamma(j-\hat{n}^*+\hat{\nu}+1/2)}{\Gamma(j-\hat{n}^*+\hat{\mu}+1/2)\Gamma(j+\hat{n}^*)} \right)^{1/2} \right)_{j=\hat{n}^*}^{n},$$

$$\mathbf{u}_2 = \left(\left((-1)^j \frac{2j\Gamma(j-\hat{n}^*+\hat{\nu}+1/2)\Gamma(j+\hat{n}^*)}{\Gamma(j-\hat{n}^*+1)\Gamma(j-\hat{n}^*+\hat{\mu}+1/2)} \right)^{1/2} \right)_{j=\hat{n}^*}^{n},$$

$$\mathbf{v}_2 = \left(\left((-1)^j \frac{2j\Gamma(j-\hat{n}^*+1)\Gamma(j-\hat{n}^*+\hat{\mu}+1/2)}{\Gamma(j-\hat{n}^*+\hat{\nu}+1/2)\Gamma(j+\hat{n}^*)} \right)^{1/2} \right)_{j=\hat{n}^*}^{n}.$$

Generalized associated Chebyshev functions of second kind

Theorem 2.36 *Let* $\{U_n^{m,m'}\}_{n\in\mathbb{N}_0,n^*\leq n}$ *and* $\{U_n^{\hat{m},\hat{m}'}\}_{n\in\mathbb{N}_0,\hat{n}^*\leq n}$ *be two sequences of generalized associated Chebyshev functions of second kind such that* $m+\hat{m}+m'+\hat{m}'$ *is an even number,* $\hat{\mu}\leq\mu$ *and* $\hat{\nu}\leq\nu$. *Then the corresponding matrix* $\mathbf{G}=(g_{i,j})_{i,j=\hat{n}^*}^{n}$ *is extended symmetric diagonal plus* $(2,2)$*-generator representable semiseparable,*

$$\begin{aligned}\mathbf{G} = \operatorname{diag}(\mathbf{d}) &+ \frac{(\mu-\hat{\mu})(\mu+\hat{\mu}+1)}{4\hat{\mu}+2}\left(\operatorname{triu}(\mathbf{u}_1\mathbf{v}_1^{\mathrm{T}},1)+\operatorname{tril}(\mathbf{v}_1\mathbf{u}_1^{\mathrm{T}},-1)\right)\\ &+ \frac{(\nu-\hat{\nu})(\nu+\hat{\nu}+1)}{4\hat{\nu}+2}\left(\operatorname{triu}(\mathbf{u}_2\mathbf{v}_2^{\mathrm{T}},1)+\operatorname{tril}(\mathbf{v}_2\mathbf{u}_2^{\mathrm{T}},-1)\right),\end{aligned}$$

with the vectors $\mathbf{d}$, $\mathbf{u}_1$, $\mathbf{v}_1$, $\mathbf{u}_2$, *and* $\mathbf{v}_2$ *given by*

$$\mathbf{d} = \left(j(j+2) + (2j+2)\left(\frac{(\mu-\hat{\mu})(\mu+\hat{\mu}+1)}{4\hat{\mu}+2} + \frac{(\nu-\hat{\nu})(\nu+\hat{\nu}+1)}{4\hat{\nu}+2} \right) \right)_{j=\hat{n}^*}^{n},$$

$$\mathbf{u}_1 = \left(\left(\frac{(2j+2)\Gamma(j-\hat{n}^*+\hat{\mu}+3/2)\Gamma(j+\hat{n}^*+2)}{\Gamma(j-\hat{n}^*+1)\Gamma(j-\hat{n}^*+\hat{\nu}+3/2)} \right)^{1/2} \right)_{j=\hat{n}^*}^{n},$$

$$\mathbf{v}_1 = \left(\left(\frac{(2j+2)\Gamma(j-\hat{n}^*+1)\Gamma(j-\hat{n}^*+\hat{\nu}+3/2)}{\Gamma(j-\hat{n}^*+\hat{\mu}+3/2)\Gamma(j+\hat{n}^*)} \right)^{1/2} \right)_{j=\hat{n}^*}^{n},$$

$$\mathbf{u}_2 = \left(\left((-1)^j \frac{(2j+2)\Gamma(j-\hat{n}^*+\hat{\nu}+3/2)\Gamma(j+\hat{n}^*+2)}{\Gamma(j-\hat{n}^*+1)\Gamma(j-\hat{n}^*+\hat{\mu}+3/2)} \right)^{1/2} \right)_{j=\hat{n}^*}^{n},$$

$$\mathbf{v}_2 = \left(\left((-1)^j \frac{(2j+2)\Gamma(j-\hat{n}^*+1)\Gamma(j-\hat{n}^*+\hat{\mu}+3/2)}{\Gamma(j-\hat{n}^*+\hat{\nu}+3/2)\Gamma(j+\hat{n}^*)} \right)^{1/2} \right)_{j=\hat{n}^*}^{n}.$$

Associated Jacobi functions

Theorem 2.37 *Let* $\{P_n^{(\alpha,\beta),m}\}_{n\in\mathbb{N}_0,\mu\leq n}$ *and* $\{P_n^{\alpha,\beta),\hat{m}}\}_{n\in\mathbb{N}_0,\hat{\mu}\leq n}$, *with* $\alpha,\beta>-1$, *be two sequences of associated Jacobi functions such that* $m+\hat{m}$ *is an even number and* $\hat{\mu}\leq\mu$.

Then the corresponding matrix $\mathbf{G} = (g_{i,j})_{i,j=\hat{\mu}}^{n}$ *is extended symmetric diagonal plus* $(2,2)$*-generator representable semiseparable,*

$$\mathbf{G} = \operatorname{diag}(\mathbf{d}) + \frac{(\mu-\hat{\mu})(2\alpha+\mu+\hat{\mu})}{4(\alpha+\hat{\mu})}\left(\operatorname{triu}(\mathbf{u}_1\,\mathbf{v}_1^{\mathrm{T}},1) + \operatorname{tril}(\mathbf{v}_1\,\mathbf{u}_1^{\mathrm{T}},-1)\right)$$
$$+ \frac{(\mu-\hat{\mu})(2\beta+\mu+\hat{\mu})}{4(\beta+\hat{\mu})}\left(\operatorname{triu}(\mathbf{u}_2\,\mathbf{v}_2^{\mathrm{T}},1) + \operatorname{tril}(\mathbf{v}_2\,\mathbf{u}_2^{\mathrm{T}},-1)\right),$$

with the vectors $\mathbf{d}$, $\mathbf{u}_1$, $\mathbf{v}_1$, $\mathbf{u}_2$, *and* $\mathbf{v}_2$ *given by*

$$\mathbf{d} = \left(j(j+\alpha+\beta+1) + (2j+\alpha+\beta+1)(\mu-\hat{\mu})\left(\frac{2\alpha+\mu+\hat{\mu}}{4(\alpha+\hat{\mu})} + \frac{2\beta+\mu+\hat{\mu}}{4(\beta+\hat{\mu})}\right)\right)_{j=\hat{\mu}}^{n},$$

$$\mathbf{u}_1 = \left(\left(\frac{(2j+\alpha+\beta+1)\Gamma(j+\alpha+1)\Gamma(j+\hat{\mu}+\alpha+\beta+1)}{\Gamma(j-\hat{\mu}+1)\Gamma(j+\beta+1)}\right)^{1/2}\right)_{j=\hat{\mu}}^{n},$$

$$\mathbf{v}_1 = \left(\left(\frac{(2j+\alpha+\beta+1)\Gamma(j-\hat{\mu}+1)\Gamma(j+\beta+1)}{\Gamma(j+\alpha+1)\Gamma(j+\hat{\mu}+\alpha+\beta+1)}\right)^{1/2}\right)_{j=\hat{\mu}}^{n},$$

$$\mathbf{u}_2 = \left(\left((-1)^j\frac{(2j+\alpha+\beta+1)\Gamma(j+\beta+1)\Gamma(j+\hat{\mu}+\alpha+\beta+1)}{\Gamma(j-\hat{\mu}+1)\Gamma(j+\alpha+1)}\right)^{1/2}\right)_{j=\hat{\mu}}^{n},$$

$$\mathbf{v}_2 = \left(\left((-1)^j\frac{(2j+\alpha+\beta+1)\Gamma(j-\hat{\mu}+1)\Gamma(j+\alpha+1)}{\Gamma(j+\beta+1)\Gamma(j+\hat{\mu}+\alpha+\beta+1)}\right)^{1/2}\right)_{j=\hat{\mu}}^{n}.$$

Associated Gegenbauer functions

Theorem 2.38 *Let* $\{C_n^{(\alpha),m}\}_{n\in\mathbb{N}_0,\mu\leq n}$ *and* $\{C_n^{(\alpha),\hat{m}}\}_{n\in\mathbb{N}_0,\hat{\mu}\leq n}$ *with* $\alpha > -1/2$ *and* $\alpha \neq 0$ *be two sequences of associated Gegenbauer functions such that* $m+\hat{m}$ *is an even number and* $\hat{\mu} \leq \mu$*. Then the corresponding matrix* $\mathbf{G} = (g_{i,j})_{i,j=\hat{\mu}}^{n}$ *is symmetric checkerboard-like diagonal plus* $(1,1)$*-generator representable semiseparable,*

$$\mathbf{G} = \operatorname{diag}(\mathbf{d}) + 2(\mu-\hat{\mu})\frac{2\alpha+\mu+\hat{\mu}-1}{2\alpha+2\hat{\mu}-1}\left(\operatorname{triuc}(\mathbf{u}\,\mathbf{v}^{\mathrm{T}},1) + \operatorname{trilc}(\mathbf{v}\,\mathbf{u}^{\mathrm{T}},-1)\right),$$

with the vectors $\mathbf{d}$, $\mathbf{u}$, *and* $\mathbf{v}$ *given by*

$$\mathbf{d} = \left(j(j+2\alpha) + 2(j+\alpha)(\mu-\hat{\mu})\frac{2\alpha+\mu+\hat{\mu}-1}{2\alpha+2\hat{\mu}-1}\right)_{j=\hat{\mu}}^{n},$$

$$\mathbf{u} = \left(\left(\frac{(j+\alpha)\Gamma(j+\hat{\mu}+2\alpha)}{\Gamma(j-\hat{\mu}+1)}\right)^{1/2}\right)_{j=\hat{\mu}}^{n},$$

$$\mathbf{v} = \left(\left(\frac{(j+\alpha)\Gamma(j-\hat{\mu}+1)}{\Gamma(j+\hat{\mu}+2\alpha)}\right)^{1/2}\right)_{j=\hat{\mu}}^{n}.$$

Associated Legendre functions

Theorem 2.39 *Let* $\{P_n^m\}_{n\in\mathbb{N}_0,\mu\leq n}$ *and* $\{P_n^{\hat{m}}\}_{n\in\mathbb{N}_0,\hat{\mu}\leq n}$ *be two sequences of associated Legendre functions such that* $m+\hat{m}$ *is an even number and* $\hat{\mu}\leq\mu$*. Then the corresponding matrix* $\mathbf{G} = (g_{i,j})_{i,j=\hat{\mu}}^{n}$ *is symmetric checkerboard-like diagonal plus* $(1,1)$*-generator representable semiseparable,*

$$\mathbf{G} = \operatorname{diag}(\mathbf{d}) + \frac{\mu^2-\hat{\mu}^2}{\hat{\mu}}\left(\operatorname{triuc}(\mathbf{u}\,\mathbf{v}^{\mathrm{T}},1) + \operatorname{trilc}(\mathbf{v}\,\mathbf{u}^{\mathrm{T}},-1)\right),$$

with the vectors $\mathbf{d}$, $\mathbf{u}$, *and* $\mathbf{v}$ *given by*

$$\mathbf{d} = \left(j(j+1) + (j+1/2)\frac{\mu^2 - \hat{\mu}^2}{\hat{\mu}} \right)_{j=\hat{\mu}}^{n},$$

$$\mathbf{u} = \left(\left(\frac{(j+1/2)\Gamma(j+\hat{\mu}+1)}{\Gamma(j-\hat{\mu}+1)} \right)^{1/2} \right)_{j=\hat{\mu}}^{n},$$

$$\mathbf{v} = \left(\left(\frac{(j+1/2)\Gamma(j-\hat{\mu}+1)}{\Gamma(j+\hat{\mu}+1)} \right)^{1/2} \right)_{j=\hat{\mu}}^{n}.$$

Associated Chebyshev functions of first kind

Theorem 2.40 *Let* $\{T_n^m\}_{n\in\mathbb{N}_0,\mu\le n}$ *and* $\{T_n^{\hat{m}}\}_{n\in\mathbb{N}_0,\hat{\mu}\le n}$ *be two sequences of associated Chebyshev functions of first kind such that* $m+\hat{m}$ *is an even number and* $\hat{\mu} \le \mu$. *Then the corresponding matrix* $\mathbf{G} = (g_{i,j})_{i,j=\hat{\mu}}^{n}$ *is symmetric checkerboard-like diagonal plus* $(1,1)$-*generator representable semiseparable,*

$$\mathbf{G} = \operatorname{diag}(\mathbf{d}) + 2(\mu - \hat{\mu})\frac{\mu + \hat{\mu} - 1}{2\hat{\mu} - 1}\left(\operatorname{triuc}(\mathbf{u}\,\mathbf{v}^{\mathrm{T}}, 1) + \operatorname{trilc}(\mathbf{v}\,\mathbf{u}^{\mathrm{T}}, -1) \right),$$

with the vectors $\mathbf{d}$, $\mathbf{u}$, *and* $\mathbf{v}$ *given by*

$$\mathbf{d} = \left(j^2 + 2j(\mu - \hat{\mu})\frac{\mu + \hat{\mu} - 1}{2\hat{\mu} - 1} \right)_{j=\hat{\mu}}^{n},$$

$$\mathbf{u} = \left(\left(\frac{j\Gamma(j+\hat{\mu})}{\Gamma(j-\hat{\mu}+1)} \right)^{1/2} \right)_{j=\mu}^{n},$$

$$\mathbf{v} = \left(\left(\frac{j\Gamma(j-\hat{\mu}+1)}{\Gamma(j+\hat{\mu})} \right)^{1/2} \right)_{j=\hat{\mu}}^{n}.$$

Associated Chebyshev functions of second kind

Theorem 2.41 *Let* $\{U_n^m\}_{n\in\mathbb{N}_0,\mu\le n}$ *and* $\{U_n^{\hat{m}}\}_{n\in\mathbb{N}_0,\hat{\mu}\le n}$ *be two sequences of associated Chebyshev functions of second kind such that* $m + \hat{m}$ *is an even number and* $\hat{\mu} \le \mu$. *Then the corresponding matrix* $\mathbf{G} = (g_{i,j})_{i,j=\hat{\mu}}^{n}$ *is symmetric checkerboard-like diagonal plus* $(1,1)$-*generator representable semiseparable,*

$$\mathbf{G} = \operatorname{diag}(\mathbf{d}) + 2(\mu - \hat{\mu})\frac{\mu + \hat{\mu} + 1}{2\hat{\mu} + 1}\left(\operatorname{triuc}(\mathbf{u}\,\mathbf{v}^{\mathrm{T}}, 1) + \operatorname{trilc}(\mathbf{v}\,\mathbf{u}^{\mathrm{T}}, -1) \right),$$

with the vectors $\mathbf{d}$, $\mathbf{u}$, *and* $\mathbf{v}$ *given by*

$$\mathbf{d} = \left(j(j+2) + 2(j+1)(\mu - \hat{\mu})\frac{\mu + \hat{\mu} + 1}{2\hat{\mu} + 1} \right)_{j=\hat{\mu}}^{n},$$

$$\mathbf{u} = \left(\left(\frac{(j+1)\Gamma(j+\hat{\mu}+2)}{\Gamma(j-\hat{\mu}+1)} \right)^{1/2} \right)_{j=\hat{\mu}}^{n},$$

$$\mathbf{v} = \left(\left(\frac{(j+1)\Gamma(j-\hat{\mu}+1)}{\Gamma(j+\hat{\mu}+2)} \right)^{1/2} \right)_{j=\hat{\mu}}^{n}.$$

2.6.3 Scaling the eigenvectors. The divide-and-conquer method for the eigendecomposition of extended symmetric diagonal plus generator representable semiseparable matrices is now ready to be applied to the matrices $\mathbf{G}$ given in the last section. But we need to ensure that the columns of the corresponding eigenvector matrix $\mathbf{Q}$ which should be identical to the connection matrix $\mathbf{K}$, are properly scaled. Recall that from the matrix $\mathbf{Q}$ only a certain number of the columns forms the actual connection matrix $\mathbf{K}$; see Remark 2.29. The rest can be ignored. Unlike the case for classical orthogonal polynomials, the divide-and-conquer method for symmetric semiseparable matrices does not allow easy control over specific entries in the connection matrix. This means that we cannot use the same procedure to achieve the desired scaling of the eigenvectors that was used in Section 2.4.5. Therefore, we propose two different methods.
Let us assume, for simplicity, that we want to compute the connection between the associated Laguerre functions $\{L_n^{(\alpha),m}\}_{n\in\mathbb{N}_0,n\geq\mu}$ and $\{L_n^{(\alpha),\hat{m}}\}_{n\in\mathbb{N}_0,n\geq\hat{\mu}}$. Then the connection matrix $\mathbf{K} = (\kappa_{i,j})$ has its entries

$$\kappa_{i,j} = \frac{1}{h_i^{(\alpha)}} \int_0^\infty L_i^{(\alpha),\hat{m}}(x) L_j^{(\alpha),m}(x)\, x^\alpha \mathrm{e}^{-x}\, \mathrm{d}x.$$

The submatrix $\hat{\mathbf{Q}}$ of the eigenvector matrix $\mathbf{Q}$, obtained from the divide-and-conquer method applied to the corresponding matrix $\mathbf{G}$, should be identical to the connection matrix $\mathbf{K}$, provided that the columns have been scaled properly. Since we can efficiently compute matrix-vector products involving the matrix $\hat{\mathbf{Q}}$, or its transpose $\hat{\mathbf{Q}}^{\mathrm{T}}$, we calculate the product

$$\mathbf{z} = \hat{\mathbf{Q}}^{\mathrm{T}}\, \mathbf{e}_1$$

where $\mathbf{e}_1$ is the first coordinate vector $\mathbf{e}_1 = (1, 0, 0, \ldots, 0)^{\mathrm{T}}$. If $\hat{\mathbf{Q}}$ is scaled correctly, then the vector $\mathbf{z}$ satisfies

$$\mathbf{z} = (\kappa_{\hat{\mu},\mu}, \kappa_{\hat{\mu},\mu+1}, \ldots)^{\mathrm{T}}.$$

Therefore, if the content of the vector $\mathbf{z}$ is known, we can obtain correction factors that must be applied to each column of the matrix $\hat{\mathbf{Q}}$ to make it identical to the connection matrix $\mathbf{K}$. This can be easily incorporated into the divide-and-conquer method at no additional cost for the actual transformation. Analogous thoughts apply to the generalized associated Jacobi functions and, thus, to all other remaining types of associated functions.
It remains to show how the connection coefficients $\kappa_{\hat{\mu},\mu}, \kappa_{\hat{\mu},\mu+1}, \ldots$ may be calculated. Here is the result for the associated Laguerre functions.

Lemma 2.42 *Let $\{L_n^{(\alpha),m}\}_{n\in\mathbb{N}_0,\mu\leq n}$ and $\{L_n^{(\alpha),\hat{m}}\}_{n\in\mathbb{N}_0,\hat{\mu}\leq n}$, with $\alpha > -1$, be two sequences of associated Laguerre functions such that $m+\hat{m}$ is an even number and $\hat{\mu} \leq \mu$. Then the connection coefficients $\kappa_{\hat{\mu},j}$ for $j = \mu, \mu+1, \ldots$ are given by*

$$\kappa_{\hat{\mu},j} = \frac{\sqrt{\Gamma(\hat{\mu}+1)}}{\Gamma(\frac{\mu-\hat{\mu}}{2})} \frac{\Gamma(\alpha+\frac{\hat{\mu}+\mu}{2}+1)}{\Gamma(\alpha+\hat{\mu}+1)} \frac{\Gamma(j-\frac{\hat{\mu}+\mu}{2})}{\sqrt{\Gamma(j+1)\Gamma(j-\mu+1)}}.$$

PROOF. Observing that even $\hat{\mu}$, μ imply $(-1)^{\hat{\mu}+\mu} = 1$, we have

$$\begin{aligned}
\kappa_{\hat{\mu},j} &= \frac{1}{h_{\hat{\mu}}^{(\alpha)}} \int_0^\infty L_{\hat{\mu}}^{(\alpha),\hat{m}}(x) L_j^{(\alpha),m}(x)\, x^\alpha \mathrm{e}^{-x}\, \mathrm{d}x \\
&= \frac{1}{h_{\hat{\mu}}^{(\alpha)}} (-1)^{\hat{\mu}+\mu} \left(\frac{\Gamma(j-\mu+1)}{\Gamma(\hat{\mu}+1)\Gamma(j+1)} \right)^{1/2} \int_0^\infty L_{j-\mu}^{(\alpha+\mu)}(x)\, x^{\alpha+\frac{\hat{\mu}+\mu}{2}} \mathrm{e}^{-x}\, \mathrm{d}x
\end{aligned}$$

$$
\begin{aligned}
&= \frac{h_{\hat{\mu}}^{(\alpha+\frac{\hat{\mu}+\mu}{2})}}{h_{\hat{\mu}}^{(\alpha)}} \left(\frac{\Gamma(j-\mu+1)}{\Gamma(\hat{\mu}+1)\Gamma(j+1)} \right)^{1/2} \kappa_{0,j-\mu}^{(\alpha+\mu)\to(\alpha+\frac{\hat{\mu}+\mu}{2})} \\
&= \frac{h_{0}^{(\alpha+\frac{\hat{\mu}+\mu}{2})}}{h_{\hat{\mu}}^{(\alpha)}} \left(\frac{\Gamma(j-\mu+1)}{\Gamma(\hat{\mu}+1)\Gamma(j+1)} \right)^{1/2} \kappa_{0,j-\mu}^{(\alpha+\mu)\to(\alpha+\frac{\hat{\mu}+\mu}{2})} \\
&= \frac{\sqrt{\Gamma(\hat{\mu}+1)}}{\Gamma(\frac{\mu-\hat{\mu}}{2})} \frac{\Gamma(\alpha+\frac{\hat{\mu}+\mu}{2}+1)}{\Gamma(\alpha+\hat{\mu}+1)} \frac{\Gamma(j-\frac{\hat{\mu}+\mu}{2})}{\sqrt{\Gamma(j+1)\Gamma(j-\mu+1)}}.
\end{aligned}
$$

□

A similar result for the generalized associated Jacobi functions can be given, but it turns out that the resulting expression still contains a sum. This makes it inefficient and possibly unstable to apply the discussed method of scaling. Therefore, we propose another method to achieve the same result. It is based on the observation that the columns of the connection matrix $\mathbf{K}$ have unit euclidean length if the corresponding sequences of associated functions are normalized. This can be seen as follows. Assume that we want to compute the connection between the two sequences of normalized generalized associated Jacobi functions $\{P_n^{(\alpha,\beta),m,m'}\}_{n\in\mathbb{N}_{0,n\geq n^*}}$ and $\{P_n^{(\alpha,\beta),\hat{m},\hat{m}'}\}_{n\in\mathbb{N}_{0,n\geq \hat{n}^*}}$. We define the vectors

$$
\mathbf{x} = (P_{n^*}^{(\alpha,\beta),m,m'}, P_{n^*+1}^{(\alpha,\beta),m,m'}, \ldots), \quad \text{and} \quad \mathbf{y} = (P_{\hat{n}^*}^{(\alpha,\beta),\hat{m},\hat{m}'}, P_{\hat{n}^*+1}^{(\alpha,\beta),\hat{m},\hat{m}'}, \ldots),
$$

and the outer product

$$
\mathbf{x} \odot \mathbf{y} := \big(\langle x_i, y_j \rangle\big),
$$

where $\langle \cdot\,, \cdot \rangle$ is the inner product with respect to which both sequences of associated functions are orthogonal. Using that the vectors $\mathbf{x}$ and $\mathbf{y}$ are related by

$$
\mathbf{y} = \mathbf{K}\mathbf{x}
$$

and that

$$
\mathbf{x} \odot \mathbf{x}^{\mathrm{T}} = \mathbf{I}, \quad \text{and} \quad \mathbf{y} \odot \mathbf{y}^{\mathrm{T}} = \mathbf{I},
$$

we find that

$$
\mathbf{K}^{\mathrm{T}}\mathbf{K} = \mathbf{K}^{\mathrm{T}}\mathbf{y}\mathbf{y}^{\mathrm{T}}\mathbf{K} = \mathbf{K}^{\mathrm{T}}\mathbf{K}\mathbf{x}\mathbf{x}^{\mathrm{T}}\mathbf{K}^{\mathrm{T}}\mathbf{K} = \mathbf{K}^{\mathrm{T}}\mathbf{K}\mathbf{K}^{\mathrm{T}}\mathbf{K}.
$$

This implies $\mathbf{K}^{\mathrm{T}}\mathbf{K} = \mathbf{I}$ and the matrix $\mathbf{K}$ must have orthogonal columns of unit euclidean length. The columns of the eigenvector matrix $\mathbf{Q}$ can be normalized at almost no cost in the divide-and-conquer algorithm by normalizing accordingly every eigenvector matrix that appears during the pre-computation phase. If the actual sequence of generalized associated Jacobi functions that we want to use is not normalized, then a final column and/or row scaling to the obtained matrix $\hat{\mathbf{Q}}$ achieves the desired result. This comes entirely at no additional cost for the actual transformation.

2.6.4 Other normalizations. It should be noted that the matrix $\mathbf{G}$ as it was defined for the classical associated functions is not affected by a change to one of the three normalizations (standard, monic, or normalized) in the involved sequences of functions. This is contrary to the case for the classical orthogonal polynomials. The reason for this is that we are only concerned with the connection between associated functions of different orders that have been derived from the same sequence of classical orthogonal polynomials. And these sequences are normalized with respect to the same norm. For example, the matrix $\bar{\mathbf{G}}$ for the monic associated Laguerre functions $\{\bar{L}_n^{(\alpha),m}\}_{n\in\mathbb{N}_0,n\geq\mu}$ and $\{\bar{L}_n^{(\beta),\hat{m}}\}_{n\in\mathbb{N}_0,n\geq\hat{\mu}}$

is identical to the matrix **G** for the associated Laguerre functions $\{L_n^{(\alpha),m}\}_{n\in\mathbb{N}_0,n\geq\mu}$ and $\{L_n^{(\beta),\hat{m}}\}_{n\in\mathbb{N}_0,n\geq\hat{\mu}}$.

Remark 2.43 Our implementation of the divide-and-conquer method for the connection between classical associated functions of different orders represents work in progress. However, the method developed in this section may be able to circumvent some numerical issues that have been observed in other methods developed for the same transformation. A more detailed discussion of this topic is beyond the scope of this text, but two applications where the newly developed method might be useful are described in Chapter 4.

3 – Techniques based on the fast multipole method

In the previous chapter techniques were introduced to efficiently compute the connection between classical orthogonal polynomials or classical associated functions. These were based on the observation that the corresponding connection matrix $\mathbf{K}$ can be represented as the properly scaled eigenvector matrix of a known (triangular) generator representable semiseparable matrix. Employing divide-and-conquer methods to compute this eigendecomposition enabled us to apply the connection matrix $\mathbf{K}$ efficiently. The reason for this is that a hierarchical representation of the connection matrix $\mathbf{K}$ is obtained that contains Cauchy-like matrices. These can be applied efficiently to any vector using the fast multipole method (FMM).

In this chapter, we provide a potentially even more efficient alternative to this technique. It is based on an FMM-like method that will be applied directly to the connection matrix $\mathbf{K}$ to compute the desired transformation. To enable and justify this procedure, smoothness results about the entries of the connection matrix $\mathbf{K}$ are needed. This will be restricted to the classical orthogonal polynomials as the explicit expressions for the connection coefficients between classical associated functions are too complicated to handle with our methods. For the classical orthogonal polynomials however, we have explicit expressions which are amenable to a detailed analysis. More precisely, when two sequences of classical orthogonal polynomials are "close enough", we will prove approximation results that will allow the FMM-like method to be used. For example, to compute the connection between the Laguerre polynomials $\{L_n^{(\alpha)}\}_{n\in\mathbb{N}_0}$ and $\{L_n^{(\beta)}\}_{n\in\mathbb{N}_0}$, we will prove that the desired approximation result holds whenever $|\alpha-\beta|<1$. Similar results will be established for Jacobi and Gegenbauer polynomials. For larger "distances", that is, when $|\alpha-\beta|>1$, we can still combine the new method with other results for the connection coefficients from Chapter 1 to maintain the efficiency of the method. This will be done by first computing a transformation from the polynomials $\{L_n^{(\alpha)}\}_{n\in\mathbb{N}_0}$ to an intermediate sequence $\{L_n^{(\alpha')}\}_{n\in\mathbb{N}_0}$. Then, the transformation from $\{L_n^{(\alpha')}\}_{n\in\mathbb{N}_0}$ to the actual target $\{L_n^{(\beta)}\}_{n\in\mathbb{N}_0}$ is calculated. The parameter α' is judiciously chosen such that $\alpha-\alpha'$ becomes an integer and, at the same time, $|\alpha'-\beta|<1$ holds.

For the first part of this two-step approach, we can use a number of results from Chapter 1 to efficiently compute the result, since the respective connection matrices will be either banded or triangular generator representable semiseparable and can thus be applied efficiently. Since the bandwidth or semiseparability rank, respectively, depend on the how large $|\alpha-\alpha'|$ is and since α' is chosen such that the quantities $\alpha-\beta$ and $\alpha-\alpha'$ are comparable, the arithmetic cost for this step will be $\mathcal{O}(|\alpha-\beta|n)$, where n is the length of the transformation.

For the second part, the approximation results let us apply the FMM-like method resulting in an $\mathcal{O}(n\log(1/\varepsilon))$ algorithm. In total, an $\mathcal{O}(n(|\alpha-\beta|+\log(1/\varepsilon)))$ method is obtained to compute both parts. With respect to the transform length n, this is asymptotically faster than the method from Chapter 2, which generally has a cost of $\mathcal{O}(|\alpha-\beta|n\log n\log(1/\varepsilon))$.

The rest of this chapter is structured as follows. In Section 3.1 a number of auxiliary results from approximation theory is given that is needed for the actual analysis. Section 3.2 shows two ways to obtain the desired approximation results for the connection coefficients $\kappa_{i,j}$ between Gegenbauer polynomials. This is motivated by historical reasons. The first technique was employed to compute the connection between Legendre and Chebyshev polynomials of first kind in [1]. Both sequences of polynomials can be seen as special cases of Gegenbauer polynomials. These results were extended by the author in [39] to

Gegenbauer polynomials. There, the original method was modified to fit the more general setting. In addition, a more powerful and arguably more elegant technique was used to obtain a stronger approximation result. While this certainly simplified the task of proving similar results for other types of classical orthogonal polynomials, the actual efficiency of the numerical method remains unaffected. The stronger theoretical results just put the numerical method on more solid ground. The analogous results for Laguerre and Jacobi polynomials are given in Section 3.3. The chapter is concluded by a number of numerical tests in Section 3.5, that mirror those conducted with the other methods on the same set of test cases in Section 2.4.6.

3.1 Preliminaries

3.1.1 Chebyshev approximation. A principal method used in this chapter is to find a good approximation to a function defined on the interval $[-1,1]$. An efficient way to do this is polynomial interpolation at the Chebyshev points.

Definition 3.1 *For $n \in \mathbb{N}_0$ and $0 \leq i \leq n$, the degree-n Chebyshev points t_i and the corresponding Lagrange polynomials $u_i(t)$ on the interval $[-1,1]$ are defined by*

$$t_i = \cos\left(\frac{(2i+1)\pi}{2(n+1)}\right), \qquad u_i(t) = \prod_{\substack{j=0\\ j\neq i}}^{n} \frac{t-t_j}{t_i-t_j}.$$

For functions on an arbitrary interval $[a,b]$, we define a linear mapping ϕ and its inverse ϕ^{-1} to transplant these functions from the interval $[a,b]$ to the interval $[-1,1]$ and vice versa. For technical reasons, the mappings are defined over the complex numbers.

Definition 3.2 *Let $[a,b]$ be an arbitrary non-empty interval on the real line. Then the linear mappings $\phi : \mathbb{C} \to \mathbb{C}$ and $\phi^{-1} : \mathbb{C} \to \mathbb{C}$ are defined by*

$$\phi(t) := \tfrac{1}{2}(b-a)(t+1)+a, \qquad \phi^{-1}(x) := 2\frac{x-a}{b-a} - 1.$$

Note that $\phi([-1,1]) = [a,b]$ and $\phi^{-1}([a,b]) = [-1,1]$. Whenever we use the mappings ϕ and ϕ^{-1} later, it will be clear from the context what the borders a and b are. We are now ready to introduce approximations based on interpolation at Chebyshev points.

Definition 3.3 *Let $f : [a,b] \to \mathbb{R}$ be an arbitrary function. Then the degree-n Chebyshev approximation $f_n : [a,b] \to \mathbb{R}$ to the function f is defined by*

$$f_n(x) = \sum_{i=0}^{n} f\big(\phi(t_i)\big)\, u_i\big(\phi^{-1}(x)\big).$$

This type of Lagrangian interpolation and the error involved are well understood. The proof of the following Theorem can be found in [5, p. 120].

Theorem 3.4 *Let $f : [a,b] \to \mathbb{R}$ be a function with $n+1$ continuous derivatives. Then*

$$\|f - f_n\|_\infty \leq \frac{2(b-a)^{n+1}}{4^{n+1}\,(n+1)!} \sup_{x\in[a,b]} \big|f^{(n+1)}(x)\big|. \tag{3.1}$$

The norm on the left hand side is taken over the interval $[a,b]$. A practical way to use Theorem 3.4 for a concrete function is provided by the following result. It is an immediate consequence of the well-known Cauchy integral formula.

Lemma 3.5 *Let $z \in \mathbb{C}$ be an arbitrary point in the complex plane and let D be a closed disc around the point z with radius r. If $f : D \to \mathbb{C}$ is continuous on D and analytic in* *its interior, then*

$$\left|f^{(n)}(z)\right| \leq \frac{\Gamma(n+1)}{r^n} \sup_{\theta\in[0,2\pi)} \left|f(z+r\mathrm{e}^{\mathrm{i}\theta})\right|. \tag{3.2}$$

The last result reveals that knowledge about the function f in the complex plane can be useful although the approximation happens on the real line. Here is a further classical result for Chebyshev approximations where the complex plane is rather essential.

Theorem 3.6 *Let $f : [a,b] \to \mathbb{R}$ be analytic in a neighbourhood of the interval $[a,b]$. Then $\|f - f_n\|_\infty = \mathcal{O}(C^n)$ for some constant $C < 1$. In particular, if $f \circ \phi$ is analytic in the ellipse with foci ± 1 and semi-major and semi-minor axis length $K \geq 1$ and $k \geq 0$, we may take $C = 1/(K+k)$.*

Proofs are found in [5, p. 121], [58, p. 297] and [7, p. 49]. In particular, if $f \circ \phi$ has poles in the complex plane, then the rate of convergence of the Chebyshev approximations is determined by the largest ellipse with foci ± 1 in which the function $f \circ \phi$ is still analytic. Figure 1(a) shows several ellipses of this type. The following corollary explains how the exponential rate C can be calculated if the poles of f are known explicitly.

Corollary 3.7 *Continuing the notation of Theorem 3.6, let $(\mu_j, \eta_j) \in [0,\infty) \times [0, 2\pi)$, for $j = 1, 2, \ldots$ be elliptical coordinates so that $z_j = \cosh(\mu_j + \mathrm{i}\,\eta_j)$ are the poles of the function $f \circ \phi$. If z^* is the point that corresponds to $\mu^* := \min_j \mu_j$ then $\|f - f_n\|_\infty = \mathcal{O}(C^n)$ and the constant C can be taken as*

$$C = \left(\alpha + \sqrt{\alpha^2 - 1}\right)^{-1}, \quad \textit{where} \quad \alpha = \tfrac{1}{2}\left(|z^*+1| + |z^*-1|\right). \tag{3.3}$$

PROOF. Exponential convergence follows directly from Theorem 3.6. The rate C is obtained from the sum of the semi-major and semi-minor axis lengths $K = \cosh\mu^*$ and $k = \sinh\mu^*$ of the convergence limiting ellipse,

$$C = 1/(K+k) = 1/(\cosh\mu^* + \sinh\mu^*)$$

Here, $\cosh\mu^*$ is obtained from z^* by using $\cosh\mu^* = (r_1 + r_2)/2s$, where $r_1 = |z^* + 1|$ and $r_2 = |z^* - 1|$ are the respective distances to the foci, and where $2s = 2$ is the distance between the foci. The semi-minor axis length $k = \sinh\mu^*$ is calculated via $\sinh\mu^* = \sqrt{\cosh^2\mu^* - 1}$. □

Theorem 3.4 and Lemma 3.5 were used in [1] to prove approximation results for the connection coefficients between Chebyshev and Legendre polynomials. Theorem 3.6 opens another way to determine the rate of convergence C of Chebyshev approximations. It was used for the first time in [39] to prove stronger approximation results for the connection coefficients between Gegenbauer polynomials.

3.1.2 An auxiliary function.

In this section, we define and analyse a function $\Lambda(z,\Delta)$ that will appear throughout the remainder of this chapter.

Definition 3.8 *For $z \in \mathbb{C}$ and $\Delta \in \mathbb{R}$, the function $\Lambda(z,\Delta)$ is defined by*

$$\Lambda(z,\Delta) = \frac{\Gamma(z+\Delta)}{\Gamma(z+1)}.$$

Unless Δ is an integer, the function $\Lambda(z,\Delta)$ is meromorphic with simple poles at $z = -\Delta, -(\Delta+1), \ldots$ and zeros at the negative integers. We will only need to evaluate $\Lambda(z,\Delta)$ in its region of analyticity. Figure 1(b) shows the function $\Lambda(z,\Delta)$ for $\Delta \in (0,1)$ which will be an important restriction in the analysis to follow.

Let us establish upper and lower bounds on the modulus $|\Lambda(z,\Delta)|$. We follow the rationale in [1] in a more general setting. First, we need the following result on the logarithm of the Gamma function $\Gamma(z)$. A proof is found in [56, p. 11].

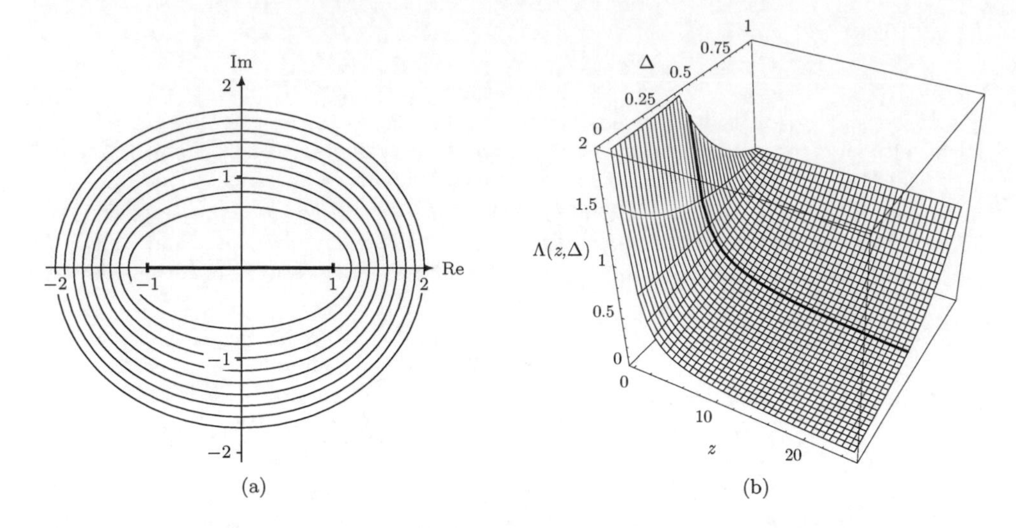

Figure 3.1: (a) Ellipses with foci ±1 enclosing the interval $[-1,1]$ in the complex plane for semi-major axis length $\cosh\mu = 1.2, 1.3, \ldots, 2.0$. (b) The function $\Lambda(z,\Delta)$ for $z \in [0,25]$ and $\Delta \in (0,1)$. The special case $\Delta = \frac{1}{2}$ is shown as a thick black line.

Lemma 3.9 *Let* $z \in \mathbb{C}$, $\operatorname{Re} z > 0$. *Then*

$$\ln\Gamma(z) = \left(z - \tfrac{1}{2}\right)\ln z - z + \frac{\ln 2\pi}{2} + I(z),$$

where

$$I(z) = \int_0^\infty \left(\frac{1}{\mathrm{e}^t - 1} - \frac{1}{t} + \frac{1}{2}\right)\frac{1}{t}\mathrm{e}^{-tz}\,\mathrm{d}t.$$

Moreover, one has $|I(z)| \le \frac{1}{6|z|}$.

The last result enables us to obtain the desired bounds on the modulus $|\Lambda(z,\Delta)|$.

Lemma 3.10 *Let* $z \in \mathbb{C}$, $\operatorname{Re} z > \frac{1}{4} - \Delta$ *with* $\Delta \in (0,1)$. *Then*

$$\frac{\mathrm{e}^{-1/(6|\operatorname{Re} z+\Delta|)}}{|z+1|^{1-\Delta}} < |\Lambda(z,\Delta)| < \frac{\mathrm{e}^{1/(6|\operatorname{Re} z+\Delta|)+1-\Delta}}{|z+1|^{1-\Delta}}. \tag{3.4}$$

Proof. We prove the equivalent inequalities

$$-\frac{1}{6|\operatorname{Re} z+\Delta|} < \ln|\Lambda(z,\Delta)| + (1-\Delta)\ln|z+1| < \frac{1}{6|\operatorname{Re} z+\Delta|} + 1 - \Delta.$$

Lemma 3.9 allows us to write the expression $\ln|\Lambda(z,\Delta)|$ as

$$\ln|\Lambda(z,\Delta)| = \operatorname{Re}(F(z,\Delta)) + (1-\Delta)(1 - \ln|z+1|) + \operatorname{Re}\big(Q(z,\Delta)\big),$$

where

$$F(z,\Delta) := \left(z + \Delta - \tfrac{1}{2}\right)\ln\left(\frac{z+\Delta}{z+1}\right), \qquad Q(z,\Delta) := I(z+\Delta) - I(z+1).$$

Taking a closer look at the function $Q(z,\Delta)$, we note that $0 < 1 - \mathrm{e}^{-t(1-\Delta)} < 1$ and $0 < \left(\frac{1}{\mathrm{e}^t-1} - \frac{1}{t} + \frac{1}{2}\right)\frac{1}{t}$ for every $t > 0$. The latter inequality is true since

$$\left(\frac{1}{\mathrm{e}^t-1} - \frac{1}{t} + \frac{1}{2}\right)\frac{1}{t} = 2\sum_{k=1}^{\infty} \frac{1}{t^2 + 4\pi^2 k^2}$$

as shown in [47, p. 378]. We conclude that

$$\begin{aligned}
\left|\operatorname{Re}\left(Q(z,\Delta)\right)\right| &\le |Q(z,\Delta)| = |I(z+\Delta) - I(z+1)| \\
&= \left|\int_0^\infty \left(\frac{1}{\mathrm{e}^t-1} - \frac{1}{t} + \frac{1}{2}\right)\frac{1}{t}\mathrm{e}^{-t(z+\Delta)}\left(1 - \mathrm{e}^{-t(1-\Delta)}\right)\,\mathrm{d}t\right| \\
&< \int_0^\infty \left(\frac{1}{\mathrm{e}^t-1} - \frac{1}{t} + \frac{1}{2}\right)\frac{1}{t}\mathrm{e}^{-t(\operatorname{Re} z+\Delta)}\,\mathrm{d}t \\
&= I\left(\operatorname{Re} z + \Delta\right) \\
&\le \frac{1}{6|\operatorname{Re} z + \Delta|}.
\end{aligned}$$

Our second goal is now the estimate $-1 + \Delta \le \operatorname{Re}(F(z,\Delta)) \le 0$. First, we write

$$\ln\left(\frac{z+\Delta}{z+1}\right) = -\ln\left(\frac{z+1}{z+\Delta}\right) = -\ln\left(1 + \frac{1-\Delta}{z+\Delta}\right) = -\int_L \frac{1}{1+w}\,\mathrm{d}w,$$

where L denotes the straight line connecting the origin with the point $z_0 := \frac{1-\Delta}{z+\Delta}$ in the complex plane. Since $\operatorname{Re} z_0 = (1-\Delta)\operatorname{Re}(z+\Delta)/|z+\Delta|^2 > 0$, we can estimate

$$\left|\left(z+\Delta-\tfrac{1}{2}\right)\ln\left(\frac{z+\Delta}{z+1}\right)\right| \le |z+\Delta|\int_L \left|\frac{1}{1+w}\right|\,\mathrm{d}|w| \le |z+\Delta|\frac{1-\Delta}{|z+\Delta|} = 1-\Delta,$$

where we have used $|z+\Delta-\frac{1}{2}| \le |z+\Delta|$, $|(1+w)^{-1}| \le 1$ and $|L| = |z_0|$. It remains to show that the real part of $F(z,\Delta)$ is negative. This follows from

$$\operatorname{Re}\left(F(z,\Delta)\right) = \underbrace{-\left(\operatorname{Re}(z) + \Delta - \tfrac{1}{2}\right)\ln|1+z_0|}_{\le 0} + \underbrace{\operatorname{Im}(z)\arg\left(1+z_0\right)}_{\le 0},$$

using that $\operatorname{sign}(\arg z) = -\operatorname{sign}\left(\arg(1+z_0)\right)$. □

Under a moderate restriction, a somewhat simpler bound is obtained.

Corollary 3.11 *Under the assumptions of Lemma* 3.10 *and if further* $\operatorname{Re} z \ge \frac{1}{2}$, *it is true that*

$$\frac{\mathrm{e}^{-\frac{1}{3}}}{|z+1|^{1-\Delta}} < |\Lambda(z,\Delta)| < \frac{\mathrm{e}^{\frac{4}{3}-\Delta}}{|z+1|^{1-\Delta}}. \tag{3.5}$$

For $\Delta = \frac{1}{2}$ the estimate is slightly tighter than Lemma 2.4 in [1, p. 161].

3.2 Gegenbauer polynomials

By Theorem 1.65 on page 32, the connection coefficients between the monic Gegenbauer polynomials $\{\bar{C}_n^{(\alpha)}\}_{n\in\mathbb{N}_0}$ and $\{\bar{C}_n^{(\beta)}\}_{n\in\mathbb{N}_0}$ are given by

$$\bar{\kappa}_{i,j}^{(\alpha)\to(\beta)} = \frac{1}{\Gamma(\alpha-\beta)}\frac{2^i}{2^j}\frac{\Gamma(i+\beta+1)}{\Gamma(i+1)}\frac{\Gamma(j+1)}{\Gamma(j+\alpha)}\frac{\Gamma\left(\frac{j-i}{2}+\alpha-\beta\right)}{\Gamma\left(\frac{j-i}{2}+1\right)}\frac{\Gamma\left(\frac{j+i}{2}+\alpha\right)}{\Gamma\left(\frac{j+i}{2}+\beta+1\right)},$$

with $i+j$ even and $i \le j$, and $\bar{\kappa}_{i,j} = 0$ otherwise. By using that

$$\bar{C}_n^{(\alpha)} = \frac{1}{k_n^{(\alpha)}}C_n^{(\alpha)} = \frac{\Gamma(\alpha)\Gamma(n+1)}{2^n\Gamma(n+\alpha)}C_n^{(\alpha)},$$

we obtain the connection coefficients $\kappa_{i,j}$ for the standard normalisation,

$$\begin{aligned}\kappa_{i,j}^{(\alpha)\to(\beta)} &= \frac{\Gamma(\beta)(j+\beta)}{\Gamma(\alpha)\Gamma(\alpha-\beta)} \frac{\Gamma(\frac{j-i}{2}+\alpha-\beta)}{\Gamma(\frac{j-i}{2}+1)} \frac{\Gamma(\frac{j+i}{2}+\alpha)}{\Gamma(\frac{j+i}{2}+\beta+1)} \\ &= \frac{\Gamma(\beta)(j+\beta)}{\Gamma(\alpha)\Gamma(\alpha-\beta)} \Lambda\left(\frac{j-i}{2}, \alpha-\beta\right) \Lambda\left(\frac{j+i}{2}+\beta, \alpha-\beta\right).\end{aligned} \tag{3.6}$$

In the following, we will analyse the expression for the Gegenbauer polynomials in the standard normalisation. This, however, does not mean that the following results in their entirety would only apply to this particular normalisation. Apart from a number of rather unimportant multiplicative constants, the results have counterparts for any other normalisation.

3.2.1 Smoothness of the connection coefficients. From (3.6) it can be seen that the connection coefficients $\kappa_{i,j}^{(\alpha)\to(\beta)}$ can be interpreted as samples at the positive integers $x = i$ and $y = j$ of a bivariate function $f^{(\alpha)\to(\beta)} : \mathbb{R}^2 \to \mathbb{R}$ that is defined by

$$f(x,y)^{(\alpha)\to(\beta)} = \frac{\Gamma(\beta)(y+\beta)}{\Gamma(\alpha)\Gamma(\alpha-\beta)} \Lambda\left(\frac{y-x}{2}, \alpha-\beta\right) \Lambda\left(\frac{y+x}{2}+\beta, \alpha-\beta\right). \tag{3.7}$$

An important observation is that the function $f^{(\alpha)\to(\beta)}$ may be split into the product of a number of expressions that either depend only on one of the two arguments, x and y, or on both. For example, in above formula, the expression

$$\frac{\Gamma(\beta)(y+\beta)}{\Gamma(\alpha)\Gamma(\alpha-\beta)}$$

only depends on y, but not on x. For reasons to become clear in the next section, the approximation results that we need are only required for the parts where both, x and y, enter together. Therefore, in the following, the function $\tilde{f}^{(\alpha)\to(\beta)} : \mathbb{R}^2 \to \mathbb{R}$ defined by

$$\tilde{f}(x,y)^{(\alpha)\to(\beta)} = \Lambda\left(\frac{y-x}{2}, \alpha-\beta\right) \Lambda\left(\frac{y+x}{2}+\beta, \alpha-\beta\right) \tag{3.8}$$

will be considered instead of the function $f^{(\alpha)\to(\beta)}$. Let us now assume that $|\alpha-\beta| < 1$. In this case, the connection matrix $\mathbf{K}^{(\alpha)\to(\beta)} = (\kappa^{(\alpha)\to(\beta)})$ has its non-vanishing entries in the upper right triangular part which do not vary much among rows and columns. More precisely, these entries $\kappa_{i,j}^{(\alpha)\to(\beta)}$ are samples of the function $f^{(\alpha)\to(\beta)}(x,y)$, which contains the part given by the function $\tilde{f}^{(\alpha)\to(\beta)}(x,y)$ that is smooth whenever $y-x$ is bounded away from zero by a certain constant. This can be quantified and exploited numerically. If the function $\tilde{f}^{(\alpha)\to(\beta)}(x,y)$ is smooth, it is well approximated by interpolation a Chebyshev nodes. To handle the smooth regions, we need the following definition.

Definition 3.12 *A square $S \subset \mathbb{R} \times \mathbb{R}$, defined by the formula $S = [x_0, x_0+c] \times [y_0, y_0+c]$ with a constant $c > 0$, is said to be* well-separated *if $y_0 - x_0 \geq 2c$.*

We prove now a generalised form of Theorem 3.2 from [1, p. 164] which Alpert and Rokhlin used as the principal tool to analyse approximations to the function $f^{(\alpha)\to(\beta)}(x,y)$ (or $\tilde{f}^{(\alpha)\to(\beta)}(x,y)$) for the case $\alpha = \frac{1}{2}$, $\beta = 0$, and vice versa. Since the cases where $\alpha = 0$ or $\beta = 0$ are undefined in the standard normalisation of Gegenbauer polynomials, this needs rescaling. Here, we stick with the standard normalisation and note that a different scaling of source and target polynomials can be incorporated later. This is also briefly explained in the next section. The assertion of the following result is that the function $\tilde{f}^{(\alpha)\to(\beta)}(x,y)$ can be well approximated by a Chebyshev approximation in x or y on any well-separated square S.

Theorem 3.13 *Let $S = [x_0, x_0+c] \times [y_0, y_0+c]$ with $c \geq 4$ be a well-separated square with $x_0, y_0 \geq 0$ and let (x, y) be an arbitrary point in S. Furthermore, denote by $\tilde{f}_n^{(\alpha)\to(\beta)}(\,\cdot\,, y)$ and $\tilde{f}_n^{(\alpha)\to(\beta)}(x,\,\cdot\,)$ the degree-n approximations to the slice functions $\tilde{f}^{(\alpha)\to(\beta)}(\,\cdot\,, y)$ and $\tilde{f}^{(\alpha)\to(\beta)}(x,\,\cdot\,)$ on the intervals $[x_0, x_0+c]$ and $[y_0, y_0+c]$ based on interpolation at Chebyshev points, respectively. If $0 < \alpha - \beta < 1$, then*

$$\|\tilde{f}^{(\alpha)\to(\beta)}(\,\cdot\,, y) - \tilde{f}_n^{(\alpha)\to(\beta)}(\,\cdot\,, y)\|_\infty < \frac{2}{3^{n+1}} \mathrm{e}^{2/3} \left(\frac{2\mathrm{e}}{3}\right)^{2(1+\beta-\alpha)},$$

$$\|\tilde{f}^{(\alpha)\to(\beta)}(x,\,\cdot\,) - \tilde{f}_n^{(\alpha)\to(\beta)}(x,\,\cdot\,)\|_\infty < \frac{2}{3^{n+1}} \mathrm{e}^{2/3} \left(\frac{2\mathrm{e}}{3}\right)^{2(1+\beta-\alpha)}.$$

Similarly, if $0 < \beta - \alpha < 1$, then

$$\|\tilde{f}^{(\alpha)\to(\beta)}(\,\cdot\,, y) - \tilde{f}_n^{(\alpha)\to(\beta)}(\,\cdot\,, y)\|_\infty < \frac{2}{3^{n+2}} \mathrm{e}^{2/3} \left(\frac{2\mathrm{e}}{3}\right)^{2(\beta-\alpha)},$$

$$\|\tilde{f}^{(\alpha)\to(\beta)}(x,\,\cdot\,) - \tilde{f}_n^{(\alpha)\to(\beta)}(x,\,\cdot\,)\|_\infty < \frac{2}{3^{n+2}} \mathrm{e}^{2/3} \left(\frac{2\mathrm{e}}{3}\right)^{2(\beta-\alpha)},$$

where the norm $\|\cdot\|_\infty$ is taken over the intervals $[x_0, x_0+c]$ and $[y_0, y_0+c]$, respectively.

PROOF. We prove only the first estimate since the reasoning for the other cases is virtually identical. Take $\theta \in [0, 2\pi)$ and consider (3.7) which gives

$$\left|\tilde{f}^{(\alpha)\to(\beta)}\left(x + \frac{3}{4}(y-x)\mathrm{e}^{\mathrm{i}\theta}, y\right)\right|$$

$$= \left|\Lambda\left(\frac{y-x}{2} - \frac{3}{8}(y-x)\mathrm{e}^{\mathrm{i}\theta}, \alpha-\beta\right)\right| \left|\Lambda\left(\frac{y+x}{2} + \frac{3}{8}(y-x)\mathrm{e}^{\mathrm{i}\theta} + \beta, \alpha-\beta\right)\right|.$$

Now, notice that since $x_0, y_0 \geq 0$, and $y_0 - x_0 \geq 2c \geq 8$, we have

$$y \geq y_0 \geq 2c \geq 8 \quad \text{and} \quad y - x \geq c \geq 4.$$

Therefore, with $\beta > -1/2$, we get

$$\operatorname{Re}\left(\frac{y-x}{2} - \frac{3}{8}(y-x)\mathrm{e}^{\mathrm{i}\theta}\right) \geq \frac{y-x}{8} \geq \frac{1}{2},$$

$$\operatorname{Re}\left(\frac{y+x}{2} + \frac{3}{8}(y-x)\mathrm{e}^{\mathrm{i}\theta} + \beta\right) \geq \frac{y+7x}{8} + \beta > \frac{1}{2}.$$

The assumption that $0 < \alpha - \beta < 1$ then allows to apply the inequality (3.5) to obtain

$$\left|\tilde{f}^{(\alpha)\to(\beta)}\left(x + \frac{3}{4}(y-x)\mathrm{e}^{\mathrm{i}\theta}, y\right)\right| < \frac{\mathrm{e}^{\frac{2}{3}+2(1+\beta-\alpha)}}{\left(\left(\frac{y-x}{8}+1\right)\left(\frac{y+7x}{8}+\beta+1\right)\right)^{1+\beta-\alpha}}.$$

$$< \mathrm{e}^{2/3}\left(\frac{4\mathrm{e}^2}{9}\right)^{1+\beta-\alpha}.$$

The Cauchy integral estimate (3.2) used in conjunction with the last expression and the fact that $y - x \geq c$ gives

$$\left|\frac{\partial^{n+1} f^{(\alpha)\to(\beta)}(\,\cdot\,, y)}{\partial x^{n+1}}\right| < \frac{(n+1)!4^{n+1}}{3^{n+1}c^{n+1}} \mathrm{e}^{2/3} \left(\frac{4\mathrm{e}^2}{9}\right)^{1+\beta-\alpha}. \tag{3.9}$$

Finally, the inequality (3.1) leads to the estimate

$$\left\| \tilde{f}^{(\alpha)\to(\beta)}(\,\cdot\,,y) - \tilde{f}_n^{(\alpha)\to(\beta)}(\,\cdot\,,y) \right\|_\infty < \frac{2}{3^{n+1}} \mathrm{e}^{2/3} \left(\frac{2\mathrm{e}}{3} \right)^{2(1+\beta-\alpha)} .$$

This proves the first inequality. Similarly, one can show that

$$\begin{aligned} &\left| \tilde{f}^{(\alpha)\to(\beta)} \left(x, y + \frac{3}{4}(y-x)\mathrm{e}^{\mathrm{i}\theta} \right) \right| \\ &< \frac{\mathrm{e}^{\frac{2}{3}+2(1+\beta-\alpha)}}{\left(\left(\frac{y-x}{8} + 1 \right) \left(\frac{y+7x}{8} + \beta + 1 \right) \right)^{1+\beta-\alpha}} \\ &< \mathrm{e}^{2/3} \left(\frac{2\mathrm{e}}{3} \right)^{2(1+\beta-\alpha)} , \end{aligned}$$

and by the same reasoning as before we get

$$\left\| \tilde{f}^{(\alpha)\to(\beta)}(x,\,\cdot\,) - \tilde{f}_n^{(\alpha)\to(\beta)}(x,\,\cdot\,) \right\|_\infty < \frac{2}{3^{n+1}} \mathrm{e}^{2/3} \left(\frac{2\mathrm{e}}{3} \right)^{2(1+\beta-\alpha)} .$$

If $0 < \beta - \alpha < 1$, we write the function $\tilde{f}^{(\alpha)\to(\beta)}(x,y)$ in the equivalent form

$$\begin{aligned} \tilde{f}^{(\alpha)\to(\beta)}(x,y) = &\frac{1}{\left(\frac{y-x}{2} + \alpha - \beta \right) \left(\frac{y+x}{2} + \beta \right)} \\ &\times \Lambda \left(\frac{y-x}{2}, 1 + \alpha - \beta \right) \Lambda \left(\frac{y+x}{2} + \beta, 1 + \alpha - \beta \right) . \end{aligned}$$

Then we get

$$\begin{aligned} \left| \tilde{f}^{(\alpha)\to(\beta)} \left(x + \frac{3}{4}(y-x)\mathrm{e}^{\mathrm{i}\theta}, y \right) \right| &< \left| \frac{1}{(2+\alpha-\beta)(4+\beta)} \right| \frac{\mathrm{e}^{\frac{2}{3}+2(\beta-\alpha)}}{\left(\left(\frac{y-x}{8} + 1 \right) \left(\frac{y+7x}{8} + \beta + 1 \right) \right)^{\beta-\alpha}} \\ &< \frac{1}{3} \mathrm{e}^{2/3} \left(\frac{2\mathrm{e}}{3} \right)^{2(\beta-\alpha)} \end{aligned}$$

and

$$\begin{aligned} \left| \tilde{f}^{(\alpha)\to(\beta)} \left(x, y + \frac{3}{4}(y-x)\mathrm{e}^{\mathrm{i}\theta} \right) \right| &< \left| \frac{1}{(2+\alpha-\beta)(4+\beta)} \right| \frac{\mathrm{e}^{\frac{2}{3}+2(\beta-\alpha)}}{\left(\left(\frac{y-x}{8} + 1 \right) \left(\frac{y+7x}{8} + \beta + 1 \right) \right)^{\beta-\alpha}} . \\ &< \frac{1}{3} \mathrm{e}^{2/3} \left(\frac{2\mathrm{e}}{3} \right)^{2(\beta-\alpha)} , \end{aligned}$$

which in conjunction with (3.1) imply the desired error estimates. □

Are the bounds of Theorem 3.13 the best we can get? Numerical results indicate that this might not be the case: Alpert and Rokhlin [1, p. 165, Remark 3.3.] state for their theorem: "Estimates (3.1)-(3.4) in Theorem 3.2 are quite pessimistic. Numerical experiments indicate that the errors in those estimates all decay approximately as 5^{-k}, as opposed to 3^{-k}." We can give a new theorem that allows us to obtain the optimal rate. The rate is optimal in the sense that a better rate cannot hold uniformly for all admissible choices of α, β, x_0, y_0 and c.

Theorem 3.14 *Let $S = [x_0, x_0+c] \times [y_0, y_0+c]$ be a well-separated square with $x_0, y_0 \geq 0$ and $c \geq 1$. Furthermore, let (x, y) be an arbitrary point in S and assume that $0 < \alpha-\beta < 1$. Then*

$$\left\|\tilde{f}^{(\alpha)\to(\beta)}(\,\cdot\,, y) - \tilde{f}_n^{(\alpha)\to(\beta)}(\,\cdot\,, y)\right\|_\infty = \mathcal{O}\Big(\big(3+\sqrt{8}\big)^{-n}\Big),$$
$$\left\|\tilde{f}^{(\alpha)\to(\beta)}(x, \,\cdot\,) - \tilde{f}_n^{(\alpha)\to(\beta)}(x, \,\cdot\,)\right\|_\infty = \mathcal{O}\Big(\big(3+\sqrt{8}\big)^{-n}\Big).$$

Similarly, if $0 < \beta - \alpha < 1$ and $c \geq 4$, then

$$\left\|\tilde{f}^{(\alpha)\to(\beta)}(\,\cdot\,, y) - \tilde{f}_n^{(\alpha)\to(\beta)}(\,\cdot\,, y)\right\|_\infty = \mathcal{O}\Big(\big(2+\sqrt{3}\big)^{-n}\Big),$$
$$\left\|\tilde{f}^{(\alpha)\to(\beta)}(x, \,\cdot\,) - \tilde{f}_n^{(\alpha)\to(\beta)}(x, \,\cdot\,)\right\|_\infty = \mathcal{O}\Big(\big(2+\sqrt{3}\big)^{-n}\Big).$$

PROOF. Let us first assume $0 < \alpha - \beta < 1$. The plan is to invoke Theorem 3.6 for the function $\tilde{f}^{(\alpha)\to(\beta)}(\,\cdot\,, y)$ on the interval $[x_0, x_0+c]$ and for the function $\tilde{f}^{(\alpha)\to(\beta)}(x, \,\cdot\,)$ on the interval $[y_0, y_0+c]$, respectively. It can be seen from (3.8) that the function $\tilde{f}^{(\alpha)\to(\beta)}(\,\cdot\,, y)$ has poles at the points

$$x = y + 2k + 2(\alpha-\beta) \quad \text{and} \quad x = -y - 2k - 2\alpha,$$

where $k \in \mathbb{N}_0$. Consequently, the function $g : [-1, 1] \to \mathbb{R}$ defined by

$$g(\,\cdot\,) = \tilde{f}^{(\alpha)\to(\beta)}\big(\phi(\,\cdot\,), y\big)$$

has poles at the points

$$z = \phi^{-1}\big(y + 2(k+\alpha-\beta)\big) = 2\frac{y + 2(k+\alpha-\beta) - x_0}{c} - 1 \geq 3 + 4\frac{k+\alpha-\beta}{c} > 3,$$
$$z = \phi^{-1}\big(-y - 2(k+\alpha)\big) = -2\frac{y + 2(k+\alpha) + x_0}{c} - 1 \leq -5 - 4\frac{k+\alpha}{c} < -3.$$

The first bound is sharp since y may be equal to y_0 and c can be taken arbitrarily large. The second bound is also sharp, since k might be zero and α can also be taken arbitrarily close to $-1/2$. Therefore, the points $z = \pm 3$ determine the convergence limiting ellipse. Using (3.3), the convergence rate C of Chebyshev approximations to the function g is then calculated as $C = 3 + \sqrt{8} \approx 5.83$. The estimate for the function $\tilde{f}^{(\alpha)\to(\beta)}(x, \,\cdot\,)$ can be proved analogously. Now we assume that $0 < \beta - \alpha < 1$ and $c \geq 4$. The poles of the function $g(\cdot)$ remain the same, but need to be estimated differently,

$$z = \phi^{-1}(y + 2(k+\alpha-\beta)) = 2\frac{y + 2(k+\alpha-\beta) - x_0}{c} - 1 \geq 3 + 4\frac{k+\alpha-\beta}{c} > 2,$$
$$z = \phi^{-1}(-y - 2(k+\alpha)) = -2\frac{y + 2(k+\alpha) + x_0}{c} - 1 \leq -5 - 4\frac{k+\alpha}{c} < -\frac{9}{2}.$$

The point $z = 2$ determines the convergence limiting ellipse and the convergence rate C of Chebyshev approximations to the function g is $C = 2 + \sqrt{3} \approx 3.73$. The estimate for the function $\tilde{f}^{(\alpha)\to(\beta)}(x, \,\cdot\,)$ can be obtained analogously. □

The bounds for the case $0 < \beta-\alpha < 1$ look worse than those for the case $0 < \alpha-\beta < 1$, but it must be realised that these depend crucially on the minimum side length c of the well-separated square S. For an actual implementation, the length c typically satisfies $c \geq 32$. The following result uses this assumption to obtain stronger approximation results.

Corollary 3.15 *Under the assumptions of Theorem 3.14 and if $0 < \beta-\alpha < 1$ and $c \geq 32$, we may take the convergence rate*

$$C = \frac{1}{8}\big(23 + \sqrt{528}\big) \approx 5.75$$

to obtain

$$\left\|\tilde{f}^{(\alpha)\to(\beta)}(\,\cdot\,,y) - \tilde{f}_n^{(\alpha)\to(\beta)}(\,\cdot\,,y)\right\|_\infty = \mathcal{O}\left(\left(\frac{1}{8}(23+\sqrt{528})\right)^{-n}\right),$$

$$\left\|\tilde{f}^{(\alpha)\to(\beta)}(x,\,\cdot\,) - \tilde{f}_n^{(\alpha)\to(\beta)}(x,\,\cdot\,)\right\|_\infty = \mathcal{O}\left(\left(\frac{1}{8}(23+\sqrt{528})\right)^{-n}\right).$$

The last result is useful because it asserts that we can get almost the same convergence rate that we found for the case $0 < \alpha - \beta < 1$, provided that the side length c of the well-separated square S is large enough. The value $c = 32$ was used in our numerical examples in Section 3.5.

3.2.2 Fast approximate matrix-vector multiplication. The results obtained in the last section can be exploited to devise a fast algorithm for computing the connection between two sequences of Gegenbauer polynomials, that is, for applying any connection matrix $\mathbf{K}^{(\alpha)\to(\beta)}$ that satisfies $|\alpha - \beta| < 1$. The method is a variant of the fast multipole method, essentially the one described in [1]. It is applicable to the more general case here "as is". The key principle is that smoothness of the function $\tilde{f}(x,y) = \tilde{f}^{(\alpha)\to(\beta)}(x,y)$ on well-seperated squares enables us to replace it by an approximation based on interpolation at Chebyshev nodes. Suppose that

$$\mathbf{K}_S^{(\alpha)\to(\beta)} = (\kappa_{i,j})_{i=i_0,j=j_0}^{i_0+c,j_0+c}$$

is a submatrix of the connection matrix $\mathbf{K}^{(\alpha)\to(\beta)}$ so that the square $S = [i_0, i_0+c] \times [j_0, j_0+c]$ is well-separated. Let us also assume for simplicity that the matrices were fully populated and that the function $\tilde{f}(x,y)$ was identical to the function $f(x,y)$, that is, the latter would not contain any parts where x and y can be separated. The standard way to compute the matrix-vector product

$$\mathbf{y} = \mathbf{K}_S^{(\alpha)\to(\beta)}\,\mathbf{x},$$

with vectors $\mathbf{x} = (x_0, \ldots, x_c)^{\mathrm{T}}$ and $\mathbf{y} = (y_0, \ldots, y_c)^{\mathrm{T}}$, is to evaluate the sums

$$y_i = \sum_{j=0}^{c} \kappa_{i_0+i,j_0+j}^{(\alpha)\to(\beta)} x_j = \sum_{j=0}^{c} f(j_0+j, i_0+i)x_j, \qquad \text{for } i = 0,1,\ldots,c. \tag{3.10}$$

This takes $\mathcal{O}(c^2)$ operations. But if the function $f^{(\alpha)\to(\beta)}(x,y)$ is well approximated by an interpolant of degree, say, p in x and y, we may justify the approximation

$$f(x,y) \approx \sum_{r=0}^{p}\sum_{s=0}^{p} f\big(\phi_1(t_r),\phi_2(t_s)\big)\, u_r\big(\phi_1^{-1}(x)\big)\, u_s\big(\phi_2^{-1}(y)\big).$$

Thus, we get

$$y_i \approx \sum_{j=0}^{c}\sum_{r=0}^{p}\sum_{s=0}^{p} f(\phi_1(t_r),\phi_2(t_s))\, u_r(\phi_1^{-1}(i_0+i))\, u_s(\phi_2^{-1}(j_0+j))\, x_j.$$

$$= \sum_{r=0}^{p}\left(u_r(\phi_1^{-1}(i_0+i))\left(\sum_{s=0}^{p} f(\phi_1(t_r),\phi_2(t_s))\left(\sum_{j=0}^{c} u_s(\phi_2^{-1}(j_0+j))\, x_j\right)\right)\right).$$

The triple sum should be evaluated in the order indicated by the parentheses. This takes $\mathcal{O}(p\,c)$ operations for the innermost sum. The cost for the remaining sums is $\mathcal{O}(p^2)$ and $\mathcal{O}(p\,c)$, respectively. If p is fixed then the total cost is $\mathcal{O}(c)$. The error can be controlled by the choice of p. To apply this principle to the whole matrix $\mathbf{K}^{(\alpha)\to(\beta)}$, a suitable decomposition of the latter into well-separated squares is needed; this is shown in Figure

2(b). The region near the diagonal that is not covered by squares is handled directly as in (3.10). The total cost is $\mathcal{O}(n \log(1/\varepsilon))$ arithmetic operations for a single application of the matrix $\mathbf{K}^{(\alpha)\to(\beta)}$. Further details can be found in [1] and [19]. This method can be seen as a variant of the original *fast multipole method (FMM)* that was introduced in [28]. It remains to explain how situations where the function $f(x,y)$ has parts that are separable, i.e., multiplicative expressions that depend only either on x or on y, but not on both, should be handled. We encountered this situation already for the Gegenbauer polynomials and proved the approximation results that enable the FMM only for the function $\tilde{f}(x,y)$ that had been freed from these parts. So let us assume that the function $f(x,y)$ can be written as the product

$$f(x,y) = f_1(x)\tilde{f}(x,y)f_2(y),$$

where the functions f_1 and f_2 depend only on x and y, respectively. To apply the FMM-like method, we only need approximation results for the function $\tilde{f}(x,y)$. This is because the functions f_1 and f_2 can be applied to the input and output coefficients of the computation directly. For the example given above, the output coefficients y_i for $i = 0, 1, \ldots, c$ are then approximated by

$$y_i \approx \sum_{r=0}^{p} \Bigg(f_2(i_0+i) u_r(\phi_1^{-1}(i_0+i)) \Bigg(\times \sum_{s=0}^{p} f(\phi_1(t_r), \phi_2(t_s)) \Bigg(\sum_{j=0}^{c} u_s(\phi_2^{-1}(j_0+j))\, f_1(j_0+j) x_j \Bigg)\Bigg)\Bigg).$$

The incorporation of the separable parts $f_1(x)$ and $f_2(y)$ of the function $f(x,y)$ can also be used to account for different normalizations of the polynomials. As seen from Lemma 1.37, rescaling the source and target polynomials is equivalent to a scaling, separable with respect to i and j, applied to the connection coefficients $\kappa_{i,j}$. This can be directly incorporated into the FMM-like method at no extra cost.

3.3 Laguerre and Jacobi polynomials

We are now ready to extend the previous results to Laguerre and Jacobi polynomials. From Theorem 1.42, we get the connection coefficients $\bar{\kappa}_{i,j}^{(\alpha)\to(\beta)}$ for the monic Laguerre polynomials,

$$\bar{\kappa}_{i,j}^{(\alpha)\to(\beta)} = \frac{(-1)^{i+j}}{\Gamma(\alpha-\beta)} \frac{\Gamma(j+1)}{\Gamma(i+1)} \frac{\Gamma(j-i+\alpha-\beta)}{\Gamma(j-i+1)}, \qquad \text{with } 0 \le i \le j.$$

Using that the leading coefficients of the Laguerre polynomials $\{L_n^{(\alpha)}\}_{n\in\mathbb{N}_0}$ are given by

$$k_n^{(\alpha)} = \frac{(-1)^n}{\Gamma(n+1)},$$

we may write the connection coefficients in the standard normalisation,

$$\kappa_{i,j}^{(\alpha)\to(\beta)} = \frac{1}{\Gamma(\alpha-\beta)} \Lambda(j-i, \alpha-\beta), \qquad \text{with } 0 \le i \le j.$$

Again, we can interpret the connection coefficients $\kappa_{i,j}^{(\alpha)\to(\beta)}$ as samples of a function $f^{(\alpha)\to(\beta)} : \mathbb{R}^2 \to \mathbb{R}$ defined by

$$f^{(\alpha)\to(\beta)}(x,y) = \frac{1}{\Gamma(\alpha-\beta)} \Lambda(y-x, \alpha-\beta).$$

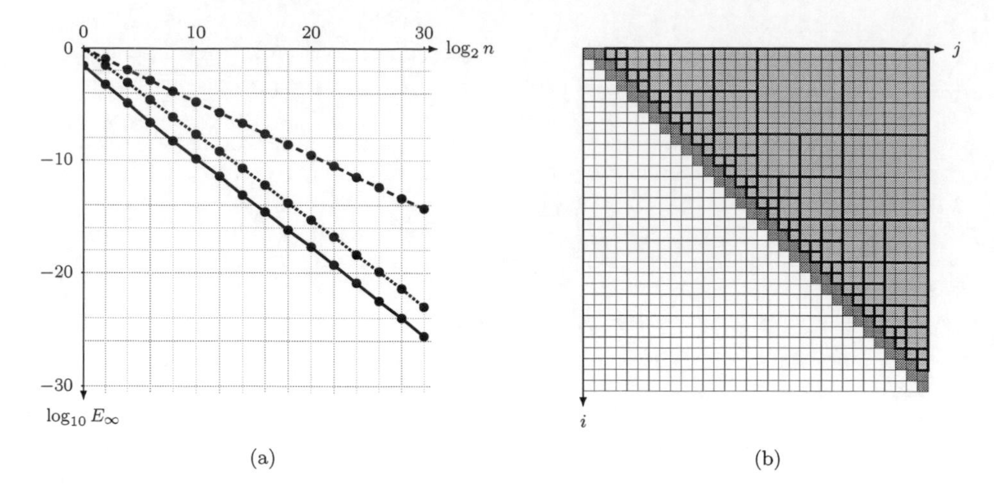

Figure 3.2: (a) Error of Chebyshev approximations to the function $\hat{f}^{(\alpha)\to(\beta)}(\,\cdot\,,192)$ for $\alpha = 4/3$ and $\beta = 1/2$ on the interval $[64, 128]$ for increasing degree n (solid) together with the bound from Theorem 3.13 (dashed) and the bound from Theorem 3.14 (dotted), where the unknown constant was assumed to be equal to one. The error E_∞ is the decimal logarithm of the maximum absolute error over 1.000 uniformly distributed random nodes in the interval $[64, 128]$. The computations were carried out in MATHEMATICA. (b) Subdivision scheme for the upper triangular connection matrix $\mathbf{K}$. The squares shown in light grey are well-separated and become smaller as they get closer to the diagonal. The darker parts near the diagonal need to be handled directly. The rest of the entries is zero.

Here is the analog of Theorem 3.14 for Laguerre polynomials. The proof is omitted because it is entirely similar.

Theorem 3.16 *Let $S = [x_0, x_0 + c] \times [y_0, y_0 + c]$ be a well-separated square with $x_0, y_0 \geq 0$ and $c \geq 1$. Furthermore, let (x, y) be a point in S and assume that $0 < \alpha - \beta < 1$. Then*

$$\left\| f^{(\alpha)\to(\beta)}(\,\cdot\,, y) - f_n^{(\alpha)\to(\beta)}(\,\cdot\,, y) \right\|_\infty = \mathcal{O}\Big(\big(3 + \sqrt{8}\big)^{-n}\Big),$$
$$\left\| f^{(\alpha)\to(\beta)}(x, \,\cdot\,) - f_n^{(\alpha)\to(\beta)}(x, \,\cdot\,) \right\|_\infty = \mathcal{O}\Big(\big(3 + \sqrt{8}\big)^{-n}\Big).$$

Similarly, if $0 < \beta - \alpha < 1$ and $c \geq 2$, then

$$\left\| f^{(\alpha)\to(\beta)}(\,\cdot\,, y) - f_n^{(\alpha)\to(\beta)}(\,\cdot\,, y) \right\|_\infty = \mathcal{O}\Big(\big(2 + \sqrt{3}\big)^{-n}\Big),$$
$$\left\| f^{(\alpha)\to(\beta)}(x, \,\cdot\,) - f_n^{(\alpha)\to(\beta)}(x, \,\cdot\,) \right\|_\infty = \mathcal{O}\Big(\big(2 + \sqrt{3}\big)^{-n}\Big).$$

The results for the case $0 < \beta - \alpha < 1$ can be made stronger by requiring a larger lower bound on the side length c of the well-separated square S.

Corollary 3.17 *Under the assumptions of Theorem 3.16 and if $0 < \beta - \alpha < 1$ and $c \geq 16$, we may take the convergence rate*

$$C = \frac{1}{8}\big(23 + \sqrt{528}\big) \approx 5.75$$

to obtain

$$\|f^{(\alpha)\to(\beta)}(\,\cdot\,,y) - f_n^{(\alpha)\to(\beta)}(\,\cdot\,,y)\|_\infty = \mathcal{O}\Big(\Big(\tfrac{1}{8}(23+\sqrt{528})\Big)^{-n}\Big),$$

$$\|f^{(\alpha)\to(\beta)}(x,\,\cdot\,) - f_n^{(\alpha)\to(\beta)}(x,\,\cdot\,)\|_\infty = \mathcal{O}\Big(\Big(\tfrac{1}{8}(23+\sqrt{528})\Big)^{-n}\Big).$$

Now for the Jacobi polynomials. We consider the connection between the Jacobi polynomials $\{P_n^{(\alpha,\beta)}\}_{n\in\mathbb{N}_0}$ and $\{P_n^{(\gamma,\beta)}\}_{n\in\mathbb{N}_0}$. Remark 3.20 below explains how this can be used for the symmetric situation where the second parameter of the Jacobi polynomials is changed. The connection coefficients $\bar{\kappa}_{i,j}^{(\alpha,\beta)\to(\gamma,\beta)}$ for monic Jacobi polynomials are obtained from Theorem 1.50,

$$\begin{aligned}\bar{\kappa}_{i,j}^{(\alpha,\beta)\to(\gamma,\beta)} &= \frac{1}{\Gamma(\alpha-\gamma)}\frac{2^j}{2^i}\frac{\Gamma(j+1)}{\Gamma(i+1)}\frac{\Gamma(j+\beta+1)}{\Gamma(i+\beta+1)}\frac{\Gamma(2i+\gamma+\beta+2)}{\Gamma(2j+\alpha+\beta+1)}\\ &\quad\times\frac{\Gamma(j+i+\alpha+\beta+1)}{\Gamma(j+i+\gamma+\beta+2)}\frac{\Gamma(j-i+\alpha-\gamma)}{\Gamma(j-i+1)},\end{aligned}$$

with $0\le i\le j$. With the leading coefficients $k_n^{(\alpha,\beta)}$ of the Jacobi polynomials $\{P_n^{(\alpha,\beta)}\}_{n\in\mathbb{N}_0}$ given by

$$k_n^{(\alpha,\beta)} = \frac{\Gamma(2n+\alpha+\beta+1)}{2^n\Gamma(n+1)\Gamma(n+\alpha+\beta+1)},$$

the connection coefficients $\kappa_{i,j}^{(\alpha,\beta)\to(\gamma,\beta)}$ for the standard normalization are calculated to

$$\begin{aligned}\kappa_{i,j}^{(\alpha,\beta)\to(\gamma,\beta)} &= \frac{2i+\gamma+\beta+1}{\Gamma(\alpha-\gamma)}\Lambda(i+\beta,1+\gamma)\Lambda(j+\alpha+\beta,1-\alpha)\\ &\quad\times\Lambda(j-i,\alpha-\gamma)\Lambda(j+i+\gamma+\beta+1,\alpha-\gamma).\end{aligned}$$

The connection coefficients $\kappa_{i,j}^{(\alpha,\beta)\to(\gamma,\beta)}$ can thus be interpreted as samples of a function $f^{(\alpha,\beta)\to(\gamma,\beta)}:\mathbb{R}^2\to\mathbb{R}$ that is defined by

$$f^{(\alpha,\beta)\to(\gamma,\beta)}(x,y) = f_1^{(\alpha,\beta)\to(\gamma,\beta)}(x)\tilde{f}^{(\alpha,\beta)\to(\gamma,\beta)}(x,y)f_2^{(\alpha,\beta)\to(\gamma,\beta)}(y),$$

with

$$f_1^{(\alpha,\beta)\to(\gamma,\beta)}(x) = \frac{2x+\gamma+\beta+1}{\Gamma(\alpha-\gamma)}\Lambda(x+\beta,1+\gamma),$$

$$\tilde{f}^{(\alpha,\beta)\to(\gamma,\beta)}(x,y) = \Lambda(y-x,\alpha-\gamma)\Lambda(y+x+\gamma+\beta+1,\alpha-\gamma),$$

$$f_2^{(\alpha,\beta)\to(\gamma,\beta)}(y) = \Lambda(y+\alpha+\beta,1-\alpha).$$

Here is the corresponding approximation result for the function $\tilde{f}^{(\alpha,\beta)\to(\gamma,\beta)}$.

Theorem 3.18 *Let $S=[x_0,x_0+c]\times[y_0,y_0+c]$ be a well-separated square with $x_0,y_0\ge 0$ and $c\ge 1$. Furthermore, let (x,y) be a point in S and assume that $0<\alpha-\gamma<1$. Then*

$$\big\|\tilde{f}^{(\alpha,\beta)\to(\gamma,\beta)}(\,\cdot\,,y) - \tilde{f}_n^{(\alpha,\beta)\to(\gamma,\beta)}(\,\cdot\,,y)\big\|_\infty = \mathcal{O}\big((3+\sqrt{8})^{-n}\big),$$

$$\big\|\tilde{f}^{(\alpha,\beta)\to(\gamma,\beta)}(x,\,\cdot\,) - \tilde{f}_n^{(\alpha,\beta)\to(\gamma,\beta)}(x,\,\cdot\,)\big\|_\infty = \mathcal{O}\big((3+\sqrt{8})^{-n}\big).$$

Similarly, if $0<\gamma-\alpha<1$ and $c\ge 2$, then

$$\big\|\tilde{f}^{(\alpha,\beta)\to(\gamma,\beta)}(\,\cdot\,,y) - \tilde{f}_n^{(\alpha,\beta)\to(\gamma,\beta)}(\,\cdot\,,y)\big\|_\infty = \mathcal{O}\big((2+\sqrt{3})^{-n}\big),$$

$$\big\|\tilde{f}^{(\alpha,\beta)\to(\gamma,\beta)}(x,\,\cdot\,) - \tilde{f}_n^{(\alpha,\beta)\to(\gamma,\beta)}(x,\,\cdot\,)\big\|_\infty = \mathcal{O}\big((2+\sqrt{3})^{-n}\big).$$

We complete the series of approximation results with the following corollary that obtains a stronger approximation result for the case $0 < \gamma - \alpha < 1$, similar to what we have seen already for Gegenbauer and Laguerre polynomials.

Corollary 3.19 *Under the assumptions of Theorem* 3.18 *and if* $0 < \gamma - \alpha < 1$ *and* $c \geq 16$, *we may take the convergence rate*

$$C = \frac{1}{8}(23 + \sqrt{528}) \approx 5.75$$

to obtain

$$\left\|\tilde{f}^{(\alpha,\beta)\to(\gamma,\beta)}(\cdot\,,y) - \tilde{f}_n^{(\alpha,\beta)\to(\gamma,\beta)}(\cdot\,,y)\right\|_\infty = \mathcal{O}\left(\left(\tfrac{1}{8}(23 + \sqrt{528})\right)^{-n}\right),$$

$$\left\|\tilde{f}^{(\alpha,\beta)\to(\gamma,\beta)}(x,\,\cdot) - \tilde{f}_n^{(\alpha,\beta)\to(\gamma,\beta)}(x,\,\cdot)\right\|_\infty = \mathcal{O}\left(\left(\tfrac{1}{8}(23 + \sqrt{528})\right)^{-n}\right).$$

Remark 3.20 We can use Lemma 1.49, that is, the identity

$$\bar{\kappa}_{i,j}^{(\alpha,\beta)\to(\gamma,\delta)} = (-1)^{i+j}\bar{\kappa}_{i,j}^{(\beta,\alpha)\to(\delta,\gamma)}$$

to obtain analogous results for the connection between the Jacobi polynomials $\{P_n^{(\alpha,\beta)}\}_{n\in\mathbb{N}_0}$ and $\{P_n^{(\alpha,\delta)}\}_{n\in\mathbb{N}_0}$. Here, the second parameter is changed. This is true since the $(-1)^{i+j} = (-1)^i(-1)^j$ part can be incorporated into the functions $f_1^{(\alpha,\beta)\to(\gamma,\beta)}$ and $f_2^{(\alpha,\beta)\to(\gamma,\beta)}$. Therefore, the approximation results for the function $\tilde{f}^{(\alpha,\beta)\to(\gamma,\beta)}$ can be used right-away to enable the FMM-like method.

The following section shows how the obtained results can be extended to efficiently compute the connection between classical orthogonal polynomials with arbitrary choices of the respective parameters.

3.4 Transforms for arbitrary indices

In this brief section, we describe how transformations between two sequences of classical orthogonal polynomials with arbitrary indices can be computed. We will explain this for two sequences of Gegenbauer polynomials $\{C_n^{(\alpha)}\}_{n\in\mathbb{N}_0}$ and $\{C_n^{(\beta)}\}_{n\in\mathbb{N}_0}$. The same idea can then also be applied to Laguerre and Jacobi polynomials.

Let us assume that the parameters α and β do not satisfy $|\alpha - \beta| < 1$, that is, the distance between α and β is too large to apply the FMM-like method on the basis of our theoretical results. To efficiently compute the transformation between the two families of Gegenbauer polynomials, we split the transformation into two steps. First, we chose the parameter α' to be the number between α and β such that $|\alpha - \alpha'|$ is an integer and as large as possible (for example, for $\alpha = \frac{1}{2}$ and $\beta = 3$, we would choose $\alpha' = \frac{5}{2}$). We can now compute the transformation from α to β by first transforming from the polynomials $\{C_n^{(\alpha)}\}_{n\in\mathbb{N}_0}$ to the proxy polynomials $\{C_n^{(\alpha')}\}_{n\in\mathbb{N}_0}$, then from there to the target $\{C_n^{(\beta)}\}_{n\in\mathbb{N}_0}$.

The first step can be handled efficiently because the results about the corresponding connection matrices $\mathbf{K}$ that were obtained in Section 1.4.2 show that these have either a banded or a semi-separable structure. The bandwidth or the semiseparability rank, respectively, are directly proportional to $|\alpha - \alpha'|$ and so are are comparable to $|\alpha - \beta|$. Consequently, for such an $n \times n$ connection matrix $\mathbf{K}$ we need $\mathcal{O}(n|\alpha - \beta|)$ arithmetic operations to calculate a matrix-vector multiplication.

The second step can also be handled efficiently because we have $|\alpha' - \beta| < 1$ and the results of the last section enable us to apply the FMM-like method to this case. The cost for this step is accordingly $\mathcal{O}\big(n\log(1/\varepsilon)\big)$. In total, the number of arithmetic operations to compute both steps together is $\mathcal{O}\big(n\big(|\alpha-\beta|+\log(1/\varepsilon)\big)\big)$.

3.5 Numerical results

The details of the implementation are described in Section 2.4.6, were the FMM was already employed. It is the same variant that was used for the tests conducted in this section. Our implementation is flexible enough to handle different kernels, checkerboard-like matrices, and also triangular matrices. We call the method that was introduced in this section the *fmm method.* The results shown in Table 3.1 to Table 3.3 correspond to the test cases seen in Section 2.4.6. It can be observed that the fmm method is slightly faster than the usmv-fmm method. This is also obvious from Figure 3.3 which compares all proposed methods, with the exception of the usmv-direct method, for selected test cases.

3.5.1 Stability issues. It was already observed in Section 2.4.6 that the usmv-fmm method can show an unacceptably large error in the results when transformations are computed with step sizes that are too large. As a remedy, we had split each transformation into many ones so that each has its step size bounded by a certain constant free of choice. The fmm method does not seem to have similar problems but the results shown so far can lead to the perception that computing the connection between any two families of classical orthogonal polynomials of the same type is a perfectly stable process in the sense of *numerically stable on a machine with the usual floating-point arithmetic.* This section is devoted to show that this is only true to a certain extent. To this end, we define in the following a test scenario that will demonstrate problematic effects that we must be aware of. This will again be done exemplarily for Gegenbauer polynomials, but also applies to the rest of the classical orthogonal polynomials as well.
Let us start with a function f that has been expanded into a finite linear combination of Gegenbauer polynomials $\{C_n^{(\alpha)}\}_{n\in\mathbb{N}_0}$, that is,

$$f \approx \sum_{j=0}^{n} x_j C_j^{(\alpha)}.$$

This is a typical situation that occurs in practice. Usually, the parameter n is large enough such that the $\approx$ sign is justified to a desired accuracy. With the connection coefficients $\kappa_{i,j}^{(\alpha)\to(\beta)}$, we can compute the coefficients y_j in the expansion

$$f \approx \sum_{j=0}^{n} y_j C_j^{(\beta)}.$$

Both representations are equivalent, but in finite precision arithmetic there can be substantial differences when it comes to the accuracy to which each value x_j or y_j can be computed. For example, if we use the connection coefficients $\kappa_{i,j}^{(\beta)\to(\alpha)}$ to recover the coefficients x_j from the computed coefficients y_j, then the result can be very different, indeed off by a large margin, compared to the original values x_j that we started with; this will be shown in detail in Section 3.5.2 below. The key observation that can be made from numerical results is that this happens whenever $|\alpha-\beta|$ is large.
An informal explanation for this behavior, without the aspirations of being an actual proof, is that whenever $|\alpha-\beta|$ is large, both sequences of Gegenbauer polynomials show such different behavior that the distribution of the expansion coefficients x_j and y_j over the different orders of magnitude will also be totally different. As an extreme example, we

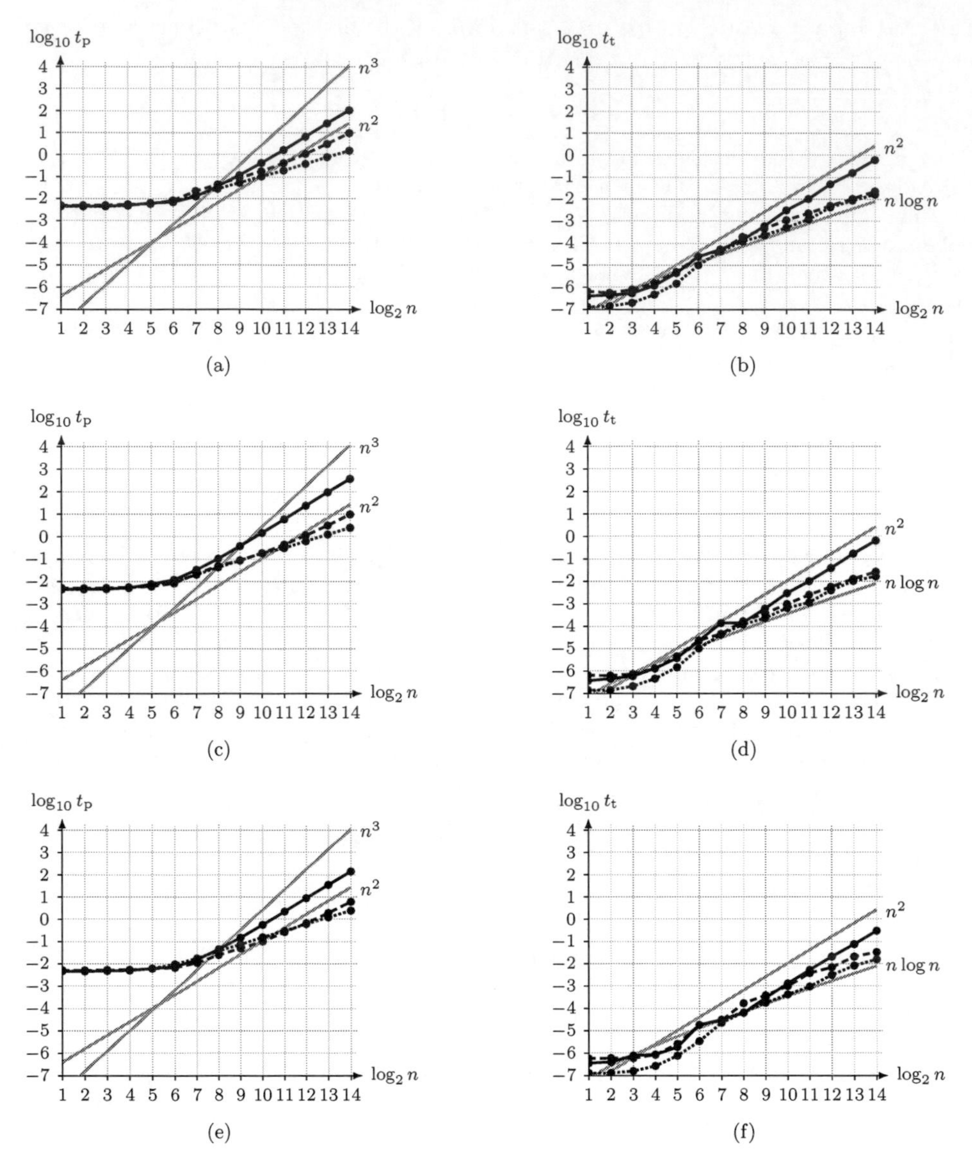

Figure 3.3: Shown from top to bottom are time measurements for Laguerre ((a) and (b)), Jacobi ((c) and (d)), and Gegenbauer polynomials ((e) and (f)), each of which correspond to the first test case reported in Tables 2.2 to 2.13 (and Tables 3.1 to 3.3), respectively. Left side: Times t_{p} for the pre-computation stage as a function of the transform size n. Shown are the direct method (solid), the usmv-fmm method (dashed), and the fmm method (dotted), with accuracy controlling parameter $p = 18$ and arbitrarily large step size s. The grey lines are to facilitate recognition of the asymptotic behavior. Right side: Times t_{t} for the computation of the actual transformation as function of the transform size n. Shown are the direct method (solid), the usmv-fmm method (dashed), and the fmm method (dotted).

α	β	n	t_{p}	t_{p}/n^2	t_{t}	t_{t}/n	E_{∞}^{c}	E_{∞}
-0.5	-0.7	256	2.8E-02	1.1E-04	1.2E-04	4.7E-07	6.4E-16	6.0E-16
-0.5	-0.7	512	5.2E-02	1.0E-04	2.3E-04	4.5E-07	6.5E-16	7.6E-16
-0.5	-0.7	1024	1.0E-01	1.0E-04	5.1E-04	4.9E-07	5.8E-16	7.6E-16
-0.5	-0.7	2048	1.9E-01	9.5E-05	1.2E-03	5.9E-07	4.9E-16	8.3E-16
-0.5	-0.7	4096	3.8E-01	9.4E-05	4.3E-03	1.0E-06	6.7E-16	1.1E-15
-0.5	-0.7	8192	7.7E-01	9.4E-05	8.8E-03	1.1E-06	8.4E-16	9.4E-16
-0.5	-0.7	16384	1.5E+00	9.3E-05	1.7E-02	1.0E-06	5.6E-16	9.8E-16
-0.5	0.2	256	3.2E-02	1.2E-04	1.1E-04	4.2E-07	6.4E-16	5.8E-16
-0.5	0.2	512	5.9E-02	1.1E-04	2.3E-04	4.5E-07	8.8E-16	5.5E-16
-0.5	0.2	1024	1.2E-01	1.1E-04	5.1E-04	5.0E-07	8.8E-16	5.3E-16
-0.5	0.2	2048	2.3E-01	1.1E-04	1.2E-03	5.7E-07	2.0E-15	6.4E-16
-0.5	0.2	4096	4.4E-01	1.1E-04	4.2E-03	1.0E-06	2.5E-15	5.0E-16
-0.5	0.2	8192	9.0E-01	1.1E-04	8.9E-03	1.1E-06	1.5E-15	6.3E-16
-0.5	0.2	16384	1.8E+00	1.1E-04	1.7E-02	1.0E-06	1.9E-15	6.2E-16
0.2	-0.5	256	2.8E-02	1.1E-04	9.7E-05	3.8E-07	3.4E-16	8.3E-16
0.2	-0.5	512	5.3E-02	1.0E-04	2.3E-04	4.5E-07	2.8E-16	7.2E-16
0.2	-0.5	1024	1.0E-01	9.8E-05	5.1E-04	5.0E-07	2.5E-16	9.2E-16
0.2	-0.5	2048	1.9E-01	9.5E-05	1.2E-03	5.6E-07	2.0E-16	1.4E-15
0.2	-0.5	4096	3.8E-01	9.3E-05	4.2E-03	1.0E-06	2.3E-16	1.1E-15
0.2	-0.5	8192	7.6E-01	9.3E-05	8.9E-03	1.1E-06	4.0E-16	1.4E-15
0.2	-0.5	16384	1.5E+00	9.3E-05	1.9E-02	1.1E-06	1.9E-16	1.7E-15
0.2	1.1	256	3.2E-02	1.2E-04	1.2E-04	4.7E-07	1.2E-15	7.0E-16
0.2	1.1	512	6.0E-02	1.2E-04	2.3E-04	4.5E-07	1.0E-15	7.3E-16
0.2	1.1	1024	1.2E-01	1.2E-04	5.1E-04	5.0E-07	1.2E-15	7.2E-16
0.2	1.1	2048	2.3E-01	1.1E-04	1.1E-03	5.6E-07	1.3E-15	9.7E-16
0.2	1.1	4096	4.5E-01	1.1E-04	4.2E-03	1.0E-06	1.4E-15	8.2E-16
0.2	1.1	8192	8.9E-01	1.1E-04	8.9E-03	1.1E-06	1.4E-15	8.2E-16
0.2	1.1	16384	1.8E+00	1.1E-04	1.6E-02	1.0E-06	1.5E-15	8.1E-16
5.6	7.8	256	3.0E-02	1.2E-04	1.0E-04	4.0E-07	6.3E-16	4.4E-16
5.6	7.8	512	5.7E-02	1.1E-04	2.4E-04	4.7E-07	6.5E-16	3.6E-16
5.6	7.8	1024	1.1E-01	1.1E-04	5.3E-04	5.2E-07	8.2E-16	4.4E-16
5.6	7.8	2048	2.2E-01	1.1E-04	1.3E-03	6.2E-07	1.5E-15	4.2E-16
5.6	7.8	4096	4.3E-01	1.1E-04	4.3E-03	1.0E-06	1.6E-15	4.1E-16
5.6	7.8	8192	8.7E-01	1.1E-04	1.0E-02	1.3E-06	1.3E-15	5.0E-16
5.6	7.8	16384	1.7E+00	1.0E-04	1.9E-02	1.1E-06	1.4E-15	5.0E-16
9.7	5.5	256	2.8E-02	1.1E-04	1.1E-04	4.3E-07	3.9E-16	3.0E-15
9.7	5.5	512	5.2E-02	1.0E-04	2.6E-04	5.1E-07	1.4E-15	3.5E-15
9.7	5.5	1024	1.0E-01	9.9E-05	5.6E-04	5.5E-07	2.3E-16	3.6E-15
9.7	5.5	2048	1.9E-01	9.4E-05	1.3E-03	6.5E-07	8.6E-16	3.8E-15
9.7	5.5	4096	3.8E-01	9.4E-05	4.1E-03	1.0E-06	6.2E-16	5.8E-15
9.7	5.5	8192	8.0E-01	9.7E-05	9.3E-03	1.1E-06	1.2E-16	5.8E-15
9.7	5.5	16384	1.5E+00	9.4E-05	2.0E-02	1.2E-06	1.2E-16	1.5E-14

Table 3.1: Test results for the connection between the Laguerre polynomials $\{L_n^{(\alpha)}\}_{n\in\mathbb{N}_0}$ and $\{L_n^{(\beta)}\}_{n\in\mathbb{N}_0}$ computed with the fmm method with accuracy controlling parameter $p = 18$ for different transform sizes n. Shown are the times for precomputation t_{p} and for the computation of the actual transform t_{t}. Both are also shown after division through the expected asymptotic expression in terms of the transform size n. Furthermore, the component-wise error E_{∞}^{c} and the relative infinity norm error E_{∞} are reported.

α	β	γ	δ	n	t_p	t_p/n^2	t_t	t_t/n	E_∞^c	E_∞
-0.7	2.0	-0.9	2.0	256	4.3E-02	1.7E-04	1.2E-04	4.7E-07	1.2E-15	1.8E-15
-0.7	2.0	-0.9	2.0	512	8.5E-02	1.7E-04	2.3E-04	4.4E-07	1.6E-15	2.2E-15
-0.7	2.0	-0.9	2.0	1024	1.9E-01	1.8E-04	6.5E-04	6.3E-07	1.7E-15	2.5E-15
-0.7	2.0	-0.9	2.0	2048	3.1E-01	1.5E-04	1.1E-03	5.6E-07	2.8E-15	5.3E-15
-0.7	2.0	-0.9	2.0	4096	6.2E-01	1.5E-04	4.0E-03	9.7E-07	3.3E-15	7.2E-15
-0.7	2.0	-0.9	2.0	8192	1.2E+00	1.5E-04	1.0E-02	1.3E-06	4.1E-15	1.1E-14
-0.7	2.0	-0.9	2.0	16384	2.5E+00	1.5E-04	1.7E-02	1.0E-06	4.4E-15	1.3E-14
-0.7	2.0	0.0	2.0	256	4.5E-02	1.8E-04	9.7E-05	3.8E-07	2.0E-15	1.5E-15
-0.7	2.0	0.0	2.0	512	8.6E-02	1.7E-04	2.3E-04	4.4E-07	2.2E-15	1.4E-15
-0.7	2.0	0.0	2.0	1024	1.7E-01	1.7E-04	5.0E-04	4.9E-07	3.7E-15	2.4E-15
-0.7	2.0	0.0	2.0	2048	3.4E-01	1.7E-04	1.1E-03	5.6E-07	5.6E-15	4.8E-15
-0.7	2.0	0.0	2.0	4096	6.9E-01	1.7E-04	3.9E-03	9.6E-07	8.7E-15	8.1E-15
-0.7	2.0	0.0	2.0	8192	1.4E+00	1.7E-04	1.0E-02	1.2E-06	1.5E-14	9.1E-15
-0.7	2.0	0.0	2.0	16384	2.8E+00	1.7E-04	1.6E-02	1.0E-06	2.1E-14	1.6E-14
0.0	2.0	-0.7	2.0	256	4.1E-02	1.6E-04	9.6E-05	3.8E-07	3.7E-16	1.1E-15
0.0	2.0	-0.7	2.0	512	7.9E-02	1.5E-04	2.3E-04	4.5E-07	5.5E-16	2.3E-15
0.0	2.0	-0.7	2.0	1024	1.6E-01	1.5E-04	5.1E-04	4.9E-07	7.2E-16	1.7E-15
0.0	2.0	-0.7	2.0	2048	3.1E-01	1.5E-04	1.2E-03	5.7E-07	3.5E-16	3.7E-15
0.0	2.0	-0.7	2.0	4096	6.2E-01	1.5E-04	3.9E-03	9.4E-07	7.2E-16	6.6E-15
0.0	2.0	-0.7	2.0	8192	1.2E+00	1.5E-04	1.1E-02	1.3E-06	4.7E-16	1.0E-14
0.0	2.0	-0.7	2.0	16384	2.4E+00	1.5E-04	1.7E-02	1.0E-06	6.2E-16	1.8E-14
0.0	2.0	0.9	2.0	256	4.7E-02	1.8E-04	1.1E-04	4.2E-07	2.1E-15	1.7E-15
0.0	2.0	0.9	2.0	512	9.2E-02	1.8E-04	2.3E-04	4.5E-07	2.9E-15	1.5E-15
0.0	2.0	0.9	2.0	1024	1.7E-01	1.6E-04	5.1E-04	5.0E-07	4.5E-15	2.9E-15
0.0	2.0	0.9	2.0	2048	3.4E-01	1.7E-04	1.2E-03	5.8E-07	6.9E-15	3.7E-15
0.0	2.0	0.9	2.0	4096	6.9E-01	1.7E-04	4.1E-03	9.9E-07	1.1E-14	7.8E-15
0.0	2.0	0.9	2.0	8192	1.4E+00	1.7E-04	1.0E-02	1.2E-06	1.5E-14	9.9E-15
0.0	2.0	0.9	2.0	16384	2.8E+00	1.7E-04	1.7E-02	1.0E-06	2.3E-14	1.2E-14
5.4	2.0	7.6	2.0	256	4.7E-02	1.8E-04	1.0E-04	4.0E-07	8.5E-15	5.6E-15
5.4	2.0	7.6	2.0	512	1.0E-01	2.0E-04	3.0E-04	5.8E-07	8.8E-15	6.4E-15
5.4	2.0	7.6	2.0	1024	1.7E-01	1.7E-04	5.3E-04	5.2E-07	9.9E-15	7.5E-15
5.4	2.0	7.6	2.0	2048	3.5E-01	1.7E-04	1.2E-03	6.0E-07	1.1E-14	8.1E-15
5.4	2.0	7.6	2.0	4096	7.0E-01	1.7E-04	4.1E-03	1.0E-06	1.3E-14	8.0E-15
5.4	2.0	7.6	2.0	8192	1.4E+00	1.7E-04	1.1E-02	1.3E-06	1.5E-14	1.1E-14
5.4	2.0	7.6	2.0	16384	2.8E+00	1.7E-04	1.7E-02	1.0E-06	2.2E-14	1.5E-14
8.6	2.0	4.3	2.0	256	4.3E-02	1.7E-04	1.1E-04	4.3E-07	7.5E-15	1.6E-15
8.6	2.0	4.3	2.0	512	8.7E-02	1.7E-04	2.6E-04	5.0E-07	5.2E-15	1.3E-14
8.6	2.0	4.3	2.0	1024	1.7E-01	1.6E-04	5.6E-04	5.4E-07	1.7E-15	8.0E-15
8.6	2.0	4.3	2.0	2048	3.3E-01	1.6E-04	1.3E-03	6.4E-07	3.4E-15	3.6E-15
8.6	2.0	4.3	2.0	4096	6.5E-01	1.6E-04	4.1E-03	1.0E-06	3.8E-15	1.3E-14
8.6	2.0	4.3	2.0	8192	1.3E+00	1.6E-04	9.9E-03	1.2E-06	1.7E-15	8.1E-15
8.6	2.0	4.3	2.0	16384	2.6E+00	1.6E-04	1.8E-02	1.1E-06	1.5E-15	1.4E-14

Table 3.2: Test results for the connection between the Jacobi polynomials $\{P_n^{(\alpha,\beta)}\}_{n\in\mathbb{N}_0}$ and $\{P_n^{(\gamma,\beta)}\}_{n\in\mathbb{N}_0}$ computed with the fmm method with accuracy controlling parameter $p = 18$ for different transform sizes n. Shown are the times for precomputation t_p and for the computation of the actual transform t_t. Both are also shown after division through the expected asymptotic expression in terms of the transform size n. Furthermore, the component-wise error E_∞^c and the relative infinity norm error E_∞ are reported.

α	β	n	t_p	t_p/n^2	t_t	t_t/n	E_∞^c	E_∞
-0.2	-0.4	256	3.8E-02	1.5E-04	6.6E-05	2.6E-07	1.8E-15	7.2E-16
-0.2	-0.4	512	7.3E-02	1.4E-04	1.8E-04	3.5E-07	1.2E-15	1.2E-15
-0.2	-0.4	1024	1.6E-01	1.5E-04	4.2E-04	4.1E-07	1.7E-15	2.3E-15
-0.2	-0.4	2048	3.0E-01	1.5E-04	9.5E-04	4.6E-07	3.3E-15	2.0E-15
-0.2	-0.4	4096	6.1E-01	1.5E-04	3.1E-03	7.7E-07	4.9E-15	3.4E-15
-0.2	-0.4	8192	1.2E+00	1.5E-04	8.3E-03	1.0E-06	4.6E-15	5.8E-15
-0.2	-0.4	16384	2.4E+00	1.5E-04	1.6E-02	9.6E-07	1.2E-14	8.3E-15
-0.2	0.5	256	3.2E-02	1.3E-04	6.6E-05	2.6E-07	1.4E-15	1.1E-15
-0.2	0.5	512	8.0E-02	1.6E-04	1.8E-04	3.5E-07	2.2E-15	8.7E-16
-0.2	0.5	1024	1.6E-01	1.5E-04	4.2E-04	4.1E-07	2.9E-15	5.6E-16
-0.2	0.5	2048	3.1E-01	1.5E-04	9.4E-04	4.6E-07	5.5E-15	2.0E-15
-0.2	0.5	4096	6.3E-01	1.5E-04	3.2E-03	7.8E-07	6.6E-15	2.5E-15
-0.2	0.5	8192	1.3E+00	1.6E-04	9.7E-03	1.2E-06	9.9E-15	1.7E-15
-0.2	0.5	16384	2.6E+00	1.6E-04	1.6E-02	9.7E-07	1.7E-14	5.7E-15
0.5	-0.2	256	3.8E-02	1.5E-04	6.6E-05	2.6E-07	2.0E-15	7.5E-16
0.5	-0.2	512	7.5E-02	1.5E-04	1.8E-04	3.5E-07	1.2E-14	1.1E-15
0.5	-0.2	1024	1.6E-01	1.5E-04	4.2E-04	4.1E-07	6.4E-15	1.2E-15
0.5	-0.2	2048	3.0E-01	1.5E-04	9.4E-04	4.6E-07	6.7E-15	2.2E-15
0.5	-0.2	4096	6.1E-01	1.5E-04	3.2E-03	7.7E-07	2.5E-14	2.8E-15
0.5	-0.2	8192	1.2E+00	1.5E-04	9.7E-03	1.2E-06	2.2E-14	3.9E-15
0.5	-0.2	16384	2.4E+00	1.5E-04	1.6E-02	9.6E-07	6.0E-14	5.4E-15
0.5	1.4	256	4.1E-02	1.6E-04	6.6E-05	2.6E-07	2.1E-15	5.5E-16
0.5	1.4	512	8.0E-02	1.6E-04	1.8E-04	3.5E-07	2.5E-15	1.0E-15
0.5	1.4	1024	1.6E-01	1.6E-04	4.3E-04	4.2E-07	3.1E-15	6.1E-16
0.5	1.4	2048	3.2E-01	1.6E-04	9.5E-04	4.6E-07	4.6E-15	5.6E-16
0.5	1.4	4096	6.5E-01	1.6E-04	3.1E-03	7.6E-07	6.6E-15	1.9E-15
0.5	1.4	8192	1.3E+00	1.6E-04	8.4E-03	1.0E-06	1.1E-14	1.8E-15
0.5	1.4	16384	2.7E+00	1.6E-04	1.5E-02	9.2E-07	1.7E-14	5.6E-15
5.9	8.1	256	4.2E-02	1.6E-04	7.0E-05	2.7E-07	7.1E-15	1.1E-14
5.9	8.1	512	8.3E-02	1.6E-04	1.9E-04	3.6E-07	7.6E-15	6.1E-15
5.9	8.1	1024	1.7E-01	1.6E-04	4.3E-04	4.2E-07	8.4E-15	6.1E-15
5.9	8.1	2048	3.3E-01	1.6E-04	9.8E-04	4.8E-07	9.7E-15	1.1E-14
5.9	8.1	4096	6.6E-01	1.6E-04	3.2E-03	7.7E-07	1.1E-14	3.3E-15
5.9	8.1	8192	1.3E+00	1.6E-04	1.0E-02	1.3E-06	1.4E-14	6.1E-15
5.9	8.1	16384	2.7E+00	1.7E-04	1.6E-02	1.0E-06	1.7E-14	4.9E-15
9.0	4.8	256	4.0E-02	1.6E-04	8.5E-05	3.3E-07	1.4E-15	1.8E-15
9.0	4.8	512	7.5E-02	1.5E-04	2.0E-04	3.9E-07	1.5E-15	1.3E-14
9.0	4.8	1024	1.6E-01	1.5E-04	4.6E-04	4.5E-07	1.4E-15	3.6E-15
9.0	4.8	2048	3.1E-01	1.5E-04	1.0E-03	5.1E-07	3.0E-15	2.8E-15
9.0	4.8	4096	6.2E-01	1.5E-04	3.6E-03	8.7E-07	2.7E-15	5.1E-15
9.0	4.8	8192	1.3E+00	1.5E-04	1.0E-02	1.2E-06	4.6E-15	4.5E-15
9.0	4.8	16384	2.5E+00	1.5E-04	1.8E-02	1.1E-06	3.7E-15	6.9E-15

Table 3.3: Test results for the connection between the Gegenbauer polynomials $\{C_n^{(\alpha)}\}_{n\in\mathbb{N}_0}$ and $\{C_n^{(\beta)}\}_{n\in\mathbb{N}_0}$ computed with the fmm method with accuracy controlling parameter $p = 18$ for different transform sizes n. Shown are the times for precomputation t_p and for the computation of the actual transform t_t. Both are also shown after division through the expected asymptotic expression in of the transform size n. Furthermore, the component-wise error E_∞^c and the relative infinity norm error E_∞ are reported.

can have all $x_j = 1$ on one hand (so one order of magnitude is spanned) and observe that the coefficients y_j span several orders of magnitude, e.g., from 10^{-10} to 10^{10} or more. This observation rings a bell to anyone familiar with the properties of floating-point arithmetic. For example, if we need to compute an expansion coefficient $y_j \approx 10^{-10}$ from other expansion coefficients $x_j = 1$, then the calculation usually incurs a large relative error in the result; see [34]. And even if the result can be computed to high relative accuracy, the recovery of the coefficients x_j from the freshly computed coefficients y_j can still fail with large errors. Let us make this more precise with a practical example.
We consider the functions $f_1 : \mathbb{R} \to \mathbb{R}$ and $f_2 : \mathbb{R} \to \mathbb{R}$ defined by

$$f_1(t) = |t - 0.1|^2 \quad \text{and} \quad f_2(t) = |t - 0.1|.$$

While the function f_1 is continuously differentiable, the function f_2 is only continuous, but not differentiable at $t = 0.1$. This choice was made to ensure that we do not compute with expansion coefficients that essentially vanish for yet moderate degrees. For both functions, we repeat the following procedure for different degrees n. First, we compute approximated expansion coefficients x_j for the respective function f such that

$$f \approx \sum_{j=0}^{n} x_j T_j.$$

This was done with a Gauss-Chebyshev quadrature rule to discretize, i.e., approximate the respective inner products. A classical result by Erdős and Turán [21] asserts that the L^2-error of the approximation is within a constant factor of the best uniform approximation by polynomials of degree at most n. Fast discrete cosine transforms can be used to compute this step efficiently. Then, from the coefficients x_j we computed the expansion coefficients y_j in

$$f \approx \sum_{j=0}^{n} y_j C_j^{(\beta)}$$

for different positive integer values β. This requires the application of different connection matrices $\mathbf{K}$. These, however, have a banded structure owing to the particular choice of the parameter β. Therefore, these matrices were applied the usual way so that errors must be caused by properties of floating-point arithmetic. After that, we tried to recover the coefficients x_j by applying the inverse of the connection matrix $\mathbf{K}$ from before. Since this matrix is always semiseparable, it was again applied cheaply without any systematic error involved. The resulting expansion is evaluated at the Chebyshev points in $[-1, 1]$ that correspond to the degree n. Finally, the result is compared in the relative infinity vector norm to the initial function values taken at the same sites. Table 3.4 shows the relative infinity error E_∞ for increasing degree n and different values β. For both functions, f_1 and f_2, it can be observed that results get more inaccurate for larger β. At the same combination of degree n and parameter β, the function f_2, which has more slowly decreasing expansion coefficients, shows results worse than for the function f_1, who has its expansion coefficients decay quicker. This behavior can be explained with catastrophic cancellations in the numerical computation as shown in the following.

3.5.2 A word on feasibility. The previous example has numerically shown that transformations between families of Gegenbauer polynomials $\{C_n^{(\alpha)}\}_{n \in \mathbb{N}_0}$ and $\{C_n^{(\beta)}\}_{n \in \mathbb{N}_0}$ can be subject to numerical instabilities. This might render the procedure infeasible in some cases. Apparently, errors in the computation grow the larger the distance between the parameters α and β gets.

Table 3.4: Combined projection and evaluation procedure applied to the functions f_1 and f_2. Shown is the relative infinity error E_∞ for increasing expansion degree n and different indices β.

		E_∞			
f	n	$\beta = 1$	$\beta = 3$	$\beta = 5$	$\beta = 7$
f_1	100	3.6E−16	3.6E−16	3.6E−16	3.6E−16
	1.000	5.5E−16	5.5E−16	5.5E−16	5.5E−16
	10.000	1.0E−15	1.0E−15	1.2E−15	1.9E−10
	100.000	5.1E−15	6.3E−15	2.5E−11	3.4E−03
f_2	100	7.0E−16	2.6E−15	1.4E−13	1.0E−11
	1.000	6.0E−16	7.2E−14	1.8E−10	1.4E−06
	10.000	1.6E−15	3.2E−12	2.7E−06	2.6E+00
	100.000	5.4E−15	6.4E−11	3.0E−03	1.2E+05

Table 3.5: The Condition number $\text{cond}_2(\mathbf{K})$ and the lower bound $\lambda_{\max}(\mathbf{K})/\lambda_{\min}(\mathbf{K})$ from (3.11) for the 51×51 matrix $\mathbf{K}^{(1)\to(\beta)}$ for different values β. As β increases, the condition number quickly grows beyond acceptable levels.

β	11	21	31	41	51
$\lambda_{\max}(\mathbf{K})/\lambda_{\min}(\mathbf{K})$	8.9E+06	8.9E+09	6.3E+11	1.2E+13	1.0E+14
$\text{cond}_2(\mathbf{K})$	2.2E+15	4.1E+23	3.1E+29	1.2E+34	6.8E+37

An often used measure for a problem's amenability to digital computation is the *condition number*. For example, the condition number associated with a system of linear equations $\mathbf{Ax} = \mathbf{b}$ gives a bound on the relative accuracy of an approximate solution when compared to relative errors in $\mathbf{A}$ and $\mathbf{b}$.

The usual 2-norm condition number $\text{cond}_2(\mathbf{A})$ for a non-singular matrix $\mathbf{A}$ is defined as $\text{cond}_2(\mathbf{A}) = \|\mathbf{A}\|_2 \|\mathbf{A}^{-1}\|_2$. A lower bound is obtained from the inequality $\lambda_{\max}(\mathbf{A}) \leq \|\mathbf{A}\|_2$, where $\lambda_{\max}(\mathbf{A})$ is the largest modulus of any eigenvalue of $\mathbf{A}$. Since $\lambda_{\max}(\mathbf{A}^{-1}) = \lambda_{\min}(\mathbf{A})^{-1}$, one obtains the inequality

$$\frac{\lambda_{\max}(\mathbf{A})}{\lambda_{\min}(\mathbf{A})} \leq \text{cond}_2(\mathbf{A}). \tag{3.11}$$

It is not trivial to work out the actual condition number $\text{cond}_2(\mathbf{K})$ for the connection matrices $\mathbf{K}$ that were used, but an explicit expression for the lower bound can be readily obtained. Since each connection matrix $\mathbf{K}$ is triangular, its eigenvalues coincide with the entries on the main diagonal. For example, when $0 < \alpha < \beta$ the largest eigenvalue is $\lambda_{\max}(\mathbf{K}) = \kappa_{0,0} = 1$ and the smallest is given by $\lambda_{\min}(\mathbf{K}) = \kappa_{n,n}$; cf. (3.6). This leads to the estimate

$$\frac{\Gamma(\alpha)\Gamma(n+\beta)}{\Gamma(\beta)\Gamma(n+\alpha)} \leq \text{cond}_2(\mathbf{K}).$$

This bound can grow quite large as $\beta - \alpha$ grows. Table 3.5 shows an example for the matrix $\mathbf{K}^{(1)\to(\beta)}$ and different values β. While the condition number is associated with solving a linear system, it is also an indicator of numerical instabilities in our situation. This is because a large condition number tell us that small errors in the computed coefficients y_j can cause the recovered coefficients x_j to contain a large relative error. For example, consider for $n \geq 1016$ the computation of the coefficient y_{1000} from the coefficients $x_j = 1$. The result, computed with a precision of 100 decimal digits is

$$\begin{aligned} x_{1000} = \quad & 3.5\ldots \times 10^{-19} - 2.8\ldots \times 10^{-18} + 9.6\ldots \times 10^{-18} - 1.9\ldots \times 10^{-17} \\ & + 2.3\ldots \times 10^{-17} - 1.9\ldots \times 10^{-17} + 9.2\ldots \times 10^{-18} - 2.6\ldots \times 10^{-18} \\ & + 3.2\ldots \times 10^{-19} \\ \approx\ & 1.6062271959049573 \times 10^{-34}. \end{aligned}$$

Computing the same sum in double precision gives a value $x_{1000} = -3.9341703542077499 \times 10^{-32}$ which is wrong in every digit. The upshot is that the above computation is correct in exact arithmetic but is subject to catastrophic cancellations in finite precision arithmetic. These occur when the result of the subtraction of two nearby numbers, say a and b, is much smaller in magnitude than a and b themselves. Cancellations become catastrophic when a and b already contain small perturbations, e.g., as the result of a previous computation. In this case, the computed result can have a large relative error compared to the correctly rounded result; see [24]. The above computation is prone to catastrophic cancellations since the connection coefficients $\kappa_{j,k}$ in $\mathbf{K}$ will already contain rounding errors in any practical situation. There does not seem to be an obvious way to circumvent this.

4 – Applications

4.1 Non-equispaced fast spherical Fourier transform

Discrete Fourier analysis in euclidean space plays an important role in a wide range of applications, for example, signal processing, image processing, computed tomography, and a lot more. However, in many fields of interest, data naturally arises on a geometry that can be identified with the surface of the two-dimensonal unit sphere

$$\mathbb{S}^2 = \left\{ \mathbf{x} \in \mathbb{R}^3 : \ \|\mathbf{x}\|_2 = 1 \right\}$$

that is embedded into the three-dimensional euclidean space $\mathbb{R}^3$. Such data occurs quite naturally when measurements are taken relative to the surface of the earth which can be roughly identified with a sphere. In similar ways to euclidean space, one considers analyzing or processing the data by expanding into a Fourier series or finite Fourier sum. For Fourier analysis on the sphere $\mathbb{S}^2$ the corresponding basis of choice are the spherical harmonics Y_ℓ^m (a precise definition is given in Section 4.1.4 below). A finite expansion of a function f in spherical harmonics in spherical coordinates (ϑ, φ) may look like

$$f(\vartheta, \varphi) = \sum_{\ell=0}^{L} \sum_{m=-\ell}^{\ell} \hat{f}_\ell^m Y_\ell^m(\vartheta, \varphi), \qquad \text{with } L \in \mathbb{N}_0, \tag{4.1}$$

where $\vartheta \in [0, \pi]$ and $\varphi \in [0, 2\pi)$ are the co-latitude and longitude, respectively.
From a computational point of view, there are two basic tasks to consider. First, if the expansion coefficients $\hat{f}_\ell^m$ are known, then one would like to evaluate the expansion (4.1) on a given finite set of nodes (ϑ_i, φ_i) for $i = 1, 2, \ldots, I$. Second, given function values $f(\vartheta_i, \varphi_i)$ at these sites, one would like to calculate the expansion coefficients $\hat{f}_l^m$. This section is devoted to propose a method that can be used to achieve both.

4.1.1 Existing algorithms for particular nodes. Discrete Fourier transforms on the sphere $\mathbb{S}^2$ have been of interest at least since the work of Driscoll and Healy [16] who introduced what is today often called the *Driscoll-Healy algorithm.* They developed an $\mathcal{O}(L^2 \log^2 L)$ algorithm to compute from $I = \mathcal{O}(L^2)$ function samples $f(\vartheta_i, \varphi_i)$, taken at specific sites, the expansion coefficients $\hat{f}_\ell^m$ for any L-bandlimited function, that is, a function that satisfies $\hat{f}_\ell^m = 0$ for $l > L$. This is more efficient than the $\mathcal{O}(L^3)$ arithmetic operations that would be needed otherwise. It is, however, later observed in [17] that the calculation of the expansion coefficients $\hat{f}_\ell^m$ may be subject to numerical problems that undermine the stability of the procedure. It is therefore noted that the algorithm might need a modification to improve the accuracy. Some modifications to the original algorithm appeared in [31], arguably without providing a totally satisfactory solution to the problems.
The first modification to the Driscoll-Healy algorithm that specifically addressed the numerical instabilities was proposed in [67] where unstable parts in the computation were identified and circumvented. The authors did this for the transposed version of the Driscoll-Healy algorithm. This is the one that allows to calculate from known expansion coefficients $\hat{f}_\ell^m$ the function values $f(\vartheta_i, \varphi_i)$ for $i = 1, 2, \ldots, I$; see (4.1). The modifications nevertheless also apply to the original variant. Similar ideas appeared later also in [32].
Another method to compute discrete spherical Fourier transforms that relies on certain matrix compression techniques was introduced in [61]. The author obtains $\mathcal{O}(L^{5/2} \log L)$ and $\mathcal{O}(L^2 \log^2 L)$ algorithms for a grid of $I = \mathcal{O}(L^2)$ specific points on the sphere.

A more recent method was proposed in [72]. It is similar to the Driscoll-Healy algorithm and the one from [67], but avoids the primary source of numerical instabilities in a new way. Instead of modifying the algorithm on a large scale, a new method for approximate polynomial multiplication, based on the fast multipole method (FMM), is used. This replaces the discrete cosine transforms that were used before and provides some flexibility in the choice of certain interpolation nodes. This way, numerically delicate calculations can be tamed. The result is an approximate $\mathcal{O}\big(L^2 \log L \log(1/\varepsilon)\big)$ algorithm, where ε is the desired accuracy.
Yet another method was developed in [70]. It uses a new approach that is entirely different than the Driscoll-Healy algorithm. The method is essentially based on the observation that the associated Legendre functions P_ℓ^m of different orders m (they appear in the definition of the spherical harmonics Y_ℓ^m below) are solutions to very similar differential equations. A matrix is constructed that has eigenvectors that appear in the matrix-representation of the conversion between associated Legendre functions of (possibly large) orders m to a set of low orders $\hat{m} \in \{1, 2\}$. This can be used to modify the original expansion (4.1) by efficiently replacing all associated Legendre functions of orders m with those of low orders $\hat{m}$. The task of evaluating the resulting expansion is reduced. Thus, an approximate $\mathcal{O}\big(L^2 \log L \log(1/\varepsilon)\big)$ method for $I = \mathcal{O}(L^2)$ specific points on the sphere is obtained. The method to be described in this section is very similar. The details are found in Section 4.1.6 below.
More recently, a modified version of these techniques appeared in [76]. The method is somewhat different in how the original expansion is modified but the key step is carried out using very similar techniques.

4.1.2 Existing algorithms for arbitrary nodes. While all algorithms mentioned so far have been described for specific sets of nodes on the sphere, an algorithm for arbitrary nodes was described in [55] for the first time. The stabilized algorithm from [67] is therein combined with the non-equispaced fast Fourier transform (NFFT), a modification to the original fast Fourier transform (FFT) for arbitrary nodes. The resulting algorithm was the first to decouple the expansion degree L and the number of nodes I to give an $\mathcal{O}\big(L^2 \log^2 L + I \log^2(1/\varepsilon)\big)$ algorithm for any set of points on the sphere. The transposed version was described and analyzed in [44] and can be seen as an analogue of the Driscoll-Healy algorithm. In other aspects, the possibility of using the FMM for polynomial multiplication is already mentioned, but not implemented, in [55]. Moreover, it should be acknowledged that it is possible to combine nearly all existing algorithms for specific nodes with the NFFT to algorithm for arbitrary nodes. This is, for example, mentioned by the authors in [70].

Remark 4.1 The term "non-equispaced" is nowadays conventional to describe any set of points free of choice, but might be a source of confusion for the casual reader. In one dimension, a set of non-equispaced points can be equivalently described as a set of arbitrary points. The NFFT algorithm provides a fast Fourier transform algorithm for these arbitrary points. The terminology becomes arguably more problematic for multi-dimensional transforms and non-euclidean manifolds. In $\mathbb{R}^3$, a set of equispaced points usually means a set of points that has been obtained by regularly sampling the coordinate axes. This enables the use of the FFT. Any other point configuration requires the NFFT or other techniques. On manifolds like the sphere, equispaced points might be equally defined as those obtained from a sampling of the axes of the spherical coordinate system. This, however, does even remotely lead to an equispaced set of points; see Figure 1(a). In the following, we will use the term *non-equispaced* as a synonym for *arbitrary*.

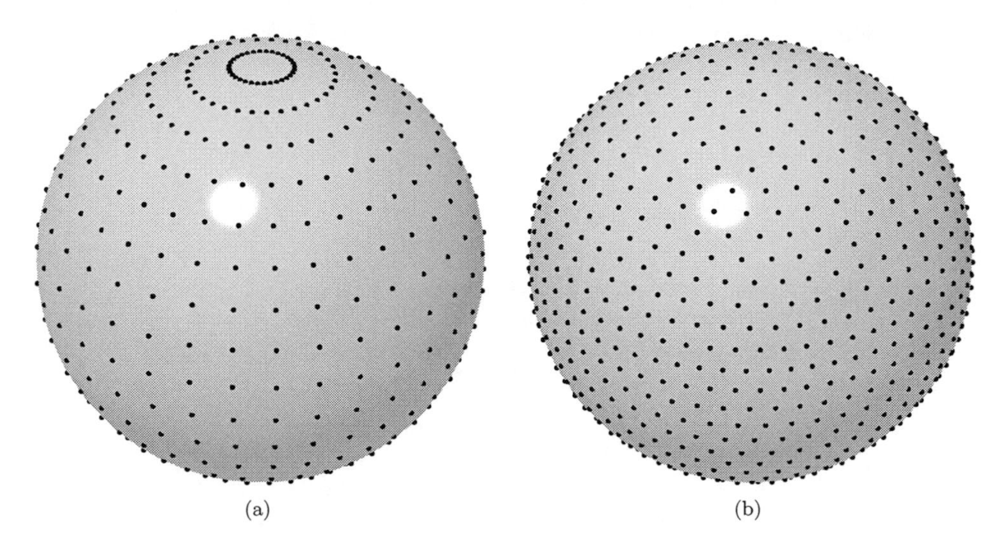

(a) (b)

Figure 4.1: (a) Clenshaw-Curtis points on the sphere. They are obtained by a regular sampling of the spherical coordinate axes and therefore cluster at the poles. (b) Almost evenly distributed points on the sphere.

4.1.3 Why arbitrary nodes are important. At first glance, the possibility to extend the fast algorithms to arbitrary nodes on the sphere seems just like a nice extra to have. But as it turns out, this is a rather essential requirement for the realization of efficient algorithms on the sphere. Some problems are just more amenable to computation if the nodes on the sphere can be distributed arbitrarily, often meaning more evenly distributed. Other algorithms that are restricted to specific point sets can fail at this point. Figure 4.1 shows two point sets on the sphere. The first one is typical for those that have been used over the years. It is obtained by sampling the sphere at regular intervals along the spherical coordinate axes (i.e., the longitudinal and latitudinal directions). It can be seen that nodes cluster near the poles. This is, in many ways, an undesirable feature. The spherical coordinate system carries some sense of arbitrariness since on a perfect sphere why would one want to distinguish two points, that is, the poles, from the rest? Most often, an approximately even distribution of nodes is desirable. Such is shown in the second image. In most cases, the distribution of the nodes, more precisely, the size of gaps between the nodes, is tightly coupled to numerical stability; see for example [42, 54] for a number of theoretical results on this subject.

The following text is structured as follows. In Section 4.1.4, we collect some basic material on Fourier analysis on the sphere $\mathbb{S}^2$. Section 4.1.5 gives a formal definition of the discrete Fourier transform on the sphere and introduces a convenient matrix-vector notation. Our fast Fourier transform algorithm for arbitrary nodes is described in Section 4.1.6. Finally, some numerical results are given in Section 4.1.7.

4.1.4 Fourier analysis on the two-sphere. In spherical coordinates, we identify each point $\mathbf{x} \in \mathbb{S}^2$ with a tuple $(\vartheta, \varphi) \in [0, \pi] \times [0, 2\pi)$ of two angles ϑ and φ. The space $\mathrm{L}^2(\mathbb{S}^2)$ is the Hilbert space of square integrable functions on the sphere $\mathbb{S}^2$ with the usual inner product given by

$$\langle f, g\rangle = \int_0^{2\pi}\int_0^{\pi} f(\vartheta,\varphi)\overline{g(\vartheta,\varphi)}\,\sin\vartheta\,\mathrm{d}\vartheta\,\mathrm{d}\varphi.$$

With the orthogonal basis of spherical harmonics Y_ℓ^m, any function f from the space $\mathrm{L}^2(\mathbb{S}^2)$ can be developed into an infinite orthogonal expansion

$$f(\vartheta,\varphi) = \sum_{\ell=0}^{\infty}\sum_{m=-\ell}^{\ell} \hat{f}_\ell^m Y_\ell^m(\vartheta,\varphi), \tag{4.2}$$

where equality holds in the L^2-sense. The expansion is finite and of degree L if the function f can be represented by an algebraic polynomial of degree at most L in three-dimensions, that has been restricted to the unit sphere. Each spherical harmonic Y_ℓ^m can be represented as an algebraic harmonic homogeneous polynomial of degree ℓ in three dimensions (restricted to the unit sphere) and is defined by

$$Y_\ell^m(\vartheta,\varphi) = P_\ell^m(\cos\vartheta)\mathrm{e}^{\mathrm{i}m\varphi}. \tag{4.3}$$

The functions P_ℓ^m are the associated Legendre functions; see Section 1.6.3.

Remark 4.2 The spherical harmonics Y_ℓ^m satisfy

$$\|Y_\ell^m\|^2 = \langle Y_\ell^m, Y_\ell^m\rangle = \frac{4\pi}{2\ell+1}.$$

It is also common to use the normalized spherical harmonics $\tilde{Y}_\ell^m$ which are defined by

$$\tilde{Y}_\ell^m = (-1)^m\sqrt{\frac{2\ell+1}{4\pi}}Y_\ell^m.$$

We have here included the factor $(-1)^m$, the so-called *Condon-Shortley phase*, which can be omitted. This is merely a matter of convention.

4.1.5 Discrete Fourier transform on the two-sphere. We recall that for $L \in \mathbb{N}_0$, our goal is the evaluation of the sums

$$f(\vartheta_i,\varphi_i) = \sum_{\ell=0}^{L}\sum_{m=-\ell}^{\ell} \hat{f}_\ell^m Y_\ell^m(\vartheta_i,\varphi_i), \qquad \text{with } i = 1,2,\ldots,I, \tag{4.4}$$

for some given points (ϑ_i,φ_i). This is called a *non-equispaced discrete spherical Fourier transform (NDSFT)* and represents a linear transformation. Therefore, we may also wish to calculate the sums

$$\tilde{f}_\ell^m = \sum_{i=1}^{I} f(\vartheta_i,\varphi_i)\overline{Y_\ell^m(\vartheta_i,\varphi_I)}, \qquad \text{with } \ell = 0,1,\ldots,L \text{ and } m = -\ell,\ldots,\ell, \tag{4.5}$$

which is the corresponding adjoint transformation, hence *adjoint NDSFT*. Note that this adjoint transform usually does not recover the coefficients $\hat{f}_\ell^m$ from (4.4), i.e., it is not the inverse transformation. To obtain these, we need weights w_i from a suitable quadrature rule, based on the given nodes, to compute

$$\hat{f}_\ell^m = \sum_{i=1}^{I} w_i f(\vartheta_i,\varphi_i)\overline{Y_\ell^m(\vartheta_i,\varphi_i)}, \quad \text{with } \ell = 0,1,\ldots,L \text{ and } m = -\ell,\ldots,\ell. \tag{4.6}$$

The weights w_i need to be chosen such the sums in (4.6) become discretized versions of the inner product integrals

$$\hat{f}_\ell^m = \int_0^{2\pi}\int_0^{\pi} f(\vartheta,\varphi)\overline{Y_\ell^m(\vartheta,\varphi)}\,\sin(\vartheta)\,\mathrm{d}\vartheta\,\mathrm{d}\varphi,$$

with $\ell = 0, 1, \ldots, L$ and $m = -\ell, \ldots, \ell$. The value $L \in \mathbb{N}_0$ is called the *bandwidth of the function* f which is likewise said to be an *L-bandlimited function* on the sphere $\mathbb{S}^2$. The complex expansion coefficients $\hat{f}_\ell^m \in \mathbb{C}$ are the *spherical Fourier coefficients* of the function f. The index ℓ is called *the degree* and m is called *the order.* A set of arbitrary nodes

$$\mathcal{X} = \left\{ (\vartheta_i, \varphi_i) \right\}_{i=1}^{I}$$

on the sphere $\mathbb{S}^2$ is called a *sampling set.* For notational convenience, we may also work with the index set

$$\mathcal{I}_L := \left\{(\ell, m) \,:\, \ell = 0, 1, \ldots, L;\; m = -\ell, \ldots, \ell\right\}$$

that specifies the allowed range for the indices l and m.
From a linear algebra point of view, evaluating the L-bandlimited function f on a sampling set $\mathcal{X}$ amounts to calculating the matrix-vector product

$$\mathbf{f}_{\mathcal{X}} = \mathbf{Y}_{L,\mathcal{X}}\, \hat{\mathbf{f}}_L$$

with

$$\begin{aligned}
\mathbf{f}_{\mathcal{X}} &:= \left(f(\vartheta_i, \varphi_i)\right)_{i=1}^{I} \in \mathbb{C}^I, \\
\mathbf{Y}_{L,\mathcal{X}} &:= \left(Y_\ell^m(\vartheta_i, \varphi_i)\right)_{i=1,\ldots,I;(\ell,m)\in\mathcal{I}_L} \in \mathbb{C}^{I\times(L+1)^2}, \\
\hat{\mathbf{f}}_L &:= \left(\hat{f}_\ell^m\right)_{(\ell,m)\in\mathcal{I}_L} \in \mathbb{C}^{(L+1)^2}.
\end{aligned}$$

We write $\mathbf{f}_{\mathcal{X}}$, $\mathbf{Y}_{L,\mathcal{X}}$ and $\hat{\mathbf{f}}_L$ to emphasize the dependence of these quantities on the concrete sampling set $\mathcal{X}$ and the bandwidth L. The computation of the adjoint transformation reads

$$\tilde{\mathbf{f}}_L = \mathbf{Y}_{L,\mathcal{X}}^{\mathrm{H}}\, \mathbf{f}_{\mathcal{X}}.$$

4.1.6 Fast Fourier transform for arbitrary nodes on the two-sphere. Before we start describing our algorithm in more detail, let us briefly discuss how an efficient algorithm for the evaluation of a finite expansion like (4.4) at a given number of nodes (ϑ_i, φ_i) can be realized in principle. The enabling idea is to perform a change of basis such that the function f in (4.4) is represented as an expansion of the form

$$f(\vartheta, \varphi) = \sum_{\ell=-L}^{L} \sum_{m=-L}^{L} \hat{c}_\ell^m \mathrm{e}^{\mathrm{i}\ell\vartheta} \mathrm{e}^{\mathrm{i}m\varphi}. \tag{4.7}$$

This is an ordinary two-dimensional Fourier sum with new coefficients $\hat{c}_\ell^m$ and it will take $\mathcal{O}\left(L^2 \log L \log(1/\varepsilon)\right)$ arithmetic operations to perform this change of basis, that is, to compute from the coefficients $\hat{f}_\ell^m$ the new coefficidents $\hat{c}_\ell^m$ up to some accuracy ε. The evaluation of the function f at the given nodes (ϑ_i, φ_i) may then be realized using the non-equispaced fast Fourier transform (NFFT) which needs another $\mathcal{O}\left(L^2 \log L + \log^2 I(1/\varepsilon)\right)$ arithmetic operations. In total, we arrive at $\mathcal{O}\left(L^2 \log L \log(1/\varepsilon) + I \log^2(1/\varepsilon)\right)$ arithmetic operations for the complete transformation.
Since this complex transformation is linear, it is possible to mechanically derive the corresponding adjoint algorithm. The first step that we have described is a real linear transformation (that may nevertheless be applied to complex data) and since the second step is a true complex linear transformation, the former has a transposed counterpart and the latter has an adjoint counterpart. For the NFFT, that is, the second part, the process of obtaining the adjoint algorithm is described, for example, in [68]; see also the references therein. For the first step, we note that while we do not describe the transposed algorithm in detail, it is easy to derive it. This is because we will see that it consists of a number of

matrix-vector multiplications that need to be performed sequentially. For these matrices the same methods as before can be used to apply their respective transpose.
We are now ready to describe the fast evaluation of the L-bandlimited function f on an arbitrary sampling set $\mathcal{X} = \big((\vartheta_i, \varphi_i) : i = 1, 2, \ldots, I\big)$ on the sphere $\mathbb{S}^2$, or equivalently, the fast evaluation of the matrix-vector product $\mathbf{f}_{\mathcal{X}} = \mathbf{Y}_{M,\mathcal{X}}\,\hat{\mathbf{f}}_M$. To this end, we define $x = \cos\vartheta$ and rearrange the spherical Fourier sum (4.4),

$$\begin{aligned} f(\vartheta,\varphi) &= \sum_{\ell=0}^{L}\sum_{m=-\ell}^{\ell} \hat{f}_\ell^m Y_\ell^m(\vartheta,\varphi) \\ &= \sum_{m=-L}^{L}\sum_{\ell=|m|}^{L} \hat{f}_\ell^m Y_\ell^m(\vartheta,\varphi) \\ &= \sum_{m=-L}^{L} \mathrm{e}^{\mathrm{i}m\phi} \sum_{\ell=|m|}^{L} \hat{f}_\ell^m P_\ell^m(\cos\vartheta) \\ &= \sum_{m=-L}^{L} \mathrm{e}^{\mathrm{i}m\phi} h_m(x), \end{aligned} \tag{4.8}$$

where

$$h_m(x) = \sum_{\ell=|m|}^{L} \hat{f}_\ell^m P_\ell^m(x).$$

The first step consists in replacing the whole expansion (4.8) with an ordinary two-dimensional Fourier sum. For this, we apply individually to each inner sum $h_m(x)$ a series of transformations. So assume that we have fixed an order m. The goal is to replace the associated Legendre functions P_ℓ^m in $h_m(x)$ with the associated Legendre functions $P_\ell^{\hat{m}}$ where $\hat{m}$ satisfies $0 \le \hat{m} \le 2$. We define in accordance with Lemma 1.93 the index

$$\hat{m} = \hat{m}(m) := \begin{cases} 0, & \text{if } m = 0, \\ 2, & \text{if } m \text{ even and } m \neq 0, \\ 1, & \text{if } m \text{ odd.} \end{cases}$$

Then each sum $h_m(x)$ may be written as

$$h_m(x) = \sum_{\ell=\hat{m}}^{L} \hat{a}_\ell^m P_\ell^{\hat{m}}(x).$$

To compute the coefficients $\hat{a}_\ell^m$, we can use the divide-and-conquer algorithm from Section 2.5 in combination with the findings in Section 2.6. For a single sum $h_m(x)$ this takes $\mathcal{O}\big(L \log L \log(1/\varepsilon)\big)$ arithmetic operations. In total, we need $\mathcal{O}\big(L^2 \log L \log(1/\varepsilon)\big)$ operations to compute for $m = -L, \ldots, L$ the new coefficients $\hat{a}_\ell^m$.
Now that we have replaced associated Legendre functions of high orders m by associated Legendre functions of low orders $\hat{m}$, the remaining sums can be manipulated further. There, we have to distinguish the three cases $\hat{m} = 0$, $\hat{m} = 1$ and $\hat{m} = 2$.

The case $\hat{m} = 0$

This is the simplest case of all three. There is only one sum $h_m(x)$ for which $\hat{m} = 0$, namely the one for $m = 0$. After the first step, we have

$$h_0(x) = \sum_{\ell=0}^{L} \hat{a}_\ell^0 P_\ell^0(x) = \sum_{\ell=0}^{L} \hat{a}_\ell^0 P_\ell(x),$$

that is, a finite linear combination of Legendre polynomials. With the fast algorithms for the connection between classical orthogonal polynomials found in Chapters 2 and 3, this can be recast into

$$h_0(x) = \sum_{\ell=0}^{L} \hat{b}_\ell^0 T_\ell(x),$$

which is a finite linear combination of Chebyshev polynomials of first kind. With the methods from Chapter 3, for example, we need $\mathcal{O}\big(L\log(1/\varepsilon)\big)$ arithmetic operations to compute the coefficients $\hat{b}_\ell^0$. To further replace the obtained linear combination with an ordinary Fourier sum, we observe that

$$T_\ell(x) = T_\ell(\cos\vartheta) = \cos(\ell\vartheta) = \frac{1}{2}\big(\mathrm{e}^{\mathrm{i}\ell\vartheta} + \mathrm{e}^{-\mathrm{i}\ell\vartheta}\big),$$

Therefore, we can easily compute the coefficients

$$\hat{c}_\ell^0 = \begin{cases} \hat{b}_0^0, & \text{if } \ell = 0, \\ \frac{1}{2}\hat{b}_{|\ell|}^0, & \text{else,} \end{cases}$$

in the Fourier sum

$$h_0(x) = \sum_{\ell=-L}^{L} \hat{c}_\ell^0 \mathrm{e}^{\mathrm{i}\ell\vartheta}$$

with $\mathcal{O}(L)$ arithmetic operations.

The case $\hat{m} = 1$

This case implies that m is an odd integer. After the first step, we have obtained the sum

$$h_m(x) = \sum_{\ell=1}^{L} \hat{a}_\ell^m P_\ell^1(x) = \sqrt{1-x^2} \sum_{\ell=0}^{L-1} \hat{a}_{\ell+1}^m \frac{1}{2}\sqrt{\frac{\ell+2}{\ell+1}} P_\ell^{(1,1)}(x).$$

This representation is obtained by using the definition of the associated Legendre functions. Similar to before, we can replace the sum with a linear combination of Chebyshev polynomials

$$h_m(x) = \sqrt{1-x^2} \sum_{\ell=0}^{L-1} \hat{b}_\ell^m T_\ell(x)$$

with $\mathcal{O}\big(L\log(1/\varepsilon)\big)$ arithmetic operations. Again, we can replace this with an ordinary Fourier sum by using that

$$\begin{aligned} \sqrt{1-x^2}T_\ell(x) &= \sin(\vartheta)T_\ell(\cos\vartheta) \\ &= \sin(\vartheta)\cos(\ell\vartheta) \\ &= \frac{1}{2\mathrm{i}}(\mathrm{e}^{\mathrm{i}\vartheta} - \mathrm{e}^{-\mathrm{i}\vartheta}) \cdot \frac{1}{2}(\mathrm{e}^{\mathrm{i}\ell\vartheta} + \mathrm{e}^{-\mathrm{i}\ell\vartheta}) \\ &= \frac{1}{4\mathrm{i}}\big(\mathrm{e}^{\mathrm{i}(\ell+1)\vartheta} - \mathrm{e}^{\mathrm{i}(\ell-1)\vartheta} + \mathrm{e}^{-\mathrm{i}(\ell-1)\vartheta} - \mathrm{e}^{-\mathrm{i}(\ell+1)\vartheta}\big). \end{aligned}$$

Thus, we compute the coefficients

$$\hat{c}_\ell^m = \operatorname{sign}(\ell) \begin{cases} 0, & \text{if } \ell = 0, \\ \frac{1}{4\mathrm{i}} \left(\hat{b}_{|\ell|-1}^m - \hat{b}_{|\ell|+1}^m \right), & \text{if } 0 < |\ell| < L-1, \\ \frac{1}{4\mathrm{i}} \hat{b}_{|\ell|-1}^m, & \text{if } |\ell| = L-1, L, \end{cases}$$

in the Fourier sum

$$h_m(x) = \sum_{\ell=-L}^{L} \hat{c}_\ell^m \mathrm{e}^{\mathrm{i}\ell\vartheta}$$

with $\mathcal{O}(L)$ arithmetic operations.

The case $\hat{m} = 2$

This case implies that m is an even integer. After the first step, we have obtained the sum

$$h_m(x) = \sum_{\ell=2}^{L} \hat{a}_\ell^m P_\ell^2(x) = (1-x^2) \sum_{\ell=0}^{L-2} \hat{a}_{\ell+2}^m \frac{1}{4} \sqrt{\frac{(\ell+3)(\ell+4)}{(\ell+1)(\ell+2)}} P_\ell^{(2,2)}(x).$$

Now, we replace this by a linear combination of Chebyshev polynomials of first kind

$$h_m(x) = (1-x^2) \sum_{\ell=0}^{L-2} \hat{b}_\ell^m T_\ell(x)$$

with $\mathcal{O}\big(L \log(1/\varepsilon)\big)$ arithmetic operations. Then with

$$\begin{aligned} (1-x^2) T_\ell(x) &= \sin^2(\vartheta) T_\ell(\cos\vartheta) \\ &= \sin^2(\vartheta) \cos(\ell\vartheta) \\ &= \frac{1}{2\mathrm{i}} \left(\mathrm{e}^{\mathrm{i}\vartheta} - \mathrm{e}^{-\mathrm{i}\vartheta} \right) \cdot \frac{1}{2\mathrm{i}} \left(\mathrm{e}^{\mathrm{i}\vartheta} - \mathrm{e}^{-\mathrm{i}\vartheta} \right) \cdot \frac{1}{2} \left(\mathrm{e}^{\mathrm{i}\ell\vartheta} + \mathrm{e}^{-\mathrm{i}\ell\vartheta} \right) \\ &= \frac{1}{8} \big(-\mathrm{e}^{\mathrm{i}(\ell+2)\vartheta} + 2\mathrm{e}^{\mathrm{i}\ell\vartheta} - \mathrm{e}^{\mathrm{i}(\ell-2)\vartheta} - \mathrm{e}^{-\mathrm{i}(\ell+2)\vartheta} + 2\mathrm{e}^{-\mathrm{i}\ell\vartheta} - \mathrm{e}^{-\mathrm{i}(\ell-2)\vartheta} \big), \end{aligned}$$

we compute the coefficients

$$\hat{c}_\ell^m = \operatorname{sign}(\ell) \begin{cases} \frac{1}{4} \left(2\hat{b}_0^m - \hat{b}_2^m \right), & \text{if } \ell = 0, \\ \frac{1}{8} \left(2\hat{b}_{|\ell|}^m - \hat{b}_{|\ell|+2}^m \right), & \text{if } |\ell| = 1, \\ \frac{1}{8} \left(-\hat{b}_{|\ell|-2}^m + 2\hat{b}_{|\ell|}^m - \hat{b}_{|\ell|+2}^m \right), & \text{if } 1 < |\ell| < L-2, \\ \frac{1}{8} \left(-\hat{b}_{|\ell|-2}^m + 2\hat{b}_{|\ell|}^m \right), & \text{if } |\ell| = L-3, L-2, \\ -\frac{1}{8} \hat{b}_{|\ell|-2}^m, & \text{if } |\ell| = L-1, L, \end{cases}$$

in the Fourier sum

$$h_m(x) = \sum_{\ell=-L}^{L} \hat{c}_\ell^m \mathrm{e}^{\mathrm{i}\ell\vartheta}$$

with $\mathcal{O}(L)$ arithmetic operations.

We have now obtained the ordinary two-dimensional Fourier sum

$$f(\vartheta, \varphi) = \sum_{m=-L}^{L} \sum_{\ell=-L}^{L} \hat{c}_\ell^m \mathrm{e}^{\mathrm{i}\ell\vartheta} \mathrm{e}^{\mathrm{i}m\varphi}.$$

This double sum can be evaluated by the NFFT algorithm on an arbitrary sampling set $\mathcal{X}$ with I nodes using $\mathcal{O}\big(L^2 \log L + I \log^2(1/\varepsilon)\big)$ arithmetic operations, where ε is the desired accuracy. We can also write complete procedure as a matrix-vector multiplication,

$$\mathbf{f}_{\mathcal{X}} = \mathbf{F}_{L,\mathcal{X}}\, \mathbf{C}_L\, \mathbf{B}_L\, \mathbf{A}_L\, \hat{\mathbf{f}}_L. \tag{4.9}$$

Here, the block-diagonal matrix $\mathbf{A}_L$ represents the step where we replace associated Legendre functions of high orders m with those of low orders $\hat{m}$, that is, we compute the coefficients $\hat{a}_\ell^m$ from the coefficients $\hat{f}_\ell^m$. Then, the matrix $\mathbf{B}_L$ replaces the several occurrences of Jacobi polynomials $P_\ell^{(0,0)}$, $P_\ell^{(1,1)}$, and $P_\ell^{(2,2)}$ with the Chebyshev polynomials of first kind T_ℓ. This is the step where we compute the coefficients $\hat{b}_\ell^m$ from the coefficients $\hat{a}_\ell^m$. The matrix $\mathbf{C}_L$ computes the coefficients $\hat{c}_\ell^m$ from the coefficients $\hat{b}_\ell^m$ to obtain the two-dimensional Fourier sum. Finally, the matrix $\mathbf{F}_{L,\mathcal{X}}$ denotes the application of the NFFT algorithm. In total, this is an $\mathcal{O}\big(L^2 \log L \log(1/\varepsilon) + I \log^2(1/\varepsilon)\big)$ algorithm.

Remark 4.3 It is important to notice that only the last step, that is, the application of the NFFT algorithm, actually depends on the nodes $\mathcal{X}$. Therefore, we might easily replace this step with a usual FFT if we are working with particular sampling sets $\mathcal{X}$. This, for example, is possible for points based on Clenshaw-Curtis quadrature rules, but not for those based on Gauss-Legendre quadrature rules.

4.1.7 Numerical results. We have implemented the described method on the same system that was used for the other tests; see 2.4.6. For the second step of the algorithm, we employed the NFFT 3.1 library [40] with the Kaiser-Bessel window function and the cut-off parameter $m = 8$. We compared our new method to the one described in [55]. An implementation of this method is already available in the NFFT library. For a set of $I = L^2$ randomly chosen nodes on the sphere $\mathbb{S}^2$, we computed the discrete spherical Fourier transform for a range of bandwidths L with both algorithms. The spherical Fourier coefficients $\hat{f}_\ell^m$ were chosen randomly from the square $[-1/2, 1/2] \times [1/2, 1/2]\imath$. We compared the obtained results against reference values computed by evaluating the double sum (4.4) via the respective three-term recurrences. Note that this was also done in double precision. We compared the results of both algorithms against the reference values in the relative infinity norm E_∞. We also compared the time in seconds needed to compute the transformation. The results are shown in Figure 4.2. From there it is evident that the method developed in this section offers comparable speed and an improved error behavior when compared to the method from [55]. Still, it should be noted that this represents work in progress.

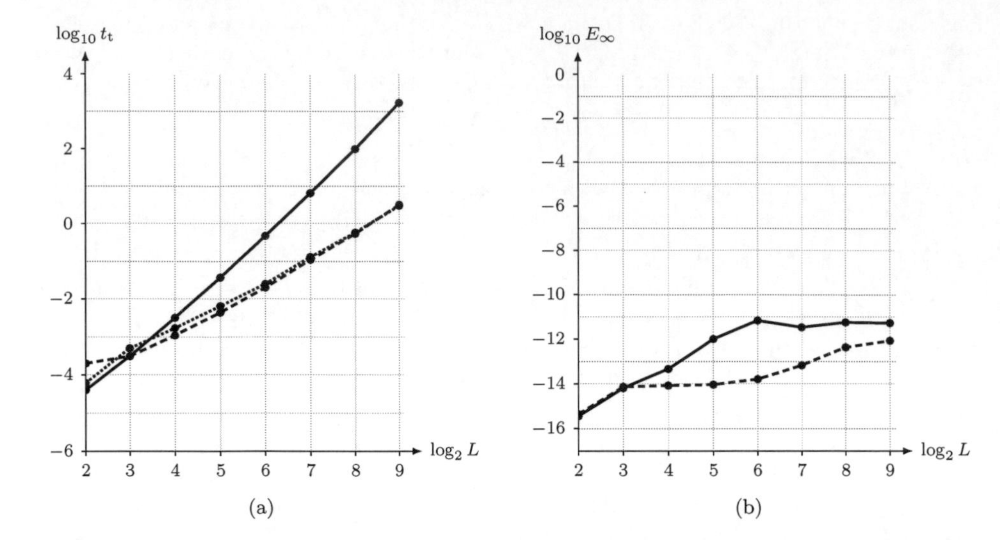

Figure 4.2: Comparison of algorithms to compute the discrete spherical Fourier transform: direct evaluation of the expansion via the three-term recurrence (solid), the method from [55] (dotted), and the method described in this section (dashed). Shown are for different degrees L and $I = L^2$ randomly chosen nodes the relative infinity error E_∞ with respect to reference values and the time in seconds needed to compute a transformation. The spherical Fourier coefficients $\hat{f}_\ell^m$ were chosen randomly.

4.2 Non-equispaced fast SO(3) Fourier transform

We have seen in the previous section how methods for the efficient conversion between expansions in different classical orthogonal polynomials or their associated functions enabled the development of an efficient algorithm for a fast Fourier transform on the sphere $\mathbb{S}^2$. This section is devoted to describing a similar algorithm for the closely related rotation group SO(3). FFT-like algorithms for the rotation group SO(3) have applications in fields like texture analysis [81], protein-protein docking [53, 69] or robot workspace generation [11]. Other related work is also found in [33]. Very similar to the spherical case, one is here interested in evaluating sums of the form

$$f(\alpha_i, \beta_i, \gamma_i) = \sum_{\ell=0}^{L} \sum_{m=-\ell}^{\ell} \sum_{m'=-\ell}^{\ell} \hat{f}_\ell^{m,m'} D_\ell^{m,m'}(\alpha_i, \beta_i, \gamma_i), \quad \text{with } i = 1, 2, \ldots, I.$$

This is a finite expansion in Wigner-D functions; they are defined in Section 4.2.2 below. The corresponding adjoint transform is the calculation of the sums

$$\tilde{f}_\ell^{m,m'} = \sum_{i=1}^{I} f(\alpha_i, \beta_i, \gamma_i) \overline{D_\ell^{m,m'}(\alpha_i, \beta_i, \gamma_i)},$$

with $\ell = 0, 1, \ldots, L$ and $m, m' = -\ell, \ldots, \ell$.

4.2.1 Existing algorithms. Apart from the approach we are going to present here, there are other algorithms for fast SO(3) Fourier transforms. Kostelec and Rockmore [52] consider an algorithm that needs $\mathcal{O}(L^4)$ arithmetic operations for a particular choice of $I = \mathcal{O}(L^3)$ points on SO(3). It is based on the Driscoll-Healy algorithm [16], although the authors do not implement the method in a way such that the theoretically attainable asymptotic cost of $\mathcal{O}(L^3 \log^2 L)$ arithmetic operations is realized. This was done in [66] where the algorithm was also generalized to arbitrary nodes using the NFFT. The result is an algorithm that needs $\mathcal{O}\big(L^3 \log^2 L + I \log(1/\varepsilon)^3\big)$ arithmetic operations.

The rest of this section is structured as follows. Section 4.2.2 covers basic material on Fourier analysis on the rotation group SO(3). In Section 4.2.3 we formally introduce discrete Fourier transforms on the rotation group SO(3) together with a convenient notation. Our new algorithm for a fast Fourier transform on the rotation group SO(3) for arbitrary nodes is developed in Section 4.2.4.

4.2.2 Fourier analysis on the rotation group SO(3). An orthogonal 3×3 matrix with unit determinant represents a rotation in $\mathbb{R}^3$. The *special orthogonal group* SO(3) is the set of all such matrices,

$$\mathrm{SO}(3) := \big\{\mathbf{R} \in \mathbb{R}^{3\times 3} : \mathbf{R}^{\mathrm{T}}\mathbf{R} = \mathbf{I},\ |\mathbf{R}| = 1\big\},$$

equipped with the usual group action, the neutral element $\mathbf{I}$, and the respective inverse elements $\mathbf{R}^{-1}$. We call the members of this group *rotations*. There are many conceivable ways to describe the group SO(3). For our purposes, we use the well-known Euler angle decomposition; see [80].

Definition 4.4 *Given angles $\alpha, \gamma \in [0, 2\pi)$, and $\beta \in [0, \pi]$, a rotation $\mathbf{R} \in \mathrm{SO}(3)$ is uniquely defined by*

$$\mathbf{R} = \mathbf{R}(\alpha, \beta, \gamma) = \mathbf{R}_Z(\alpha)\, \mathbf{R}_Y(\beta) \mathbf{R}_Z(\gamma),$$

with the Y-axis and Z-axis rotation matrices

$$\mathbf{R}_Y(\theta) := \begin{pmatrix} \cos\theta & 0 & \sin\theta \\ 0 & 1 & 0 \\ -\sin\theta & 0 & \cos\theta \end{pmatrix}, \qquad \mathbf{R}_Z(\theta) := \begin{pmatrix} \cos\theta & -\sin\theta & 0 \\ \sin\theta & \cos\theta & 0 \\ 0 & 0 & 1 \end{pmatrix}.$$

This representation is called Euler angle decomposition.

We denote by $\mathrm{L}^2\big(\mathrm{SO}(3)\big)$ the space of square integrable functions on the rotation group SO(3). Any of these functions $f : \mathrm{SO}(3) \to \mathbb{C}$ can be represented in Euler angles, hence we may write

$$f(\alpha,\beta,\gamma) := f\big(\mathbf{R}(\alpha,\beta,\gamma)\big).$$

The standard inner product in $\mathrm{L}^2\big(\mathrm{SO}(3)\big)$ is given by

$$\langle f, g\rangle = \int_{\mathrm{SO}(3)} f(\mathbf{R})\overline{g(\mathbf{R})}\,\mathrm{d}\mu(\mathbf{R}) = \int_0^{2\pi}\int_0^{\pi}\int_0^{2\pi} f(\alpha,\beta,\gamma)\overline{g(\alpha,\beta,\gamma)}\sin\beta\,\mathrm{d}\gamma\,\mathrm{d}\beta\,\mathrm{d}\alpha.$$

A convenient orthogonal basis for $\mathrm{L}^2\big(\mathrm{SO}(3)\big)$ are the Wigner-D functions $D_\ell^{m,m'}$ which can be represented in Euler angles by

$$D_\ell^{m,m'}(\alpha,\beta,\gamma) = \mathrm{e}^{-\mathrm{i}m\alpha}\, d_\ell^{m,m'}(\cos\beta)\,\mathrm{e}^{-\mathrm{i}m'\gamma},$$

where $d_n^{m,m'}$ are the *Wigner-d functions* (see below). Any function f from the space SO(3) can be developed into an infinite expansion

$$f(\alpha,\beta,\gamma) = \sum_{\ell=0}^{L}\sum_{m=-\ell}^{\ell}\sum_{m'=-\ell}^{\ell} \hat{f}_\ell^{m,m'} D_\ell^{m,m'}(\alpha,\beta,\gamma).$$

Wigner-D functions satisfy

$$\|D_\ell^{m,m'}\|^2 \langle D_\ell^{m,m'}, D_\ell^{m,m'}\rangle = \frac{8\pi^2}{2\ell+1},$$

hence the normalized Wigner-D functions $\tilde{D}_\ell^{m,m'}$ are given by

$$\tilde{D}_\ell^{m,m'} = \frac{1}{2\pi}\sqrt{\tfrac{2\ell+1}{2}}\,D_\ell^{m,m'}.$$

The Wigner-d functions are defined by

$$d_\ell^{m,m'}(x) = \omega\, 2^{-\ell^*}\sqrt{\frac{\Gamma(\ell-\ell^*+1)\Gamma(\ell+\ell^*+1)}{\Gamma(\ell-\ell^*+\mu+1)\Gamma(\ell-\ell^*+\nu+1)}}(1-x)^{\frac{\mu}{2}}(1+x)^{\frac{\nu}{2}}P_{\ell-\ell^*}^{(\mu,\nu)}(x),$$

where

$$\omega = \omega(m,m') := \begin{cases} 1, & \text{if } m > m', \\ (-1)^{\ell-m}, & \text{if } m \le m', \end{cases}$$

and

$$\mu := |m'-m|, \quad \nu := |m'+m|, \quad \ell^* := \max\{|m|,|m'|\} = \frac{\mu+\nu}{2}.$$

This shows that these are, up to the factor ω, identical to the generalized associated Legendre functions $P_\ell^{m,m'}$; see (1.48).

4.2.3 Discrete Fourier transform on the rotation group SO(3)**.** We recall that for $L \in \mathbb{N}_0$ our goal is the evaluation of the sums

$$f(\alpha_i, \beta_i, \gamma_i) = \sum_{\ell=0}^{L} \sum_{m=-\ell}^{\ell} \sum_{m'=-\ell}^{\ell} \hat{f}_\ell^{m,m'} D_\ell^{m,m'}(\alpha_i, \beta_i, \gamma_i), \quad \text{with } i = 1, 2, \ldots, I. \tag{4.10}$$

We call this the *non-equispaced discrete* SO(3) *Fourier transform (NDSOFT)*. The *adjoint NDSOFT* is accordingly defined as the calculation of the sums

$$\tilde{f}_\ell^{m,m'} = \sum_{i=1}^{I} f(\alpha_i, \beta_i, \gamma_i) \overline{\tilde{D}_\ell^{m,m'}(\alpha_i, \beta_i, \gamma_i)},$$

with $\ell = 0, 1, \ldots, L$ and $m, m' = -\ell, \ldots, \ell$. To recover the SO(3) Fourier coefficients $\hat{f}_\ell^{m,m'}$, we again need a suitable quadrature rule with weights w_i to discretize the inner products

$$\hat{f}_\ell^{m,n} = \int_0^{2\pi} \int_0^{\pi} \int_0^{2\pi} f(\alpha, \beta, \gamma) \overline{D_\ell^{m,m'}(\alpha, \beta, \gamma)} \sin\beta \, \mathrm{d}\gamma \, \mathrm{d}\beta \, \mathrm{d}\alpha.$$

Similar to the spherical case, we work with sampling sets of arbitrary nodes

$$\mathcal{X} = \big\{(\alpha_i, \beta_i, \gamma_i)\big\}_{i=1}^{I},$$

and define the index set

$$\mathcal{I}_L := \big\{(\ell, m, m') \,:\, \ell = 0, 1, \ldots, L; m, m' = -\ell, \ldots, \ell\big\}.$$

In matrix-vector notation, we can write the calculation of an NDSOFT as

$$\mathbf{f}_{\mathcal{X}} = \mathbf{D}_{L,\mathcal{X}} \,\hat{\mathbf{f}}_L$$

with

$$\begin{aligned}
\mathbf{f}_{\mathcal{X}} &:= \big(f(\alpha_i, \beta_i, \gamma_i)\big)_{i=1}^{I} \in \mathbb{C}^I, \\
\mathbf{D}_{L,\mathcal{X}} &:= \big(D_\ell^{m,m'}(\alpha_i, \beta_i, \gamma_i)\big)_{i=1,2,\ldots,I;(\ell,m,m')\in\mathcal{I}_L} \in \mathbb{C}^{I\times\frac{1}{3}(L+1)(2L+1)(2L+3)}, \\
\hat{\mathbf{f}}_{\mathbf{L}} &:= \big(\hat{f}_\ell^{m,m'}\big) \in \mathbb{C}^{\frac{1}{3}(L+1)(2L+1)(2L+3)}.
\end{aligned}$$

Then the adjoint NDSOFT reads

$$\tilde{\mathbf{f}}_L = \mathbf{D}_{L,\mathcal{X}}^{\mathrm{H}} \,\mathbf{f}_{\mathcal{X}}.$$

4.2.4 Fast Fourier transform for arbitrary nodes on SO(3)**.** The chief idea for our fast Fourier transform algorithm on the rotation group SO(3) is very similar to that for the sphere $\mathbb{S}^2$. We first rewrite (4.10) as a three-dimensional Fourier sum,

$$f(\alpha, \beta, \gamma) = \sum_{\ell=-L}^{L} \sum_{m=-L}^{L} \sum_{m'=-L}^{L} \hat{c}_\ell^{m,m'} \mathrm{e}^{-\mathrm{i}m\alpha} \mathrm{e}^{\mathrm{i}\ell\beta} \mathrm{e}^{-\mathrm{i}m'\gamma}. \tag{4.11}$$

The coefficients $\hat{c}_\ell^{m,m'}$ therein will be computed with $\mathcal{O}\big(L^3 \log L \log(1/\varepsilon)\big)$ arithmetic operations. Then we can use the NFFT to evaluate the obtained expansion with another $\mathcal{O}\big(L^3 \log L + I \log(1/\varepsilon)^3\big)$ arithmetic operations. In total, this is an $\mathcal{O}\big(L^3 \log L \log(1/\varepsilon) + I \log(1/\varepsilon)^3\big)$ algorithm to calculate the whole transformation.

Equation (4.10) can be rearranged to

$$\begin{aligned} f(\alpha,\beta,\gamma) &= \sum_{\ell=0}^{L}\sum_{m=-\ell}^{\ell}\sum_{m'=-\ell}^{\ell} \hat{f}_\ell^{m,m'} D_\ell^{m,m'}(\alpha,\beta,\gamma) \\ &= \sum_{m=-L}^{L}\sum_{m'=-L}^{L}\sum_{\ell=\ell^*}^{L} \hat{f}_\ell^{m,m'} D_\ell^{m,m'}(\alpha,\beta,\gamma) \\ &= \sum_{m=-L}^{L}\sum_{m'=-L}^{L} \mathrm{e}^{-\mathrm{i}m\alpha}\,\mathrm{e}^{-\mathrm{i}m'\gamma} \sum_{\ell=l^*}^{L} \hat{f}_\ell^{m,m'} d_\ell^{m,m'}(\cos\beta) \\ &= \sum_{m=-L}^{L}\sum_{m'=-L}^{L} \mathrm{e}^{-\mathrm{i}m\alpha}\,\mathrm{e}^{-\mathrm{i}m'\gamma}\, h_{m,m'}(x), \end{aligned}$$

where

$$h_{m,m'}(x) = \sum_{\ell=l^*}^{L} \hat{f}_\ell^{m,m'} d_\ell^{m,m'}(x).$$

The first step is to replace the rearranged expansion by an ordinary three-dimensional Fourier sum. Similar to the spherical case, we apply to each individual sum $h_{m,m'}(x)$ a number of transformations. We therefore assume now that a pair of orders m,m' has been fixed. First, we want to replace the Wigner-d functions $d_\ell^{m,m'}$ by those of low orders $d_\ell^{\hat{m},\hat{m}'}$. Lemma 1.94 motivates us to define

$$\hat{m} = \hat{m}(m,m') := \begin{cases} 0, & \text{if } m = m' = 0, \\ 1, & \text{if } m = \pm m, \\ 0, & \text{else,} \end{cases} \qquad \hat{m}' := \begin{cases} 0, & \text{if } m = m' = 0, \\ \pm 1, & \text{if } m = \pm n, \\ 2, & \text{if } m + m' \text{ even,} \\ 1, & \text{if } m + m' \text{ odd.} \end{cases} \tag{4.12}$$

With the help of the divide-and-conquer algorithm from Section 2.5 and the results found in Section 2.6, we can write the sum $h_{m,m'}(x)$ as

$$h_{m,m'}(x) = \sum_{\ell=\hat{l}^*}^{L} \hat{a}_\ell^{m,m'} d_\ell^{\hat{m},\hat{m}'}(x).$$

For the computation of the coefficients $\hat{a}_\ell^{m,m'}$ from the known coefficients $\hat{f}_\ell^{m,m'}$ we need $\mathcal{O}\big(L\log L\log(1/\varepsilon)\big)$ arithmetic operations, where ε is the desired accuracy. In total, we need $\mathcal{O}\big(L^3\log L\log(1/\varepsilon)\big)$ arithmetic operations to calculate all coefficients $\hat{a}_\ell^{m,m'}$ for all orders m and m'.
It remains to manipulate the obtained sum $h_{m,m'}(x)$ until we have obtained an ordinary Fourier sum. To this end, we must distinguish five different cases with respect to the orders $\hat{m}$ and $\hat{m}'$. Three of them, namely the cases $(\hat{m},\hat{m}') = (0,0),(0,1),(0,2)$, are entirely equivalent to the three cases that have been described for the fast spherical Fourier transform; see Section 4.1.6. Therefore, we only describe the remaining two cases here, which might be succinctly handled together as follows.

The case $\hat{m} = 1$, $\hat{m}' = \pm 1$

This case implies that $m = \pm m'$. After the first step, we obtain the sum

$$h_m(x) = \sum_{\ell=1}^{L} \hat{a}_\ell^{m,m'} P_\ell^{1,\pm 1}(x) = (1 \pm x)\sum_{\ell=0}^{L-1} \hat{a}_{\ell+1}^{m,m'}\frac{1}{2}P_\ell^{(1\mp 1,1\pm 1)}(x).$$

Now, we replace this by a linear combination of Chebyshev polynomials of first kind

$$h_m(x) = (1+x)\sum_{\ell=0}^{L-1} \hat{b}_\ell^{m,m'} T_\ell(x),$$

and with

$$\begin{aligned}(1 \pm x)T_\ell(x) &= \pm \frac{1+\delta_{0,\ell}}{2} T_{\ell+1}(\cos\beta) + T_\ell(\cos\beta) \pm \frac{1-\delta_{0,\ell}}{2} T_{\ell-1}(\cos\beta)\\ &= \pm \frac{1+\delta_{0,\ell}}{2}\cos((\ell+1)\beta) + \cos(\ell\beta) \pm \frac{1-\delta_{0,\ell}}{2}\cos((\ell-1)\beta)\\ &= \pm \frac{1+\delta_{0,\ell}}{4}(\mathrm{e}^{(\ell+1)\beta} + \mathrm{e}^{-(\ell+1)\beta}) + \frac{1}{2}(\mathrm{e}^{\ell\beta} + \mathrm{e}^{-\ell\beta})\\ &\quad \pm \frac{1-\delta_{0,\ell}}{4}(\mathrm{e}^{(\ell-1)\beta} + \mathrm{e}^{-(\ell-1)\beta}).\end{aligned}$$

we compute the coefficients

$$\hat{c}_\ell^m = \begin{cases} \frac{1}{2}(2\hat{b}_0^{m,m'} \pm \hat{b}_1^{m,m'}), & \text{if } \ell = 0,\\ \frac{1}{4}(\pm 2\hat{b}_0^{m,m'} + 2\hat{b}_1^{m,m'} \pm \hat{b}_2^{m,m'}), & \text{if } |\ell| = 1,\\ \frac{1}{4}(\pm \hat{b}_{|\ell|-1}^{m,m'} + 2\hat{b}_{|\ell|}^{m,m'} \pm \hat{b}_{|\ell|+1}^{m,m'}), & \text{if } 1 < |\ell| < L-1,\\ \frac{1}{4}(\pm \hat{b}_{L-2}^{m,m'} + 2\hat{b}_{L-1}^{m,m'}), & \text{if } |\ell| = L-1,\\ \pm\frac{1}{4}\hat{b}_{L-1}^{m,m'}, & \text{if } |\ell| = L, \end{cases}$$

in the Fourier sum

$$h_m(x) = \sum_{\ell=-L}^{L} \hat{c}_\ell^{m,m'} \mathrm{e}^{\mathrm{i}\ell\gamma}$$

with $\mathcal{O}(L)$ arithmetic operations. After these transformations, we obtain the ordinary three-dimensional Fourier sum

$$f(\alpha,\beta,\gamma) = \sum_{m=-L}^{L}\sum_{m'=-L}^{L}\sum_{\ell=\ell^*}^{L} \hat{c}_\ell^{m,m'} \mathrm{e}^{-\mathrm{i}m\alpha}\mathrm{e}^{\mathrm{i}\ell\beta}\mathrm{e}^{-\mathrm{i}m'\gamma}. \tag{4.13}$$

This triple sum can be evaluated with the NFFT algorithm on an arbitrary sampling set $\mathcal{X}$ with I nodes using $\mathcal{O}(L^3 \log L + I \log^3(1/\varepsilon))$ arithmetic operations, where ε is the desired accuracy. We can also write complete procedure as a matrix-vector multiplication,

$$\mathbf{f}_{\mathcal{X}} = \mathbf{F}_{L,\mathcal{X}}\, \mathbf{C}_L\, \mathbf{B}_L\, \mathbf{A}_L\, \hat{\mathbf{f}}_L. \tag{4.14}$$

Here, the block-diagonal matrix $\mathbf{A}_L$ represents the step where we replace generalized associated Legendre functions of high orders m and m' with those of low orders $\hat{m}$ and $\hat{m}'$, that is, we compute the coefficients $\hat{a}_\ell^{m,m'}$ from the coefficients $\hat{f}_\ell^{m,m'}$. Then, the matrix $\mathbf{B}_L$ replaces the several occurrences of Jacobi polynomials $P_\ell^{(0,0)}$, $P_\ell^{(1,1)}$, $P_\ell^{(2,2)}$, $P_\ell^{(0,2)}$, and $P_\ell^{(2,0)}$ with the Chebyshev polynomials of first kind T_ℓ. This is the step where we compute the coefficients $\hat{b}_\ell^{m,m'}$ from the coefficients $\hat{a}_\ell^{m.m'}$. The matrix $\mathbf{C}_L$ computes the coefficients $\hat{c}_\ell^{m,m'}$ from the coefficients $\hat{b}_\ell^{m,m'}$ to obtain the three-dimensional Fourier sum (4.13). Finally, the matrix $\mathbf{F}_{L,\mathcal{X}}$ denotes the application of the NFFT algorithm. In total, this is an $\mathcal{O}(L^3 \log L \log(1/\varepsilon) + I \log^3(1/\varepsilon))$ algorithm.

The implementation of the described method is work in progress, but first tests indicate that it does not suffer from some negative effects that were observed in [66] for a competing method based on a different algorithm. Future work includes more detailed investigations about this subject.

Appendix A – Formula reference

A.1 Classical orthogonal polynomials

A.1.1 Laguerre polynomials.

Symbol

$$L_n^{(\alpha)} : \mathbb{R} \to \mathbb{R}, \quad n \in \mathbb{N}_0, \quad -1 < \alpha.$$

Inner product and weight function

$$\langle f, g \rangle = \int_I f(x) g(x)\, w(x)\, \mathrm{d}x, \quad I = [-1, 1], \quad w(x) = x^\alpha \mathrm{e}^{-x}.$$

Differential equation

$$\sigma(x) y''(x) + \tau(x) y'(x) + \lambda_n y(x) = 0, \quad y = L_n^{(\alpha)},$$
$$\sigma(x) = x, \quad \tau(x) = -x + \alpha + 1, \quad \lambda_n = n.$$

Differential operator

$$\mathcal{D} = -x \frac{\mathrm{d}^2}{\mathrm{d}x^2} + (x - \alpha - 1) \frac{\mathrm{d}}{\mathrm{d}x}.$$

Rodrigues formula

$$L_n^{(\alpha)}(x) = \frac{x^{-\alpha} \mathrm{e}^x}{\Gamma(n+1)} \frac{\mathrm{d}^n}{\mathrm{d}x^n} \left(x^{n+\alpha} \mathrm{e}^{-x} \right),$$

$$\frac{\mathrm{d}^m}{\mathrm{d}x^m} L_n^{(\alpha)}(x) = (-1)^m \frac{x^{-(\alpha+m)} \mathrm{e}^x}{\Gamma(n-m+1)} \frac{\mathrm{d}^{n-m}}{\mathrm{d}x^{n-m}} \left(x^{n+\alpha} \mathrm{e}^{-x} \right) = (-1)^m L_{n-m}^{(\alpha+m)}(x).$$

Leading coefficients

$$k_n = \frac{(-1)^n}{\Gamma(n+1)}, \qquad \bar{k}_n = 1, \qquad \tilde{k}_n = \frac{(-1)^n}{\sqrt{\Gamma(n+1)\Gamma(n+\alpha+1)}}.$$

Squared norms

$$h_n = \frac{\Gamma(n+\alpha+1)}{\Gamma(n+1)}, \qquad \bar{h}_n = \Gamma(n+1)\Gamma(n+\alpha+1), \qquad \tilde{h}_n = 1.$$

Three-term recurrence

$$a_n = -\frac{1}{n+1}, \qquad a'_n = -n-1,$$
$$b_n = -\frac{2n+\alpha+1}{n+1}, \qquad b'_n = 2n+\alpha+1,$$
$$c_n = \frac{n+\alpha}{n+1}, \qquad c'_n = -n-\alpha,$$

$$\bar{a}_n = 1, \qquad \bar{a}'_n = 1,$$
$$\bar{b}_n = 2n+\alpha+1, \qquad \bar{b}'_n = 2n+\alpha+1,$$
$$\bar{c}_n = n(n+\alpha), \qquad \bar{c}'_n = n(n+\alpha),$$

$$\tilde{a}_n = -\frac{1}{\sqrt{(n+1)(n+\alpha+1)}}, \qquad \tilde{a}'_n = -\sqrt{(n+1)(n+\alpha+1)},$$
$$\tilde{b}_n = -\frac{2n+\alpha+1}{\sqrt{(n+1)(n+\alpha+1)}}, \qquad \tilde{b}'_n = \quad 2n+\alpha+1,$$
$$\tilde{c}_n = \quad \sqrt{\frac{n}{n+1}\frac{n+\alpha}{n+\alpha+1}}, \qquad \tilde{c}'_n = -\sqrt{n(n+\alpha)}.$$

Derivative identities

$$\frac{\mathrm{d}}{\mathrm{d}x}L_n^{(\alpha)}(x) = -L_{n-1}^{(\alpha+1)}(x),$$
$$\frac{\mathrm{d}^m}{\mathrm{d}x^m}L_n^{(\alpha)}(x) = (-1)^m L_{n-m}^{(\alpha+m)}(x),$$
$$\frac{\mathrm{d}}{\mathrm{d}x}L_n^{(\alpha)}(x) = A_n \sum_{j=0}^{n-1} B_j L_n^{(\alpha)}(x), \quad A_n = -1, \quad B_j = 1,$$

$$\frac{\mathrm{d}}{\mathrm{d}x}\bar{L}_n^{(\alpha)}(x) = n\bar{L}_{n-1}^{(\alpha+1)}(x),$$
$$\frac{\mathrm{d}^m}{\mathrm{d}x^m}\bar{L}_n^{(\alpha)}(x) = \frac{\Gamma(n+1)}{\Gamma(n-m+1)}\bar{L}_{n-m}^{(\alpha+m)}(x),$$
$$\frac{\mathrm{d}}{\mathrm{d}x}\bar{L}_n^{(\alpha)}(x) = \bar{A}_n \sum_{j=0}^{n-1} \bar{B}_j \bar{L}_j^{(\alpha)}(x), \quad \bar{A}_n = (-1)^{n+1}\Gamma(n+1), \quad \bar{B}_j = \frac{(-1)^j}{\Gamma(j+1)},$$

$$\frac{\mathrm{d}}{\mathrm{d}x}\tilde{L}_n^{(\alpha)}(x) = -\sqrt{n}\tilde{L}_{n-1}^{(\alpha+1)}(x),$$
$$\frac{\mathrm{d}^m}{\mathrm{d}x^m}\tilde{L}_n^{(\alpha)}(x) = (-1)^m\sqrt{\frac{\Gamma(n+1)}{\Gamma(n-m+1)}}\tilde{L}_{n-m}^{(\alpha+m)}(x),$$
$$\frac{\mathrm{d}}{\mathrm{d}x}\tilde{L}_n^{(\alpha)}(x) = \tilde{A}_n \sum_{j=0}^{n-1} \tilde{B}_j \tilde{L}_j^{(\alpha)}(x), \quad \tilde{A}_n = -\sqrt{\frac{\Gamma(n+1)}{\Gamma(n+\alpha+1)}}, \quad \tilde{B}_j = \sqrt{\frac{\Gamma(j+\alpha+1)}{\Gamma(j+1)}}.$$

Connection coefficients

$\{L_n^{(\alpha)}\}_{n\in\mathbb{N}_0} \to \{L_n^{(\beta)}\}_{n\in\mathbb{N}_0}$:

$$\kappa_{i,j} = \frac{1}{\Gamma(\alpha-\beta)}\frac{\Gamma(j-i+\alpha-\beta)}{\Gamma(j-i+1)},$$
$$\bar{\kappa}_{i,j} = \frac{(-1)^{i+j}}{\Gamma(\alpha-\beta)}\frac{\Gamma(j+1)}{\Gamma(i+1)}\frac{\Gamma(j-i+\alpha-\beta)}{\Gamma(j-i+1)},$$
$$\tilde{\kappa}_{i,j} = \frac{(-1)^{i+j}}{\Gamma(\alpha-\beta)}\sqrt{\frac{\Gamma(j+1)}{\Gamma(i+1)}\frac{\Gamma(i+\beta+1)}{\Gamma(j+\alpha+1)}}\frac{\Gamma(j-i+\alpha-\beta)}{\Gamma(j-i+1)}.$$

$\{L_n^{(\alpha)}\}_{n\in\mathbb{N}_0} \to \{L_n^{(\beta)}\}_{n\in\mathbb{N}_0}$, $\beta-\alpha\in\mathbb{N}_0$:

$$\kappa_{i,j} = \begin{cases} (-1)^{i+j}\binom{\beta-\alpha}{j-i}, & \text{if } i\le j\le i+\beta-\alpha,\\ 0, & \text{else,}\end{cases}$$

$$\bar\kappa_{i,j} = \begin{cases} \frac{\Gamma(j+1)}{\Gamma(i+1)}\binom{\beta-\alpha}{j-i}, & \text{if } i\le j\le i+\beta-\alpha,\\ 0, & \text{else,}\end{cases}$$

$$\tilde\kappa_{i,j} = \begin{cases} \sqrt{\frac{\Gamma(j+1)}{\Gamma(i+1)}\frac{\Gamma(i+\beta+1)}{\Gamma(j+\alpha+1)}}\binom{\beta-\alpha}{j-i}, & \text{if } i\le j\le i+\beta-\alpha,\\ 0, & \text{else.}\end{cases}$$

$\{L_n^{(\alpha)}\}_{n\in\mathbb{N}_0} \to \{L_n^{(\alpha+1)}\}_{n\in\mathbb{N}_0}$:

$$\kappa_{i,j} = \begin{cases} 1, & \text{if } j=i,\\ (-1)^{i+j}, & \text{if } j=i+1,\\ 0, & \text{else,}\end{cases}$$

$$\bar\kappa_{i,j} = \begin{cases} 1 & , \text{ if } j=i,\\ j & , \text{ if } j=i+1,\\ 0 & , \text{ else,}\end{cases}$$

$$\tilde\kappa_{i,j} = \begin{cases} \sqrt{j+\alpha+1}, & \text{if } j=i,\\ j\sqrt{\frac{\Gamma(i+1)}{\Gamma(j+1)}}, & \text{if } j=i+1,\\ 0, & \text{else.}\end{cases}$$

$\{L_n^{(\alpha)}\}_{n\in\mathbb{N}_0} \to \{L_n^{(\alpha-1)}\}_{n\in\mathbb{N}_0}$:

$$\kappa_{i,j}=1, \qquad \bar\kappa_{i,j} = (-1)^{i+j}\frac{\Gamma(j+1)}{\Gamma(i+1)}, \qquad \tilde\kappa_{i,j} = (-1)^{i+j}\sqrt{\frac{\Gamma(j+1)}{\Gamma(i+1)}\frac{\Gamma(i+\alpha)}{\Gamma(j+\alpha+1)}}.$$

The matrix $\mathbf{G}$

$\{L_n^{(\alpha)}\}_{n=0}^N \to \{L_n^{(\beta)}\}_{n=0}^N$:

$$\mathbf{G} = \operatorname{diag}(\mathbf{d}) + (\alpha-\beta)\operatorname{triu}(\mathbf{u}\,\mathbf{v}^{\mathrm{T}},1) \in \mathbb{R}^{(N+1)\times(N+1)},$$

$$\begin{aligned} \mathbf{d} &= (d_j)_{j=0}^N, & d_j &= j,\\ \mathbf{u} &= (u_j)_{j=0}^N, & u_j &= 1,\\ \mathbf{v} &= (v_j)_{j=0}^N, & v_j &= 1. \end{aligned}$$

$$\bar{\mathbf{G}} = \operatorname{diag}(\bar{\mathbf{d}}) + (\alpha - \beta)\operatorname{triu}(\bar{\mathbf{u}}\,\bar{\mathbf{v}}^{\mathrm{T}}, 1) \in \mathbb{R}^{(N+1)\times(N+1)},$$

$$\begin{aligned}
\bar{\mathbf{d}} &= (\bar{d}_j)_{j=0}^{N}, & \bar{d}_j &= j,\\
\bar{\mathbf{u}} &= (\bar{u}_j)_{j=0}^{N}, & \bar{u}_j &= (-1)^j \frac{1}{\Gamma(j+1)},\\
\bar{\mathbf{v}} &= (\bar{v}_j)_{j=0}^{N}, & \bar{v}_j &= (-1)^j \Gamma(j+1).
\end{aligned}$$

$$\tilde{\mathbf{G}} = \operatorname{diag}(\tilde{\mathbf{d}}) + (\alpha - \beta)\operatorname{triu}(\tilde{\mathbf{u}}\,\tilde{\mathbf{v}}^{\mathrm{T}}, 1) \in \mathbb{R}^{(N+1)\times(N+1)},$$

$$\begin{aligned}
\tilde{\mathbf{d}} &= (\tilde{d}_j)_{j=0}^{N}, & \tilde{d}_j &= j,\\
\tilde{\mathbf{u}} &= (\tilde{u}_j)_{j=0}^{N}, & \tilde{u}_j &= \sqrt{\frac{\Gamma(j+\beta+1)}{\Gamma(i+1)}},\\
\tilde{\mathbf{v}} &= (\tilde{v}_j)_{j=0}^{N}, & \tilde{v}_j &= \sqrt{\frac{\Gamma(j+1)}{\Gamma(j+\beta+1)}}.
\end{aligned}$$

A.1.2 Jacobi polynomials.

Symbol

$$P_n^{(\alpha,\beta)} : \mathbb{R} \to \mathbb{R}, \quad n \in \mathbb{N}_0, \quad -1 < \alpha, \beta.$$

Inner product and weight function

$$\langle f, g\rangle = \int_I f(x)g(x)\, w(x)\, \mathrm{d}x, \quad I = [-1,1], \quad w(x) = (1-x)^\alpha (1+x)^\beta.$$

Differential equation

$$\sigma(x)y''(x) + \tau(x)y'(x) + \lambda_n y(x) = 0, \quad y = P_n^{(\alpha,\beta)},$$
$$\sigma(x) = 1 - x^2, \quad \tau(x) = -(\alpha+\beta+2)x + \beta - \alpha, \quad \lambda_n = n(n+\alpha+\beta+1).$$

Differential operator

$$\mathcal{D} = -(1-x^2)\frac{\mathrm{d}^2}{\mathrm{d}x^2} + \big((\alpha+\beta+2)x + \beta - \alpha\big)\frac{\mathrm{d}}{\mathrm{d}x}.$$

Rodrigues formula

$$P_n^{(\alpha,\beta)}(x) = \frac{(-1)^n}{2^n\Gamma(n+1)}(1-x)^{-\alpha}(1+x)^{-\beta}\frac{\mathrm{d}^n}{\mathrm{d}x^n}\left((1-x)^{n+\alpha}(1+x)^{n+\beta}\right),$$

$$\begin{aligned}\frac{\mathrm{d}^m}{\mathrm{d}x^m}P_n^{(\alpha,\beta)}(x) &= \frac{(-1)^{n-m}\Gamma(n+m+\alpha+\beta+1)}{2^n\Gamma(n+\alpha+\beta+1)\Gamma(n-m+1)}(1-x)^{-(\alpha+m)}(1+x)^{-(\beta+m)}\\ &\quad\times\frac{\mathrm{d}^{n-m}}{\mathrm{d}x^{n-m}}\left((1-x)^{n+\alpha}(1+x)^{n+\beta}\right)\\ &= \frac{\Gamma(n+m+\alpha+\beta+1)}{2^m\Gamma(n+\alpha+\beta+1)}P_{n-m}^{(\alpha+m,\beta+m)}(x).\end{aligned}$$

Leading coefficients

$$\begin{aligned}k_n &= \frac{\Gamma(2n+\alpha+\beta+1)}{2^n\Gamma(n+1)\Gamma(n+\alpha+\beta+1)},\\ \bar{k}_n &= 1,\\ \tilde{k}_n &= \frac{\Gamma(2n+\alpha+\beta+1)}{2^{n+(\alpha+\beta+1)/2}}\sqrt{\frac{(2n+\alpha+\beta+1)}{\Gamma(n+1)\Gamma(n+\alpha+1)\Gamma(n+\beta+1)\Gamma(n+\alpha+\beta+1)}}.\end{aligned}$$

Squared norms

$$\begin{aligned}h_n &= \frac{2^{\alpha+\beta+1}\Gamma(n+\alpha+1)\Gamma(n+\beta+1)}{(2n+\alpha+\beta+1)\Gamma(n+1)\Gamma(n+\alpha+\beta+1)},\\ \bar{h}_n &= 2^{2n+\alpha+\beta+1}\frac{\Gamma(n+1)\Gamma(n+\alpha+1)\Gamma(n+\beta+1)\Gamma(n+\alpha+\beta+1)}{\Gamma(2n+\alpha+\beta+1)\Gamma(2n+\alpha+\beta+2)},\\ \tilde{h}_n &= 1.\end{aligned}$$

Three-term recurrence

$$a_n = \frac{(2n+\alpha+\beta+1)(2n+\alpha+\beta+2)}{2(n+1)(n+\alpha+\beta+1)}, \qquad a'_n = \frac{2(n+1)(n+\alpha+\beta+1)}{(2n+\alpha+\beta+1)_2},$$

$$b_n = \frac{(\beta^2-\alpha^2)(2n+\alpha+\beta+1)}{2(n+1)(n+\alpha+\beta+1)(2n+\alpha+\beta)}, \qquad b'_n = \frac{\beta^2-\alpha^2}{(2n+\alpha+\beta)(2n+\alpha+\beta+2)},$$

$$c_n = \frac{(n+\alpha)(n+\beta)(2n+\alpha+\beta+2)}{(n+1)(n+\alpha+\beta+1)(2n+\alpha+\beta)}, \qquad c'_n = \frac{2(n+\alpha)(n+\beta)}{(2n+\alpha+\beta)(2n+\alpha+\beta+1)},$$

$$\bar{a}_n = \bar{a}'_n = 1,$$

$$\bar{b}_n = \bar{b}'_n = \frac{\beta^2-\alpha^2}{(2n+\alpha+\beta)(2n+\alpha+\beta+2)},$$

$$\bar{c}_n = \bar{c}'_n = \frac{4n(n+\alpha+\beta)(n+\alpha)(n+\beta)}{(2n+\alpha+\beta-1)(2n+\alpha+\beta)(2n+\alpha+\beta)(2n+\alpha+\beta+1)},$$

$$\tilde{a}_n = \frac{(2n+\alpha+\beta+2)}{2}\sqrt{\frac{(2n+\alpha+\beta+1)(2n+\alpha+\beta+3)}{(n+1)(n+\alpha+\beta+1)(n+\alpha+1)(n+\beta+1)}},$$

$$\tilde{b}_n = \frac{\beta^2-\alpha^2}{2(2n+\alpha+\beta)}\sqrt{\frac{(2n+\alpha+\beta+1)(2n+\alpha+\beta+3)}{(n+1)(n+\alpha+\beta+1)(n+\alpha+1)(n+\beta+1)}},$$

$$\tilde{c}_n = \frac{2n+\alpha+\beta+2}{2n+\alpha+\beta}\sqrt{\frac{n(n+\alpha)(n+\beta)(n+\alpha+\beta)(2n+\alpha+\beta+3)}{(n+1)(n+\alpha+1)(n+\beta+1)(n+\alpha+\beta+1)(2n+\alpha+\beta-1)}},$$

$$\tilde{a}'_n = \frac{2}{(2n+\alpha+\beta+2)}\sqrt{\frac{(n+1)(n+\alpha+\beta+1)(n+\alpha+1)(n+\beta+1)}{(2n+\alpha+\beta+1)(2n+\alpha+\beta+3)}},$$

$$\tilde{b}'_n = \frac{\beta^2-\alpha^2}{(2n+\alpha+\beta)(2n+\alpha+\beta+2)},$$

$$\tilde{c}'_n = \frac{2}{2n+\alpha+\beta}\sqrt{\frac{n(n+\alpha)(n+\beta)(n+\alpha+\beta)}{(2n+\alpha+\beta-1)(n+\alpha+\beta+1)}}.$$

Derivative identities

$$\frac{\mathrm{d}}{\mathrm{d}x}P_n^{(\alpha,\beta)}(x) = \frac{n+\alpha+\beta+1}{2}P_{n-1}^{(\alpha+1,\beta+1)}(x),$$

$$\frac{\mathrm{d}^m}{\mathrm{d}x^m}P_n^{(\alpha,\beta)}(x) = \frac{\Gamma(n+m+\alpha+\beta+1)}{2^m\Gamma(n+\alpha+\beta+1)}P_{n-m}^{(\alpha+m,\beta+m)}(x),$$

$$\frac{\mathrm{d}}{\mathrm{d}x}P_n^{(\alpha,\beta)}(x) = A_n\sum_{j=0}^{n-1}B_jP_j^{(\alpha,\beta)}(x) + C_n\sum_{j=0}^{n-1}D_jP_j^{(\alpha,\beta)}(x),$$

$$A_n = (-1)^{n+1}\frac{\Gamma(n+\alpha+1)}{2\Gamma(n+\alpha+\beta+1)}, \qquad B_j = (-1)^j(2j+\alpha+\beta+1)\frac{\Gamma(j+\alpha+\beta+1)}{\Gamma(j+\alpha+1)},$$

$$C_n = \frac{\Gamma(n+\beta+1)}{2\Gamma(n+\alpha+\beta+1)}, \qquad D_j = (2j+\alpha+\beta+1)\frac{\Gamma(j+\alpha+\beta+1)}{\Gamma(j+\beta+1)},$$

$$\frac{\mathrm{d}}{\mathrm{d}x}\bar{P}_n^{(\alpha,\beta)}(x) = n\bar{P}_{n-1}^{(\alpha+1,\beta+1)}(x),$$

$$\frac{\mathrm{d}^m}{\mathrm{d}x^m}\bar{P}_n^{(\alpha,\beta)}(x) = \frac{\Gamma(n+1)}{\Gamma(n-m+1)}\bar{P}_{n-m}^{(\alpha+m,\beta+m)}(x),$$

$$\frac{\mathrm{d}}{\mathrm{d}x}\bar{P}_n^{(\alpha,\beta)}(x) = \bar{A}_n\sum_{j=0}^{n-1}\bar{B}_j\bar{P}_j^{(\alpha,\beta)}(x) + \bar{C}_n\sum_{j=0}^{n-1}\bar{D}_j\bar{P}_j^{(\alpha,\beta)}(x),$$

$$\bar{A}_n = (-1)^{n+1}\frac{2^{n-1}\Gamma(n+1)\Gamma(n+\alpha+1)}{\Gamma(2n+\alpha+\beta+1)}, \qquad \bar{B}_j = (-1)^j\frac{2^j\Gamma(2j+\alpha+\beta+1)}{\Gamma(j+1)\Gamma(j+\alpha+1)},$$

$$\bar{C}_n = \frac{2^{n-1}\Gamma(n+1)\Gamma(n+\beta+1)}{\Gamma(2n+\alpha+\beta+1)}, \qquad \bar{D}_j = \frac{2^j\Gamma(2j+\alpha+\beta+1)}{\Gamma(j+1)\Gamma(j+\beta+1)},$$

$$\frac{\mathrm{d}}{\mathrm{d}x}\tilde{P}_n^{(\alpha,\beta)}(x) = \sqrt{n(n+\alpha+\beta+1)}\tilde{P}_{n-1}^{(\alpha+1,\beta+1)}(x),$$

$$\frac{\mathrm{d}^m}{\mathrm{d}x^m}\tilde{P}_n^{(\alpha,\beta)}(x) = \sqrt{\frac{\Gamma(n+1)\Gamma(n+m+\alpha+\beta+1)}{\Gamma(n-m+1)\Gamma(n+\alpha+\beta+1)}}\tilde{P}_{n-m}^{(\alpha+m,\beta+m)}(x),$$

$$\frac{\mathrm{d}}{\mathrm{d}x}\tilde{P}_n^{(\alpha,\beta)}(x) = \tilde{A}_n\sum_{j=0}^{n-1}\tilde{B}_j\tilde{P}_j^{(\alpha,\beta)}(x) + \tilde{C}_n\sum_{j=0}^{n-1}\tilde{D}_j\tilde{P}_j^{(\alpha,\beta)}(x),$$

$$\tilde{A}_n = (-1)^{n+1}\frac{1}{2}\sqrt{2n+\alpha+\beta+1}\tilde{E}_n\tilde{F}_n, \qquad \tilde{B}_j = (-1)^j\sqrt{2j+\alpha+\beta+1}\tilde{E}_j^{-1}\tilde{F}_j^{-1},$$

$$\tilde{C}_n = \frac{1}{2}\sqrt{2n+\alpha+\beta+1}\tilde{E}_n\tilde{F}_n^{-1}, \qquad \tilde{D}_j = \sqrt{2j+\alpha+\beta+1}\tilde{E}_j^{-1}\tilde{F}_j,$$

$$\tilde{E}_n = \sqrt{\frac{\Gamma(n+1)}{\Gamma(n+\alpha+\beta+1)}}, \qquad \tilde{F}_n = \sqrt{\frac{\Gamma(n+\alpha+1)}{\Gamma(n+\beta+1)}}.$$

Connection coefficients

$\{P_n^{(\alpha,\beta)}\}_{n\in\mathbb{N}_0} \to \{P_n^{(\gamma,\beta)}\}_{n\in\mathbb{N}_0}$:

$$\kappa_{i,j} = \frac{(2i+\gamma+\beta+1)}{\Gamma(\alpha-\gamma)}\frac{\Gamma(j+\beta+1)}{\Gamma(i+\beta+1)}\frac{\Gamma(i+\gamma+\beta+1)}{\Gamma(j+\alpha+\beta+1)}\frac{\Gamma(j-i+\alpha-\gamma)}{\Gamma(j-i+1)} \times\frac{\Gamma(j+i+\alpha+\beta+1)}{\Gamma(j+i+\gamma+\beta+2)},$$

$$\bar{\kappa}_{i,j} = \frac{2^{j-i}}{\Gamma(\alpha-\gamma)}\frac{\Gamma(j+1)}{\Gamma(i+1)}\frac{\Gamma(j+\beta+1)}{\Gamma(i+\beta+1)}\frac{\Gamma(2i+\gamma+\beta+2)}{\Gamma(2j+\alpha+\beta+1)}\frac{\Gamma(j-i+\alpha-\gamma)}{\Gamma(j-i+1)} \times\frac{\Gamma(j+i+\alpha+\beta+1)}{\Gamma(j+i+\gamma+\beta+2)},$$

$$\tilde{\kappa}_{i,j} = \frac{2^{(\gamma-\alpha)/2}}{\Gamma(\alpha-\gamma)}\frac{\Gamma(2i+\gamma+\beta+2)}{\Gamma(2i+\gamma+\beta+1)}\frac{\Gamma(j-i+\alpha-\gamma)}{\Gamma(j-i+1)}\frac{\Gamma(j+i+\alpha+\beta+1)}{\Gamma(j+i+\gamma+\beta+2)} \times\sqrt{\frac{2j+\alpha+\beta+1}{2i+\gamma+\beta+1}\frac{\Gamma(j+1)}{\Gamma(i+1)}\frac{\Gamma(j+\beta+1)}{\Gamma(i+\beta+1)}\frac{\Gamma(i+\gamma+1)}{\Gamma(j+\alpha+1)}\frac{\Gamma(i+\gamma+\beta+1)}{\Gamma(j+\alpha+\beta+1)}}.$$

$\{P_n^{(\alpha,\beta)}\}_{n\in\mathbb{N}_0} \to \{P_n^{(\alpha,\delta)}\}_{n\in\mathbb{N}_0}$:

$$\kappa_{i,j} = (-1)^{i+j}\frac{(2i+\alpha+\delta+1)}{\Gamma(\beta-\delta)}\frac{\Gamma(j+\alpha+1)}{\Gamma(i+\alpha+1)}\frac{\Gamma(i+\alpha+\delta+1)}{\Gamma(j+\alpha+\beta+1)}\frac{\Gamma(j-i+\beta-\delta)}{\Gamma(j-i+1)} \times \frac{\Gamma(i+j+\alpha+\beta+1)}{\Gamma(i+j+\alpha+\delta+2)},$$

$$\bar{\kappa}_{i,j} = (-1)^{i+j}\frac{2^{j-i}}{\Gamma(\beta-\delta)}\frac{\Gamma(j+1)}{\Gamma(i+1)}\frac{\Gamma(j+\alpha+1)}{\Gamma(i+\alpha+1)}\frac{\Gamma(2i+\alpha+\delta+2)}{\Gamma(2j+\alpha+\beta+1)}\frac{\Gamma(j-i+\beta-\delta)}{\Gamma(j-i+1)} \times \frac{\Gamma(j+i+\alpha+\beta+1)}{\Gamma(j+i+\alpha+\delta+2)},$$

$$\tilde{\kappa}_{i,j} = (-1)^{i+j}\frac{2^{(\delta-\beta)/2}}{\Gamma(\beta-\delta)}\frac{\Gamma(2i+\alpha+\delta+2)}{\Gamma(2i+\alpha+\delta+1)}\frac{\Gamma(j-i+\beta-\delta)}{\Gamma(j-1+1)}\frac{\Gamma(j+i+\alpha+\beta+1)}{\Gamma(j+i+\alpha+\delta+2)} \times \sqrt{\frac{(2j+\alpha+\beta+1)}{(2i+\alpha+\delta+1)}\frac{\Gamma(j+1)}{\Gamma(i+1)}\frac{\Gamma(j+\alpha+1)}{\Gamma(i+\alpha+1)}\frac{\Gamma(i+\delta+1)}{\Gamma(j+\beta+1)}\frac{\Gamma(i+\alpha+\delta+1)}{\Gamma(j+\alpha+\beta+1)}}.$$

$\{P_n^{(\alpha,\beta)}\}_{n\in\mathbb{N}_0} \to \{P_n^{(\gamma,\beta)}\}_{n\in\mathbb{N}_0}$, $\gamma-\alpha\in\mathbb{N}$:

$$\kappa_{i,j} = \begin{cases} (-1)^{i+j}(2i+\gamma+\beta+1)\binom{\gamma-\alpha}{j-i} \\ \quad\times\frac{\Gamma(j+\beta+1)}{\Gamma(i+\beta+1)}\frac{\Gamma(i+\gamma+\beta+1)}{\Gamma(j+\alpha+\beta+1)} \\ \quad\times\frac{\Gamma(j+i+\alpha+\beta+1)}{\Gamma(j+i+\gamma+\beta+2)}, & \text{if } i\le j\le i+\gamma-\alpha, \\ 0, & \text{else,} \end{cases}$$

$$\bar{\kappa}_{i,j} = \begin{cases} (-1)^{i+j}2^{j-i}\binom{\gamma-\alpha}{j-i}\frac{\Gamma(j+1)}{\Gamma(i+1)}\frac{\Gamma(j+\beta+1)}{\Gamma(i+\beta+1)} \\ \quad\times\frac{\Gamma(j+i+\alpha+\beta+1)}{\Gamma(j+i+\gamma+\beta+2)},\frac{\Gamma(2i+\gamma+\beta+2)}{\Gamma(2j+\alpha+\beta+1)}, & \text{if } i\le j\le i+\gamma-\alpha, \\ 0, & \text{else,} \end{cases}$$

$$\tilde{\kappa}_{i,j} = \begin{cases} (-1)^{j+i}2^{(\gamma-\alpha)/2}\binom{\gamma-\alpha}{j-i}\sqrt{\frac{\Gamma(j+\beta+1)}{\Gamma(i+\beta+1)}} \\ \quad\times\sqrt{\frac{\Gamma(i+\gamma+1)}{\Gamma(j+\alpha+1)}\frac{\Gamma(j+1)}{\Gamma(i+1)}} \\ \quad\times\sqrt{\frac{(2j+\alpha+\beta+1)}{(2i+\gamma+\beta+1)}\frac{\Gamma(i+\gamma+\beta+1)}{\Gamma(j+\alpha+\beta+1)}} \\ \quad\times\frac{\Gamma(j+i+\alpha+\beta+1)}{\Gamma(j+i+\gamma+\beta+2)}\frac{\Gamma(2i+\gamma+\beta+2)}{\Gamma(2i+\gamma+\beta+1)}, & \text{if } i\le j\le i+\gamma-\alpha, \\ 0, & \text{else.} \end{cases}$$

$\{P_n^{(\alpha,\beta)}\}_{n\in\mathbb{N}_0} \to \{P_n^{(\alpha,\delta)}\}_{n\in\mathbb{N}_0}$, $\delta-\beta\in\mathbb{N}$:

$$\kappa_{i,j} = \begin{cases} (2i+\alpha+\delta+1)\binom{\delta-\beta}{j-i}\frac{\Gamma(j+\alpha+1)}{\Gamma(i+\alpha+1)} \\ \quad\times\frac{\Gamma(i+\alpha+\delta+1)}{\Gamma(j+\alpha+\beta+1)}\frac{\Gamma(j+i+\alpha+\beta+1)}{\Gamma(j+i+\alpha+\delta+2)}, & \text{if } i\le j\le i+\delta-\beta, \\ 0, & \text{else,} \end{cases}$$

$$\bar{\kappa}_{i,j} = \begin{cases} 2^{j-i}\binom{\delta-\beta}{j-i}\frac{\Gamma(j+1)}{\Gamma(i+1)}\frac{\Gamma(j+\alpha+1)}{\Gamma(i+\alpha+1)} \\ \quad\times\frac{\Gamma(2i+\alpha+\delta+2)}{\Gamma(2j+\alpha+\beta+1)}\frac{\Gamma(i+j+\alpha+\beta+1)}{\Gamma(i+j+\alpha+\delta+2)}, & \text{if } i\le j\le i+\delta-\beta, \\ 0, & \text{else,} \end{cases}$$

$$\tilde{\kappa}_{i,j} = \begin{cases} 2^{\delta-\beta/2}\binom{\delta-\beta}{j-i}\sqrt{\frac{\Gamma(j+\alpha+1)}{\Gamma(i+\alpha+1)}} \\ \quad\times\sqrt{\frac{\Gamma(i+\delta+1)}{\Gamma(j+\beta+1)}\frac{\Gamma(j+1)}{\Gamma(i+1)}} \\ \quad\times\sqrt{\frac{(2j+\alpha+\beta+1)}{(2i+\alpha+\delta+1)}\frac{\Gamma(i+\alpha+\delta+1)}{\Gamma(j+\alpha+\beta+1)}} \\ \quad\times\frac{\Gamma(i+j+\alpha+\beta+1)}{\Gamma(j+i+\alpha+\delta+2)}\frac{\Gamma(2i+\alpha+\delta+2)}{\Gamma(2i+\alpha+\delta+1)}, & \text{if } i\le j\le i+\delta-\beta, \\ 0, & \text{else.} \end{cases}$$

$\{P_n^{(\alpha,\beta)}\}_{n\in\mathbb{N}_0} \to \{P_n^{(\alpha+1,\beta)}\}_{n\in\mathbb{N}_0}$:

$$\kappa_{i,j} = \begin{cases} \frac{j+\alpha+\beta+1}{2j+\alpha+\beta+1}, & \text{if } i=j, \\ \frac{-(j+\beta)}{2j+\alpha+\beta+1}, & \text{if } j=i+1, \\ 0, & \text{else,} \end{cases}$$

$$\bar{\kappa}_{i,j} = \begin{cases} 1, & \text{if } j=i, \\ \frac{-2j(j+\beta)}{(2j+\alpha+\beta)(2j+\alpha+\beta+1)}, & \text{if } j=i+1, \\ 0, & \text{else,} \end{cases}$$

$$\tilde{\kappa}_{i,j} = \begin{cases} \sqrt{\frac{2(j+\alpha+1)(j+\alpha+\beta+1)}{(2j+\alpha+\beta+1)(2j+\alpha+\beta+2)}}, & \text{if } i=j, \\ -\sqrt{\frac{2j(j+\beta)\Gamma(i+1)}{(2j+\alpha+\beta)(2j+\alpha+\beta+1)\Gamma(j)}}, & \text{if } j=i+1, \\ 0, & \text{else.} \end{cases}$$

$\{P_n^{(\alpha,\beta)}\}_{n\in\mathbb{N}_0} \to \{P_n^{(\alpha,\beta+1)}\}_{n\in\mathbb{N}_0}$:

$$\kappa_{i,j} = \begin{cases} \dfrac{j+\alpha+\beta+1}{2j+\alpha+\beta+1}, & \text{if } i=j, \\ \dfrac{j+\alpha}{2j+\alpha+\beta+1}, & \text{if } j=i+1, \\ 0, & \text{else,} \end{cases}$$

$$\bar{\kappa}_{i,j} = \begin{cases} 1, & \text{if } j=i, \\ \dfrac{2j(j+\alpha)}{(2j+\alpha+\beta)(2j+\alpha+\beta+1)}, & \text{if } j=i+1, \\ 0, & \text{else,} \end{cases}$$

$$\tilde{\kappa}_{i,j} = \begin{cases} \sqrt{\dfrac{2(j+\beta+1)(j+\alpha+\beta+1)}{(2j+\alpha+\beta+1)(2j+\alpha+\beta+2)}}, & \text{if } i=j, \\ \sqrt{\dfrac{2j(j+\alpha)\Gamma(i+1)}{(2j+\alpha+\beta)(2j+\alpha+\beta+1)\Gamma(j)}}, & \text{if } j=i+1, \\ 0, & \text{else.} \end{cases}$$

$\{P_n^{(\alpha,\beta)}\}_{n\in\mathbb{N}_0} \to \{P_n^{(\alpha+1,\beta+1)}\}_{n\in\mathbb{N}_0}$:

$$\kappa_{i,j} = 2^{i-j}\frac{\Gamma(i+1)}{\Gamma(j+1)}\frac{\Gamma(i+\alpha+\beta+3)}{\Gamma(j+\alpha+\beta+1)}\frac{\Gamma(2j+\alpha+\beta+1)}{\Gamma(2i+\alpha+\beta+3)}$$

$$\times \begin{cases} 1, & \text{if } j=i, \\ \dfrac{2j(\alpha-\beta)}{(2j+\alpha+\beta)(2j+\alpha+\beta+2)}, & \text{if } j=i+1, \\ \dfrac{-4(j-1)j(j+\alpha)(j+\beta)}{(2j+\alpha+\beta-1)(2j+\alpha+\beta)^2(2j+\alpha+\beta+1)}, & \text{if } j=i+2, \end{cases}$$

$$\bar{\kappa}_{i,j} = \begin{cases} 1, & \text{if } j=i, \\ \dfrac{2j(\alpha-\beta)}{(2j+\alpha+\beta)(2j+\alpha+\beta+2)}, & \text{if } j=i+1, \\ \dfrac{-4(j-1)j(j+\alpha)(j+\beta)}{(2j+\alpha+\beta-1)(2j+\alpha+\beta)^2(2j+\alpha+\beta+1)}, & \text{if } j=i+2, \end{cases}$$

$$\tilde{\kappa}_{i,j} = \frac{2^{i+1}}{2^j}\frac{\Gamma(2j+\alpha+\beta+1)}{\Gamma(2i+\alpha+\beta+3)}\sqrt{\frac{\Gamma(i+1)}{\Gamma(j+1)}\frac{\Gamma(i+\alpha+2)}{\Gamma(j+\alpha+1)}\frac{\Gamma(i+\beta+2)}{\Gamma(j+\beta+1)}}$$

$$\times \sqrt{\frac{\Gamma(i+\alpha+\beta+3)}{\Gamma(j+\alpha+\beta+1)}\frac{(2j+\alpha+\beta+1)}{(2i+\alpha+\beta+3)}}$$

$$\times \begin{cases} 1 & \text{if } i=j, \\ \dfrac{2j(\alpha-\beta)}{(2j+\alpha+\beta)(2j+\alpha+\beta+2)}, & \text{if } j=i+1, \\ \dfrac{-4(j-1)j(j+\alpha)(j+\beta)}{(2j+\alpha+\beta-1)(2j+\alpha+\beta)^2(2j+\alpha+\beta+1)}, & \text{if } j=i+2. \end{cases}$$

$\{P_n^{(\alpha,\beta)}\}_{n\in\mathbb{N}_0} \to \{P_n^{(\alpha-1,\beta)}\}_{n\in\mathbb{N}_0}$:

$$\kappa_{i,j} = (2i+\alpha+\beta)\frac{\Gamma(j+\beta+1)}{\Gamma(i+\beta+1)}\frac{\Gamma(i+\alpha+\beta)}{\Gamma(j+\alpha+\beta+1)},$$

$$\bar{\kappa}_{i,j} = \frac{2^j}{2^i}\frac{\Gamma(j+1)}{\Gamma(i+1)}\frac{\Gamma(j+\beta+1)}{\Gamma(i+\beta+1)}\frac{\Gamma(2i+\alpha+\beta+1)}{\Gamma(2j+\alpha+\beta+1)},$$

$$\tilde{\kappa}_{i,j} = \frac{2i+\alpha+\beta}{\sqrt{2}}\sqrt{\frac{\Gamma(j+1)}{\Gamma(i+1)}\frac{\Gamma(i+\alpha)}{\Gamma(j+\alpha+1)}\frac{\Gamma(j+\beta+1)}{\Gamma(i+\beta+1)}\frac{\Gamma(i+\alpha+\beta)}{\Gamma(j+\alpha+\beta+1)}\frac{(2j+\alpha+\beta+1)}{(2i+\alpha+\beta)}}.$$

$\{P_n^{(\alpha,\beta)}\}_{n\in\mathbb{N}_0} \to \{P_n^{(\alpha,\beta-1)}\}_{n\in\mathbb{N}_0}$:

$$\kappa_{i,j} = (-1)^{i+j}(2i+\alpha+\beta)\frac{\Gamma(j+\alpha+1)}{\Gamma(i+\alpha+1)}\frac{\Gamma(i+\alpha+\beta)}{\Gamma(j+\alpha+\beta+1)},$$

$$\bar{\kappa}_{i,j} = (-1)^{i+j}\frac{2^j}{2^i}\frac{\Gamma(j+1)}{\Gamma(i+1)}\frac{\Gamma(j+\alpha+1)}{\Gamma(i+\alpha+1)}\frac{\Gamma(2i+\alpha+\beta+1)}{2j+\alpha+\beta+1)},$$

$$\tilde{\kappa}_{i,j} = (-1)^{i+j}\frac{2i+\alpha+\beta+1}{\sqrt{2}}\sqrt{\frac{\Gamma(j+1)}{\Gamma(i+1)}\frac{\Gamma(j+\alpha+1)}{\Gamma(i+\alpha+1)}\frac{\Gamma(i+\beta)}{\Gamma(j+\beta+1)}\frac{\Gamma(i+\alpha+\beta)}{\Gamma(j+\alpha+\beta+1)}}$$
$$\times\sqrt{\frac{(2j+\alpha+\beta+1)}{(2i+\alpha+\beta)}}.$$

$(\alpha,\beta)\to(\alpha-1,\beta-1)$:

$$\kappa_{i,j} = \left((-1)^{i+j}\Gamma(j+\alpha+1)\Gamma(i+\beta)+\Gamma(i+\alpha)\Gamma(j+\beta+1)\right)$$
$$\times\frac{(2i+\alpha+\beta-1)\Gamma(i+\alpha+\beta-1)}{\Gamma(i+\alpha)\Gamma(i+\beta)\Gamma(j+\alpha+\beta+1)},$$

$$\bar{\kappa}_{i,j} = \left((-1)^{i+j}\frac{\Gamma(j+\alpha+1)}{\Gamma(i+\alpha)}+\frac{\Gamma(j+\beta+1)}{\Gamma(i+\beta)}\right)\frac{2^j}{2^i}\frac{\Gamma(j+1)}{\Gamma(i+1)}\frac{\Gamma(2i+\alpha+\beta)}{\Gamma(2j+\alpha+\beta+1)},$$

$$\tilde{\kappa}_{i,j} = \left((-1)^{i+j}\sqrt{\frac{\Gamma(j+\alpha+1)}{\Gamma(j+\beta+1)}\frac{\Gamma(i+\beta)}{\Gamma(i+\alpha)}}+\sqrt{\frac{\Gamma(j+\beta+1)}{\Gamma(j+\alpha+1)}\frac{\Gamma(i+\alpha)}{\Gamma(i+\beta)}}\right)$$
$$\times\sqrt{\frac{\Gamma(j+1)}{\Gamma(i+1)}\frac{(2j+\alpha+\beta+1)}{(2i+\alpha+\beta-1)}\frac{\Gamma(i+\alpha+\beta-1)}{\Gamma(j+\alpha+\beta+1)}\frac{\Gamma(2i+\alpha+\beta)}{2\Gamma(2i+\alpha+\beta-1)}}.$$

The matrix $\mathbf{G}$

$\{P_n^{(\alpha,\beta)}\}_{n=0}^N \to \{P_n^{(\gamma,\delta)}\}_{n=0}^N$:

$$\mathbf{G} = \operatorname{diag}(\mathbf{d}) + (\alpha-\gamma)\operatorname{triu}(\mathbf{u}\,\mathbf{v}^{\mathrm{T}},1) + (\beta-\delta)\operatorname{triu}(\mathbf{w}\,\mathbf{z}^{\mathrm{T}},1) \in \mathbb{R}^{(N+1)\times(N+1)},$$

$$\begin{aligned}
\mathbf{d} &= (d_j)_{j=0}^N, & d_j &= j(j+\alpha+\beta+1),\\
\mathbf{u} &= (u_j)_{j=0}^N, & u_j &= (2j+\gamma+\delta+1)\frac{\Gamma(j+\gamma+\delta+1)}{\Gamma(j+\delta+1)},\\
\mathbf{v} &= (v_j)_{j=0}^N, & v_j &= \frac{\Gamma(j+\delta+1)}{\Gamma(j+\gamma+\delta+1)},\\
\mathbf{w} &= (w_j)_{j=0}^N, & w_j &= (-1)^j(2j+\gamma+\delta+1)\frac{\Gamma(j+\gamma+\delta+1)}{\Gamma(j+\gamma+1)},\\
\mathbf{z} &= (z_j)_{j=0}^N, & z_j &= (-1)^j\frac{\Gamma(j+\gamma+1)}{\Gamma(j+\gamma+\delta+1)}.
\end{aligned}$$

$$\bar{\mathbf{G}} = \operatorname{diag}(\bar{\mathbf{d}}) + (\alpha-\gamma)\operatorname{triu}(\bar{\mathbf{u}}\,\bar{\mathbf{v}}^{\mathrm{T}},1) + (\beta-\delta)\operatorname{triu}(\bar{\mathbf{w}}\,\bar{\mathbf{z}}^{\mathrm{T}},1) \in \mathbb{R}^{(N+1)\times(N+1)},$$

$$\begin{aligned}
\bar{\mathbf{d}} &= (\bar{d}_j)_{j=0}^N, & \bar{d}_j &= j(j+\alpha+\beta+1),\\
\bar{\mathbf{u}} &= (\bar{u}_j)_{j=0}^N, & \bar{u}_j &= \frac{\Gamma(2j+\gamma+\delta+2)}{2^j\Gamma(j+1)\Gamma(j+\delta+1)},\\
\bar{\mathbf{v}} &= (\bar{v}_j)_{j=0}^N, & \bar{v}_j &= \frac{2^j\Gamma(j+1)\Gamma(j+\delta+1)}{\Gamma(2j+\gamma+\delta+1)},\\
\bar{\mathbf{w}} &= (\bar{w}_j)_{j=0}^N, & \bar{w}_j &= (-1)^j\frac{\Gamma(2j+\gamma+\delta+2)}{2^j\Gamma(j+1)\Gamma(j+\gamma+1)},\\
\bar{\mathbf{z}} &= (\bar{z}_j)_{j=0}^N, & \bar{z}_j &= (-1)^j\frac{2^j\Gamma(j+1)\Gamma(j+\gamma+1)}{\Gamma(2j+\gamma+\delta+1)}.
\end{aligned}$$

$$\tilde{\mathbf{G}} = \operatorname{diag}(\tilde{\mathbf{d}}) + (\alpha-\gamma)\operatorname{triu}(\tilde{\mathbf{u}}\,\tilde{\mathbf{v}}^{\mathrm{T}},1) + (\beta-\delta)\operatorname{triu}(\tilde{\mathbf{w}}\,\tilde{\mathbf{z}}^{\mathrm{T}},1) \in \mathbb{R}^{(N+1)\times(N+1)},$$

$$\begin{aligned}
\tilde{\mathbf{d}} &= (\tilde{d}_j)_{j=0}^N, & \tilde{d}_j &= j(j+\alpha+\beta+1),\\
\tilde{\mathbf{u}} &= (\tilde{u}_j)_{j=0}^N, & \tilde{u}_j &= \sqrt{(2j+\gamma+\delta+1)\frac{\Gamma(j+\gamma+\delta+1)\Gamma(j+\gamma+1)}{\Gamma(j+1)\Gamma(j+\delta+1)}},\\
\tilde{\mathbf{v}} &= (\tilde{v}_j)_{j=0}^N, & \tilde{v}_j &= \sqrt{(2j+\gamma+\delta+1)\frac{\Gamma(j+1)\Gamma(j+\delta+1)}{\Gamma(j+\gamma+\delta+1)\Gamma(j+\gamma+1)}},\\
\tilde{\mathbf{w}} &= (\tilde{w}_j)_{j=0}^N, & \tilde{w}_j &= (-1)^j\sqrt{(2j+\gamma+\delta+1)\frac{\Gamma(j+\gamma+\delta+1)\Gamma(j+\delta+1)}{\Gamma(j+1)\Gamma(j+\gamma+1)}},\\
\tilde{\mathbf{z}} &= (\tilde{z}_j)_{j=0}^N, & \tilde{z}_j &= (-1)^j\sqrt{(2j+\gamma+\delta+1)\frac{\Gamma(j+1)\Gamma(j+\gamma+1)}{\Gamma(j+\gamma+\delta+1)\Gamma(j+\delta+1)}}.
\end{aligned}$$

A.1.3 Gegenbauer polynomials.

Symbol

$$C_n^{(\alpha)} : \mathbb{R} \to \mathbb{R}, \quad n \in \mathbb{N}_0, \quad -1/2 < \alpha.$$

Inner product and weight function

$$\langle f, g \rangle = \int_I f(x) g(x)\, w(x)\, \mathrm{d}x, \quad I = [-1, 1], \quad w(x) = (1 - x^2)^{\alpha - 1/2}.$$

Differential equation

$$\sigma(x) y''(x) + \tau(x) y'(x) + \lambda_n y(x) = 0, \quad y = C_n^{(\alpha)}$$
$$\sigma(x) = 1 - x^2, \quad \tau(x) = -(2\alpha + 1)x, \quad \lambda_n = n(n + 2\alpha).$$

Differential operator

$$\mathcal{D} = -(1 - x^2) \frac{\mathrm{d}^2}{\mathrm{d}x^2} + (2\alpha + 1) x \frac{\mathrm{d}}{\mathrm{d}x}.$$

Rodrigues formula

$$C_n^{(\alpha)}(x) = \frac{(-1)^n \Gamma(\alpha + 1/2) \Gamma(n + 2\alpha)}{2^n \Gamma(n + 1) \Gamma(n + \alpha + 1/2) \Gamma(2\alpha)} (1 - x^2)^{1/2 - \alpha} \frac{\mathrm{d}^n}{\mathrm{d}x^n} \left((1 - x^2)^{n + \alpha - 1/2} \right),$$

$$\begin{aligned} \frac{\mathrm{d}^m}{\mathrm{d}x^m} C_n^{(\alpha)}(x) &= \frac{(-1)^{n-m} \Gamma(\alpha + 1/2) \Gamma(n + 2\alpha + m)}{2^n \Gamma(n - m + 1) \Gamma(n + \alpha + 1/2) \Gamma(2\alpha)} (1 - x^2)^{1 - 2\alpha - m} \\ &\quad \times \frac{\mathrm{d}^{n-m}}{\mathrm{d}x^{n-m}} \left((1 - x^2)^{n + \alpha - 1/2} \right) \\ &= 2^m \frac{\Gamma(\alpha + m)}{\Gamma(\alpha)} C_{n-m}^{(\alpha + m)}(x). \end{aligned}$$

Leading coefficients

$$k_n = \frac{2^n}{\Gamma(\alpha)} \frac{\Gamma(n + \alpha)}{\Gamma(n + 1)}, \qquad \bar{k}_n = 1, \qquad \tilde{k}_n = \frac{2^{n+\alpha}}{\sqrt{2\pi}} \frac{\Gamma(n + \alpha + 1)}{\sqrt{\Gamma(n + 1)(n + \alpha)\Gamma(n + 2\alpha)}}.$$

Squared norms

$$h_n = \frac{2^{1-2\alpha}\pi}{\Gamma(\alpha)^2} \frac{\Gamma(n + 2\alpha)}{(n + \alpha)\Gamma(n + 1)}, \qquad \bar{h}_n = \frac{2\pi}{2^{2(n+\alpha)}} \frac{\Gamma(n + 1)\Gamma(n + 2\alpha)}{\Gamma(n + \alpha)\Gamma(n + \alpha + 1)}, \qquad \tilde{h}_n = 1.$$

Three-term recurrence

$$a_n = \frac{2(n + \alpha)}{n + 1}, \qquad a'_n = \frac{n + 1}{2(n + \alpha)},$$
$$b_n = 0, \qquad b'_n = 0,$$
$$c_n = \frac{n + 2\alpha - 1}{n + 1}, \qquad c'_n = \frac{n + 2\alpha - 1}{2(n + \alpha)},$$

$$\bar{a}_n = 1, \qquad \bar{a}'_n = 1,$$
$$\bar{b}_n = 0, \qquad \bar{b}'_n = 0,$$
$$\bar{c}_n = \frac{n(n + 2\alpha - 1)}{4(n + \alpha - 1)(n + \alpha)}, \qquad \bar{c}'_n = \frac{n(n + 2\alpha - 1)}{4(n + \alpha - 1)(n + \alpha)},$$

$$\tilde{a}_n = 2\sqrt{\frac{(n+\alpha)(n+\alpha+1)}{(n+1)(n+2\alpha)}}, \qquad \tilde{a}'_n = \frac{1}{2}\sqrt{\frac{(n+1)(n+2\alpha)}{(n+\alpha)(n+\alpha+1)}},$$

$$\tilde{b}_n = 0, \qquad \tilde{b}'_n = 0,$$

$$\tilde{c}_n = \sqrt{\frac{n(n+\alpha+1)(n+2\alpha-1)}{(n+1)(n+\alpha-1)(n+2\alpha)}}, \qquad \tilde{c}'_n = \frac{1}{2}\sqrt{\frac{n(n+2\alpha-1)}{(n+\alpha-1)(n+\alpha)}}.$$

Derivative identities

$$\frac{\mathrm{d}}{\mathrm{d}x}C_n^{(\alpha)}(x) = 2\alpha C_{n-1}^{(\alpha+1)}(x),$$

$$\frac{\mathrm{d}^m}{\mathrm{d}x^m}C_n^{(\alpha)}(x) = 2^m\frac{\Gamma(\alpha+m)}{\Gamma(\alpha)}C_{n-m}^{(\alpha+m)}(x),$$

$$\frac{\mathrm{d}}{\mathrm{d}x}C_n^{(\alpha)}(x) = A_n\sum_{j=0}^{n-1}B_jC_j^{(\alpha)}(x) + C_n\sum_{j=0}^{n-1}D_jC_j^{(\alpha)}(x) = A'_n\sum_{j=0}^{\lfloor\frac{n-1}{2}\rfloor}B'_{2j+\chi}C_{2j+\chi}^{(\alpha)}(x),$$

$$A_n = (-1)^{n+1}, \qquad B_j = (-1)^j(j+\alpha),$$

$$C_n = 1, \qquad D_j = j+\alpha,$$

$$A'_n = 2, \qquad B'_j = j+\alpha,$$

$$\frac{\mathrm{d}}{\mathrm{d}x}\bar{C}_n^{(\alpha)}(x) = n\bar{C}_{n-1}^{(\alpha+1)}(x),$$

$$\frac{\mathrm{d}^m}{\mathrm{d}x^m}\bar{C}_n^{(\alpha)}(x) = \frac{\Gamma(n+1)}{\Gamma(n-m+1)}\bar{C}_{n-m}^{(\alpha+m)}(x),$$

$$\frac{\mathrm{d}}{\mathrm{d}x}\bar{C}_n^{(\alpha)}(x) = \bar{A}_n\sum_{j=0}^{n-1}\bar{B}_j\bar{C}_j^{(\alpha)}(x) + \bar{C}_n\sum_{j=0}^{n-1}\bar{D}_j\bar{C}_j^{(\alpha)}(x) = \bar{A}'_n\sum_{j=0}^{\lfloor\frac{n-1}{2}\rfloor}\bar{B}'_{2j+\chi}\bar{C}_{2j+\chi}^{(\alpha)}(x),$$

$$\bar{A}_n = (-1)^{n+1}2^{-n}\frac{\Gamma(n+1)}{\Gamma(n+\alpha)}, \qquad \bar{B}_j = (-1)^j2^j\frac{\Gamma(j+\alpha+1)}{\Gamma(j+1)},$$

$$\bar{C}_n = 2^{-n}\frac{\Gamma(n+1)}{\Gamma(n+\alpha)}, \qquad \bar{D}_j = 2^j\frac{\Gamma(j+\alpha+1)}{\Gamma(j+1)},$$

$$\bar{A}'_n = 2^{1-n}\frac{\Gamma(n+1)}{\Gamma(n+\alpha)}, \qquad \bar{B}'_j = 2^j\frac{\Gamma(j+\alpha+1)}{\Gamma(j+1)},$$

$$\frac{\mathrm{d}}{\mathrm{d}x}\tilde{C}_n^{(\alpha)}(x) = \frac{\sqrt{n(n+2\alpha)}}{\mathrm{sign}(\alpha)}\tilde{C}_{n-1}^{(\alpha+1)}(x),$$

$$\frac{\mathrm{d}^m}{\mathrm{d}x^m}\tilde{C}_n^{(\alpha)}(x) = \mathrm{sign}(\alpha)\sqrt{\frac{\Gamma(n+1)\Gamma(n+m+2\alpha)}{\Gamma(n-m+1)\Gamma(n+2\alpha)}}\tilde{C}_{n-m}^{(\alpha+m)}(x),$$

$$\frac{\mathrm{d}}{\mathrm{d}x}\tilde{C}_n^{(\alpha)}(x) = \tilde{A}_n\sum_{j=0}^{n-1}\tilde{B}_j\tilde{C}_j^{(\alpha)}(x) + \tilde{C}_n\sum_{j=0}^{n-1}\tilde{D}_j\tilde{C}_j^{(\alpha)}(x) = \tilde{A}'_n\sum_{j=0}^{\lfloor\frac{n-1}{2}\rfloor}\tilde{B}'_{2j+\chi}\tilde{C}_{2j+\chi}^{(\alpha)}(x),$$

$$\tilde{A}_n = (-1)^{n+1}\sqrt{\frac{(n+\alpha)\Gamma(n+1)}{\Gamma(n+2\alpha)}}, \qquad \tilde{B}_j = (-1)^j\sqrt{\frac{(j+\alpha)\Gamma(j+2\alpha)}{\Gamma(j+1)}},$$

$$\tilde{C}_n = \sqrt{\frac{(n+\alpha)\Gamma(n+1)}{\Gamma(n+2\alpha)}}, \qquad \tilde{D}_j = \sqrt{\frac{(j+\alpha)\Gamma(j+2\alpha)}{\Gamma(j+1)}},$$

$$\tilde{A}'_n = 2\sqrt{\frac{(n+\alpha)\Gamma(n+1)}{\Gamma(n+2\alpha)}}, \qquad \tilde{B}'_j = \sqrt{\frac{(j+\alpha)\Gamma(j+2\alpha)}{\Gamma(j+1)}}.$$

Connection coefficients

$\{C_n^{(\alpha)}\}_{n\in\mathbb{N}_0} \to \{C_n^{(\beta)}\}_{n\in\mathbb{N}_0}$:

$$\kappa_{i,j} = \frac{\Gamma(\beta)(i+\beta)}{\Gamma(\alpha)\Gamma(\alpha-\beta)} \frac{\Gamma\left(\frac{j-i}{2}+\alpha-\beta\right)}{\Gamma\left(\frac{j-i}{2}+1\right)} \frac{\Gamma\left(\frac{j+i}{2}+\alpha\right)}{\Gamma\left(\frac{j+i}{2}+\beta+1\right)},$$

$$\bar{\kappa}_{i,j} = \frac{1}{\Gamma(\alpha-\beta)} \frac{2^i}{2^j} \frac{\Gamma(j+1)}{\Gamma(i+1)} \frac{\Gamma(i+\beta+1)}{\Gamma(j+\alpha)} \frac{\Gamma\left(\frac{j-i}{2}+\alpha-\beta\right)}{\Gamma\left(\frac{j-i}{2}+1\right)} \frac{\Gamma\left(\frac{j+i}{2}+\alpha\right)}{\Gamma\left(\frac{j+i}{2}+\beta+1\right)},$$

$$\tilde{\kappa}_{i,j} = \frac{\sqrt{(i+\beta)(j+\alpha)}}{\Gamma(\alpha-\beta)} \sqrt{\frac{\Gamma(j+1)\Gamma(i+2\hat{\alpha})}{\Gamma(i+1)\Gamma(j+2\alpha)}} \frac{2^\alpha}{2^\beta} \frac{\Gamma\left(\frac{j-i}{2}+\alpha-\beta\right)}{\Gamma\left(\frac{j-i}{2}+1\right)} \frac{\Gamma\left(\frac{j+i}{2}+\alpha\right)}{\Gamma\left(\frac{j+i}{2}+\beta+1\right)}$$

$\{C_n^{(\alpha)}\}_{n\in\mathbb{N}_0} \to \{C_n^{(\beta)}\}_{n\in\mathbb{N}_0}$, $\beta-\alpha \in \mathbb{N}$:

$$\kappa_{i,j} = \begin{cases} (-1)^{(i+j)/2} \dfrac{\Gamma(\beta)(i+\beta)}{\Gamma(\alpha)} \dbinom{\beta-\alpha}{\frac{j-i}{2}} \dfrac{\Gamma\left(\frac{j+i}{2}+\alpha\right)}{\Gamma\left(\frac{j+i}{2}+\beta+1\right)}, & \text{if } i \le j \le i+2(\beta-\alpha), \\ 0, & \text{else,} \end{cases}$$

$$\bar{\kappa}_{i,j} = \begin{cases} (-1)^{(i+j)/2} \dfrac{2^i}{2^j} \dfrac{\Gamma(j+1)}{\Gamma(i+1)} \dfrac{\Gamma(i+\beta+1)}{\Gamma(j+\alpha)} \dbinom{\beta-\alpha}{\frac{j-i}{2}} & \\ \times \dfrac{\Gamma\left(\frac{j+i}{2}+\alpha\right)}{\Gamma\left(\frac{j+i}{2}+\beta+1\right)}, & \text{if } i \le j \le i+2(\beta-\alpha), \\ 0, & \text{else,} \end{cases}$$

$$\tilde{\kappa}_{i,j} = \begin{cases} (-1)^{(i+j)/2} \dfrac{2^\alpha}{2^\beta} \sqrt{(i+\beta)(j+\alpha)} \dbinom{\beta-\alpha}{\frac{j-i}{2}} & \\ \times \dfrac{\Gamma\left(\frac{j+i}{2}+\alpha\right)}{\Gamma\left(\frac{j+i}{2}++\beta+1\right)} \sqrt{\dfrac{\Gamma(j+1)}{\Gamma(i+1)} \dfrac{\Gamma(i+2\beta)}{\Gamma(j+2\alpha)}}, & \text{if } i \le j \le i+2(\beta-\alpha), \\ 0, & \text{else.} \end{cases}$$

$\{C_n^{(\alpha)}\}_{n\in\mathbb{N}_0} \to \{C_n^{(\alpha+1)}\}_{n\in\mathbb{N}_0}$:

$$\kappa_{i,j} = \begin{cases} \dfrac{\alpha}{j+\alpha}, & \text{if } j = i, \\ -\dfrac{\alpha}{j+\alpha}, & \text{if } j = i+2, \\ 0, & \text{else}, \end{cases}$$

$$\bar{\kappa}_{i,j} = \begin{cases} 1, & \text{if } j = i, \\ -\dfrac{1}{4}\dfrac{j-1}{j+\alpha-1}\dfrac{j}{j+\alpha}, & \text{if } j = i+2, \\ 0, & \text{else}, \end{cases}$$

$$\tilde{\kappa}_{i,j} = \begin{cases} \dfrac{1}{2}\sqrt{\dfrac{j+2\alpha}{j+\alpha}\dfrac{j+2\alpha+1}{j+\alpha+1}}, & \text{if } j = i, \\ -\dfrac{1}{2}\dfrac{1}{\sqrt{(j+\alpha-1)(j+\alpha)}}, & \text{if } j = i+2, \\ 0, & \text{else}. \end{cases}$$

$\{C_n^{(\alpha)}\}_{n\in\mathbb{N}_0} \to \{C_n^{(\alpha-1)}\}_{n\in\mathbb{N}_0}$:

$$\kappa_{i,j} = \frac{i+\alpha-1}{\alpha-1},$$

$$\bar{\kappa}_{i,j} = \frac{2^i}{2^j}\frac{\Gamma(j+1)}{\Gamma(i+1)}\frac{\Gamma(i+\alpha)}{\Gamma(j+\alpha)},$$

$$\tilde{\kappa}_{i,j} = \frac{\sqrt{(i+\alpha-1)(j+\alpha)}}{2}\sqrt{\frac{\Gamma(j+1)}{\Gamma(i+1)}\frac{\Gamma(i+2\alpha-2)}{\Gamma(j+2\alpha)}}.$$

The matrix $\mathbf{G}$

$\{C_n^{(\alpha)}\}_{n=0}^N \to \{C_n^{(\beta)}\}_{n=0}^N$:

$$\mathbf{G} = \operatorname{diag}(\mathbf{d}) + 4(\alpha-\beta)\operatorname{triuc}(\mathbf{u}\,\mathbf{v}^{\mathrm{T}}, 1) \in \mathbb{R}^{(N+1)\times(N+1)},$$

$$\mathbf{d} = (d_j)_{j=0}^N, \qquad d_j = j(j+2\alpha),$$

$$\mathbf{u} = (u_j)_{j=0}^N, \qquad u_j = j+\beta,$$

$$\mathbf{v} = (v_j)_{j=0}^N, \qquad v_j = 1.$$

$$\bar{\mathbf{G}} = \operatorname{diag}(\bar{\mathbf{d}}) + 4(\alpha-\beta)\operatorname{triuc}(\bar{\mathbf{u}}\,\bar{\mathbf{v}}^{\mathrm{T}}, 1) \in \mathbb{R}^{(N+1)\times(N+1)},$$

$$\bar{\mathbf{d}} = (\bar{d}_j)_{j=0}^N, \qquad \bar{d}_j = j(j+2\alpha),$$

$$\bar{\mathbf{u}} = (\bar{u}_j)_{j=0}^N, \qquad \bar{u}_j = \frac{2^j\Gamma(j+\beta+1)}{\Gamma(j+1)},$$

$$\bar{\mathbf{v}} = (\bar{v}_j)_{j=0}^N, \qquad \bar{v}_j = \frac{\Gamma(j+1)}{2^j\Gamma(j+\beta)}.$$

$$
\begin{aligned}
\tilde{\mathbf{G}} &= \operatorname{diag}(\tilde{\mathbf{d}}) + 4(\alpha-\beta)\operatorname{triu}(\tilde{\mathbf{u}}\,\tilde{\mathbf{v}}^{\mathrm{T}}, 1) \in \mathbb{R}^{(N+1)\times(N+1)},\\
&\tilde{\mathbf{d}} = (\tilde{d}_j)_{j=0}^N, \qquad \tilde{d}_j = j(j+2\alpha),\\
&\tilde{\mathbf{u}} = (\tilde{u}_j)_{j=0}^N, \qquad \tilde{u}_j = \sqrt{(j+\beta)\frac{\Gamma(j+2\beta)}{\Gamma(j+1)}},\\
&\tilde{\mathbf{v}} = (\tilde{v}_j)_{j=0}^N, \qquad \tilde{v}_j = \sqrt{(j+\beta)\frac{\Gamma(j+1)}{\Gamma(j+2\beta)}}.
\end{aligned}
$$

A.1.4 Legendre polynomials.

Symbol

$$P_n : \mathbb{R} \to \mathbb{R}, \quad n \in \mathbb{N}_0.$$

Inner product and weight function

$$\langle f, g \rangle = \int_I f(x) g(x)\, w(x)\, \mathrm{d}x, \quad I = [-1, 1], \quad w(x) = 1.$$

Differential equation

$$\sigma(x) y''(x) + \tau(x) y'(x) + \lambda_n y(x) = 0, \quad y = P_n,$$
$$\sigma(x) = 1 - x^2, \quad \tau(x) = -2x, \quad \lambda_n = n(n+1).$$

Differential operator

$$\mathcal{D} = -(1 - x^2)\frac{\mathrm{d}^2}{\mathrm{d}x^2} + 2x\frac{\mathrm{d}}{\mathrm{d}x}.$$

Rodrigues formula

$$P_n(x) = \frac{(-1)^n}{2^n n!}\frac{\mathrm{d}^n}{\mathrm{d}x^n}\left((1 - x^2)^n\right),$$

$$\frac{\mathrm{d}^m}{\mathrm{d}x^m} P_n(x) = \frac{(-1)^{n-m}(n+1)_m}{2^n (n-m)!(1-x^2)^m}\frac{\mathrm{d}^{n-m}}{\mathrm{d}x^{n-m}}\left((1 - x^2)^n\right) = \frac{\Gamma(n+m+1)}{2^m \Gamma(n+1)} P_{n-m}^{(m,m)}(x).$$

Leading coefficients

$$k_n = \frac{2^n}{\sqrt{\pi}}\frac{\Gamma(n+1/2)}{\Gamma(n+1)}, \qquad \bar{k}_n = 1, \qquad \tilde{k}_n = \sqrt{\frac{2n+1}{2}}\frac{2^n}{\sqrt{\pi}}\frac{\Gamma(n+1/2)}{\Gamma(n+1)}.$$

Squared norms

$$h_n = \frac{2}{2n+1}, \qquad \bar{h}_n = \frac{\Gamma(n+1)^2}{2^{2n}\Gamma(n+1/2)\Gamma(n+3/2)}, \qquad \tilde{h}_n = 1.$$

Three-term recurrence

$$a_n = \frac{2n+1}{n+1}, \qquad a'_n = \frac{n+1}{2n+1},$$
$$b_n = 0, \qquad b'_n = 0,$$
$$c_n = \frac{n}{n+1}, \qquad c'_n = \frac{n}{2n+1}.$$

$$\bar{a}_n = 1, \qquad \bar{a}'_n = 1,$$
$$\bar{b}_n = 0, \qquad \bar{b}'_n = 0,$$
$$\bar{c}_n = \frac{n^2}{(2n-1)(2n+1)}, \qquad \bar{c}'_n = \frac{n^2}{(2n-1)(2n+1)}.$$

$$\tilde{a}_n = \frac{\sqrt{(2n+1)(2n+3)}}{n+1}, \qquad \tilde{a}'_n = \frac{n+1}{\sqrt{(2n+1)(2n+3)}}$$
$$\tilde{b}_n = 0, \qquad \tilde{b}'_n = 0,$$
$$\tilde{c}_n = \frac{n}{n+1}\sqrt{\frac{2n+3}{2n-1}}, \qquad \tilde{c}'_n = \frac{n}{\sqrt{(2n-1)(n+1)}}.$$

Derivative identities

$$\frac{\mathrm{d}}{\mathrm{d}x}P_n(x) = \frac{n+1}{2}P^{(1,1)}_{n-1}(x),$$

$$\frac{\mathrm{d}^m}{\mathrm{d}x^m}P_n(x) = \frac{\Gamma(n+m+1)}{2^m\Gamma(n+1)}P^{(m,m)}_{n-m}(x),$$

$$\frac{\mathrm{d}}{\mathrm{d}x}P_n(x) = A_n\sum_{j=0}^{n-1}B_jP_j(x) + C_n\sum_{j=0}^{n-1}D_jP_j(x) = A'_n\sum_{j=0}^{\lfloor\frac{n-1}{2}\rfloor}B'_{2j+\chi}P_{2j+\chi}(x),$$

$$A_n = (-1)^{n+1}\frac{1}{2}, \qquad B_j = (-1)^j(2j+1),$$

$$C_n = \frac{1}{2}, \qquad D_j = 2j+1,$$

$$A'_n = 1, \qquad B'_j = 2j+1,$$

$$\frac{\mathrm{d}}{\mathrm{d}x}\bar{P}_n(x) = n\bar{P}^{(1,1)}_{n-1}(x),$$

$$\frac{\mathrm{d}^m}{\mathrm{d}x^m}\bar{P}_n(x) = \frac{\Gamma(n+1)}{\Gamma(n-m+1)}\bar{P}^{(m,m)}_{n-m}(x),$$

$$\frac{\mathrm{d}}{\mathrm{d}x}\bar{P}_n(x) = \bar{A}_n\sum_{j=0}^{n-1}\bar{B}_j\bar{P}_j(x) + \bar{C}_n\sum_{j=0}^{n-1}\bar{D}_j\bar{P}_j(x) = \bar{A}'_n\sum_{j=0}^{\lfloor\frac{n-1}{2}\rfloor}\bar{B}'_{2j+\chi}\bar{P}_{2j+\chi}(x),$$

$$\bar{A}_n = (-1)^{n+1}\frac{1}{2^{n+1}}\frac{\Gamma(n+1)}{\Gamma(n+1/2)}, \qquad \bar{B}_j = (-1)^j2^j\frac{\Gamma(j+1/2)}{\Gamma(j+1)},$$

$$\bar{C}_n = \frac{1}{2^{n+1}}\frac{\Gamma(n+1)}{\Gamma(n+1/2)}, \qquad \bar{D}_j = 2^j\frac{\Gamma(j+1/2)}{\Gamma(j+1)},$$

$$\bar{A}'_n = \frac{1}{2^n}\frac{\Gamma(n+1)}{\Gamma(n+1/2)}, \qquad \bar{B}'_j = 2^j\frac{\Gamma(j+1/2)}{\Gamma(j+1)},$$

$$\frac{\mathrm{d}}{\mathrm{d}x}\tilde{P}_n(x) = \sqrt{n(n+1)}\tilde{P}^{(1,1)}_{n-1}(x),$$

$$\frac{\mathrm{d}^m}{\mathrm{d}x^m}\tilde{P}_n(x) = \sqrt{\frac{\Gamma(n+1)\Gamma(n+m+1)}{\Gamma(n-m+1)\Gamma(n+1)}}\tilde{P}^{(m,m)}_{n-m}(x),$$

$$\frac{\mathrm{d}}{\mathrm{d}x}\tilde{P}_n(x) = \tilde{A}_n\sum_{j=0}^{n-1}\tilde{B}_j\tilde{P}_j(x) + \tilde{C}_n\sum_{j=0}^{n-1}\tilde{D}_j\tilde{P}_j(x) = \tilde{A}'_n\sum_{j=0}^{\lfloor\frac{n-1}{2}\rfloor}\tilde{B}'_{2j+\chi}\tilde{P}_{2j+\chi}(x),$$

$$\tilde{A}_n = (-1)^{n+1}\frac{1}{2}\sqrt{2n+1}, \qquad \tilde{B}_j = (-1)^j\sqrt{2j+1},$$

$$\tilde{C}_n = \frac{1}{2}\sqrt{2n+1}, \qquad \tilde{D}_j = \sqrt{2j+1},$$

$$\tilde{A}'_n = \sqrt{2n+1}, \qquad \tilde{B}'_j = \sqrt{2j+1}.$$

Connection coefficients

$\{P_n\}_{n\in\mathbb{N}_0} \to \{T_n\}_{n\in\mathbb{N}_0}$:

$$\kappa_{i,j} = \frac{2-\delta_{i,0}}{\pi}\frac{\Gamma\left(\frac{j-i}{2}+\frac{1}{2}\right)}{\Gamma\left(\frac{j-i}{2}+1\right)}\frac{\Gamma\left(\frac{j+i}{2}+\frac{1}{2}\right)}{\Gamma\left(\frac{j+i}{2}+1\right)},$$

$$\bar{\kappa}_{i,j} = \frac{1}{\sqrt{\pi}}\frac{2^i}{2^j}\frac{\Gamma(j+1)}{\Gamma\left(j+\frac{1}{2}\right)}\frac{\Gamma\left(\frac{j-i}{2}+\frac{1}{2}\right)}{\Gamma\left(\frac{j-i}{2}+1\right)}\frac{\Gamma\left(\frac{j+i}{2}+\frac{1}{2}\right)}{\Gamma\left(\frac{j+i}{2}+1\right)},$$

$$\tilde{\kappa}_{i,j} = \sqrt{\frac{2-\delta_{i,0}}{2}\frac{2j+1}{\pi}}\frac{\Gamma\left(\frac{j-i}{2}+\frac{1}{2}\right)}{\Gamma\left(\frac{j-i}{2}+1\right)}\frac{\Gamma\left(\frac{j+i}{2}+\frac{1}{2}\right)}{\Gamma\left(\frac{j+i}{2}+1\right)}.$$

$\{P_n\}_{n\in\mathbb{N}_0} \to \{U_n\}_{n\in\mathbb{N}_0}$:

$$\kappa_{i,j} = -\frac{i+1}{2\pi}\frac{\Gamma\left(\frac{j-i}{2}-\frac{1}{2}\right)}{\Gamma\left(\frac{j-i}{2}+1\right)}\frac{\Gamma\left(\frac{j+i}{2}+\frac{1}{2}\right)}{\Gamma\left(\frac{j+i}{2}+2\right)},$$

$$\bar{\kappa}_{i,j} = -\frac{i+1}{2\sqrt{\pi}}\frac{2^i}{2^j}\frac{\Gamma(j+1)}{\Gamma\left(j+\frac{1}{2}\right)}\frac{\Gamma\left(\frac{j-i}{2}-\frac{1}{2}\right)}{\Gamma\left(\frac{j-i}{2}+1\right)}\frac{\Gamma\left(\frac{j+i}{2}+\frac{1}{2}\right)}{\Gamma\left(\frac{j+i}{2}+2\right)},$$

$$\tilde{\kappa}_{i,j} = -\frac{i+1}{4}\sqrt{\frac{2j+1}{\pi}}\frac{\Gamma\left(\frac{j-i}{2}-\frac{1}{2}\right)}{\Gamma\left(\frac{j-i}{2}+1\right)}\frac{\Gamma\left(\frac{j+i}{2}+\frac{1}{2}\right)}{\Gamma\left(\frac{j+i}{2}+2\right)}.$$

A.1.5 Chebyshev polynomials of first kind.

Symbol

$$T_n : \mathbb{R} \to \mathbb{R}, \quad n \in \mathbb{N}_0.$$

Inner product and weight function

$$\langle f, g \rangle = \int_I f(x) g(x)\, w(x)\, \mathrm{d}x, \quad I = [-1, 1], \quad w(x) = \frac{1}{\sqrt{1-x^2}}.$$

Differential equation

$$\sigma(x) y''(x) + \tau(x) y'(x) + \lambda_n y(x) = 0, \quad y = T_n,$$
$$\sigma(x) = 1 - x^2, \quad \tau(x) = -x, \quad \lambda_n = n^2.$$

Differential operator

$$\mathcal{D} = -(1-x^2)\frac{\mathrm{d}^2}{\mathrm{d}x^2} + x\frac{\mathrm{d}}{\mathrm{d}x}.$$

Rodrigues formula

$$T_n(x) = \frac{(-1)^n}{2^n} \frac{\sqrt{\pi(1-x^2)}}{\Gamma\left(n+\frac{1}{2}\right)} \frac{\mathrm{d}^n}{\mathrm{d}x^n}\left((1-x^2)^{n-1/2}\right),$$

$$\frac{\mathrm{d}^m}{\mathrm{d}x^m} T_n(x) = \frac{(-1)^{n-m}\sqrt{\pi}n\Gamma(n+m)(1-x^2)^{\frac{1}{2}-m}}{2^n\Gamma(n-m+1)\Gamma(n+1/2)} \frac{\mathrm{d}^{n-m}}{\mathrm{d}x^{n-m}}\left((1-x^2)^{n-\frac{1}{2}}\right).$$

Leading coefficients

$$k_n = \begin{cases} 1, & \text{if } n = 0, \\ 2^{n-1}, & \text{else,} \end{cases} \qquad \bar{k}_n = 1, \qquad \tilde{k}_n = \begin{cases} \pi^{-1/2}, & \text{if } n = 0, \\ \pi^{-1/2} 2^{n-1/2}, & \text{else.} \end{cases}$$

Squared norms

$$h_n = \begin{cases} \pi, & \text{if } n = 0, \\ \pi/2, & \text{else,} \end{cases} \qquad \bar{h}_n = \begin{cases} \pi, & \text{if } n = 0, \\ 2^{1-2n}\pi, & \text{else,} \end{cases} \qquad \tilde{h}_n = 1.$$

Three-term recurrence

$$a_n = \begin{cases} 1, & \text{if } n = 0, \\ 2, & \text{else,} \end{cases} \qquad a'_n = \begin{cases} 1, & \text{if } n = 0, \\ 1/2, & \text{else,} \end{cases}$$
$$b_n = 0, \qquad b'_n = 0,$$
$$c_n = 1, \qquad c'_n = 1/2,$$

$$\bar{a}_n = 1, \qquad \bar{a}'_n = 1,$$
$$\bar{b}_n = 0, \qquad \bar{b}'_n = 0,$$
$$\bar{c}_n = \begin{cases} 1/2, & \text{if } n = 1, \\ 1/4, & \text{else,} \end{cases} \qquad \bar{c}'_n = \begin{cases} 1/2, & \text{if } n = 1, \\ 1/4, & \text{else,} \end{cases}$$

$$\tilde{a}_n = \begin{cases} 2, & \text{if } n = 0, \\ \sqrt{2}, & \text{else,} \end{cases} \qquad \tilde{a}'_n = \begin{cases} 1/2, & \text{if } n = 0, \\ 1/\sqrt{2}, & \text{else,} \end{cases}$$

$$\tilde{b}_n = 0, \qquad \tilde{b}'_n = 0,$$

$$\tilde{c}_n = \begin{cases} \sqrt{2}, & \text{if } n = 1, \\ 1, & \text{else,} \end{cases} \qquad \tilde{c}'_n = \begin{cases} 1/\sqrt{2}, & \text{if } n = 1, \\ 1/2, & \text{else.} \end{cases}$$

Derivative identities

$$\frac{\mathrm{d}}{\mathrm{d}x} T_n(x) = \frac{\sqrt{\pi}}{2} \frac{\Gamma(n+2)}{\Gamma(n+1/2)} P_{n-1}^{(1/2,1/2)}(x) = nU_{n-1}(x),$$

$$\frac{\mathrm{d}^m}{\mathrm{d}x^m} T_n(x) = \frac{\sqrt{\pi}}{2^m} \frac{n\Gamma(n+m)}{\Gamma(n+1/2)} P_{n-m}^{(m-1/2,m-1/2)}(x),$$

$$\frac{\mathrm{d}}{\mathrm{d}x} T_n(x) = A_n \sum_{j=0}^{n-1} B_j T_j(x) + C_n \sum_{j=0}^{n-1} D_j T_j(x) = A'_n \sum_{j=0}^{\lfloor \frac{n-1}{2} \rfloor} B'_{2j+\chi} T_{2j+\chi}(x),$$

$$A_n = (-1)^{n+1} \begin{cases} 1/2, & \text{if } n = 0, \\ n, & \text{else,} \end{cases} \qquad B_j = (-1)^j,$$

$$C_n = \begin{cases} 1/2, & \text{if } n = 0, \\ n, & \text{else,} \end{cases} \qquad D_j = 1,$$

$$A'_n = \begin{cases} 1, & \text{if } n = 0, \\ 2n, & \text{else,} \end{cases} \qquad B'_j = 1,$$

$$\frac{\mathrm{d}}{\mathrm{d}x} \bar{T}_n(x) = n\bar{P}_{n-1}^{(1/2,1/2)}(x) = n\bar{U}_{n-1}(x),$$

$$\frac{\mathrm{d}^m}{\mathrm{d}x^m} \bar{T}_n(x) = \frac{\Gamma(n+1)}{\Gamma(n-m+1)} \bar{P}_{n-m}^{(m-1/2,m-1/2)}(x),$$

$$\frac{\mathrm{d}}{\mathrm{d}x} \bar{T}_n(x) = \bar{A}_n \sum_{j=0}^{n-1} \bar{B}_j \bar{T}_j(x) + \bar{C}_n \sum_{j=0}^{n-1} \bar{D}_j \bar{T}_j(x) = \tilde{A}'_n \sum_{j=0}^{\lfloor \frac{n-1}{2} \rfloor} \tilde{B}'_{2j+\chi} \bar{T}_{2j+\chi}(x),$$

$$\bar{A}_n = (-1)^{n+1} \frac{n}{2^n}, \qquad \bar{B}_j = (-1)^j 2^j,$$

$$\bar{C}_n = \frac{n}{2^n}, \qquad \bar{D}_j = 2^j,$$

$$\bar{A}'_n = \frac{2n}{2^n}, \qquad \bar{B}'_j = 2^j,$$

$$\frac{\mathrm{d}}{\mathrm{d}x}\tilde{T}_n(x) = n\tilde{P}_{n-1}^{(1/2,1/2)}(x) = n\tilde{U}_{n-1}(x),$$

$$\frac{\mathrm{d}^m}{\mathrm{d}x^m}\tilde{T}_n(x) = \sqrt{\frac{n\Gamma(n+m)}{\Gamma(n-m+1)}}\tilde{P}_{n-m}^{(m-1/2,m-1/2)}(x),$$

$$\frac{\mathrm{d}}{\mathrm{d}x}\tilde{T}_n(x) = \tilde{A}_n\sum_{j=0}^{n-1}\tilde{B}_j\tilde{T}_j(x) + \tilde{C}_n\sum_{j=0}^{n-1}\tilde{D}_j\tilde{T}_j(x) = \tilde{A}'_n\sum_{j=0}^{\lfloor\frac{n-1}{2}\rfloor}\tilde{B}'_{2j+\chi}\tilde{T}_{2j+\chi}(x),$$

$$\tilde{A}_n = (-1)^{n+1}\begin{cases}1/\sqrt{2}, & \text{if } n=0,\\ n, & \text{else,}\end{cases} \qquad \tilde{B}_j = (-1)^j,$$

$$\tilde{C}_n = \begin{cases}1/\sqrt{2}, & \text{if } n=0,\\ n, & \text{else,}\end{cases} \qquad \tilde{D}_j = 1,$$

$$\tilde{A}'_n = \begin{cases}\sqrt{2}, & \text{if } n=0,\\ 2n, & \text{else,}\end{cases} \qquad \tilde{B}'_j = 1.$$

Connection coefficients

$\{T_n\}_{n\in\mathbb{N}_0} \to \{U_n\}_{n\in\mathbb{N}_0}$:

$$\kappa_{i,j} = \begin{cases}1, & \text{if } j=i=0,\\ 1/2, & \text{if } j=i,\\ -1/2, & \text{if } j=i+2,\end{cases}$$

$$\bar{\kappa}_{i,j} = \begin{cases}1, & \text{if } j=i,\\ -1/4, & \text{if } j=i+2,\end{cases}$$

$$\tilde{\kappa}_{i,j} = \begin{cases}1/\sqrt{2}, & \text{if } j=i=0,\\ 1/2, & \text{if } j=i,\\ -1/2, & \text{if } j=i+2.\end{cases}$$

$\{T_n\}_{n\in\mathbb{N}_0} \to \{P_n\}_{n\in\mathbb{N}_0}$:

$$\kappa_{i,j} = -\frac{j(2i+1)}{8}\frac{\Gamma\left(\frac{j-i}{2}-\frac{1}{2}\right)}{\Gamma\left(\frac{j-i}{2}+1\right)}\frac{\Gamma\left(\frac{j+i}{2}\right)}{\Gamma\left(\frac{j+i}{2}+\frac{3}{2}\right)},$$

$$\bar{\kappa}_{i,j} = -\frac{j}{2\sqrt{\pi}}\frac{2^i}{2^j}\frac{\Gamma\left(i+\frac{3}{2}\right)}{\Gamma(i+1)}\frac{\Gamma\left(\frac{j-i}{2}-\frac{1}{2}\right)}{\Gamma\left(\frac{j-i}{2}+1\right)}\frac{\Gamma\left(\frac{j+i}{2}\right)}{\Gamma\left(\frac{j+i}{2}+\frac{3}{2}\right)},$$

$$\tilde{\kappa}_{i,j} = -\frac{j}{4}\sqrt{\frac{2i+1}{\pi}}\frac{\Gamma\left(\frac{j-i}{2}-\frac{1}{2}\right)}{\Gamma\left(\frac{j-i}{2}+1\right)}\frac{\Gamma\left(\frac{j+i}{2}\right)}{\Gamma\left(\frac{j+i}{2}+\frac{3}{2}\right)}.$$

A.1.6 Chebyshev polynomials of second kind.

Symbol

$$U_n : \mathbb{R} \to \mathbb{R}, \quad n \in \mathbb{N}_0.$$

Inner product and weight function

$$\langle f, g\rangle = \int_I f(x)g(x)\, w(x)\, \mathrm{d}x, \quad I = [-1,1], \quad w(x) = \sqrt{1-x^2}.$$

Differential equation

$$\sigma(x)y''(x) + \tau(x)y'(x) + \lambda_n y(x) = 0, \quad y = U_n,$$
$$\sigma(x) = 1-x^2, \quad \tau(x) = -3x, \quad \lambda_n = n(n+2).$$

Differential operator

$$\mathcal{D} = -(1-x^2)\frac{\mathrm{d}^2}{\mathrm{d}x^2} + 3x\frac{\mathrm{d}}{\mathrm{d}x}.$$

Rodrigues formula

$$U_n(x) = \frac{(-1)^n}{2^{n+1}}\frac{(n+1)}{\Gamma\left(n+\frac{3}{2}\right)}\sqrt{\frac{\pi}{1-x^2}}\frac{\mathrm{d}^n}{\mathrm{d}x^n}\left((1-x^2)^{n+1/2}\right),$$

$$\frac{\mathrm{d}^m}{\mathrm{d}x^m}U_n(x) = \frac{(-1)^{n-m}\sqrt{\pi}\Gamma(n+m+2)}{2^{n+1}\Gamma(n-m+1)\Gamma\left(n+\frac{3}{2}\right)(1-x^2)^{m+1/2}}\frac{\mathrm{d}^{n-m}}{\mathrm{d}x^{n-m}}\left((1-x^2)^{n+\frac{1}{2}}\right).$$

Leading coefficients

$$k_n = 2^n, \qquad \bar{k}_n = 1, \qquad \tilde{k}_n = \sqrt{\frac{2}{\pi}}2^n.$$

Squared norms

$$h_n = \frac{\pi}{2}, \qquad \bar{h}_n = \frac{\pi}{2}2^{-2n}, \qquad \tilde{h}_n = 1.$$

Three-term recurrence

$$a_n = 2, \qquad a'_n = 1/2,$$
$$b_n = 0, \qquad b'_n = 0,$$
$$c_n = 1, \qquad c'_n = 1/2,$$

$$\bar{a}_n = 1, \qquad \bar{a}'_n = 1,$$
$$\bar{b}_n = 0, \qquad \bar{b}'_n = 0,$$
$$\bar{c}_n = 1/4, \qquad \bar{c}'_n = 1/4,$$

$$\tilde{a}_n = 2, \qquad \tilde{a}'_n = 1/2,$$
$$\tilde{b}_n = 0, \qquad \tilde{b}'_n = 0,$$
$$\tilde{c}_n = 1, \qquad \tilde{c}'_n = 1/2.$$

Derivative identities

$$\frac{\mathrm{d}}{\mathrm{d}x}U_n(x) = \frac{\sqrt{\pi}}{4}\frac{\Gamma(n+3)}{\Gamma\left(n+\frac{3}{2}\right)}P_{n-1}^{(3/2,3/2)}(x),$$

$$\frac{\mathrm{d}^m}{\mathrm{d}x^m}U_n(x) = \frac{\sqrt{\pi}}{2^{m+1}}\frac{\Gamma(n+m+2)}{\Gamma\left(n+\frac{3}{2}\right)}P_{n-m}^{(m+1/2,m+1/2)}(x),$$

$$\frac{\mathrm{d}}{\mathrm{d}x}U_n(x) = A_n\sum_{j=0}^{n-1}B_jT_j(x) + C_n\sum_{j=0}^{n-1}D_jT_j(x) = A'_n\sum_{j=0}^{\left\lfloor\frac{n-1}{2}\right\rfloor}B'_{2j+\chi}U_{2j+\chi}(x),$$

$$A_n = (-1)^{n+1}, \qquad B_j = (-1)^j(j+1),$$

$$C_n = 1, \qquad D_j = j+1$$

$$A'_n = 2, \qquad B'_j = j+1,$$

$$\frac{\mathrm{d}}{\mathrm{d}x}\bar{U}_n(x) = n\bar{P}_{n-1}^{(3/2,3/2)}(x),$$

$$\frac{\mathrm{d}^m}{\mathrm{d}x^m}\bar{U}_n(x) = \frac{\Gamma(n+1)}{\Gamma(n-m+1)}\bar{P}_{n-m}^{(m+1/2,m+1/2)}(x),$$

$$\frac{\mathrm{d}}{\mathrm{d}x}\bar{U}_n(x) = \bar{A}_n\sum_{j=0}^{n-1}\bar{B}_j\bar{U}_j(x) + \bar{C}_n\sum_{j=0}^{n-1}\bar{B}_j\bar{U}_j(x) = \bar{A}'_n\sum_{j=0}^{\left\lfloor\frac{n-1}{2}\right\rfloor}\bar{B}'_{2j+\chi}\bar{U}_{2j+\chi}(x),$$

$$\bar{A}_n = (-1)^{n+1}2^{-n}, \qquad \bar{B}_j = (-1)^j2^j(j+1),$$

$$\bar{C}_n = 2^{-n}, \qquad \bar{D}_j = 2^j(j+1),$$

$$\bar{A}'_n = 2^{1-n}, \qquad \bar{B}'_j = 2^j(j+1),$$

$$\frac{\mathrm{d}}{\mathrm{d}x}\tilde{U}_n(x) = \sqrt{n(n+2)}\tilde{P}_{n-1}^{(3/2,3/2)}(x),$$

$$\frac{\mathrm{d}^m}{\mathrm{d}x^m}\tilde{U}_n(x) = \sqrt{\frac{\Gamma(n+m+2)}{(n+1)\Gamma(n-m+1)}}\tilde{P}_{n-m}^{(m+1/2,m+1/2)}(x),$$

$$\frac{\mathrm{d}}{\mathrm{d}x}\tilde{U}_n(x) = \tilde{A}_n\sum_{j=0}^{n-1}\tilde{B}_j\tilde{U}_j(x) + \tilde{C}_n\sum_{j=0}^{n-1}\tilde{D}_j\tilde{U}_j(x) = \tilde{A}'_n\sum_{j=0}^{\left\lfloor\frac{n-1}{2}\right\rfloor}\tilde{B}'_j\tilde{U}_{2j+\chi}(x),$$

$$\tilde{A}_n = (-1)^{n+1}, \qquad \tilde{B}_j = (-1)^j(j+1),$$

$$\tilde{C}_n = 1, \qquad \tilde{D}_j = j+1,$$

$$\tilde{A}'_n = 2, \qquad \tilde{B}'_j = j+1.$$

Connection coefficients

$\{U_n\}_{n\in\mathbb{N}_0} \to \{T_n\}_{n\in\mathbb{N}_0}$:

$$\kappa_{i,j} = \begin{cases} 1, & \text{if } i = 0 \text{ and } j \text{ even}, \\ 2, & \text{if } i + j \text{ even}, \\ 0, & \text{else}, \end{cases}$$

$$\bar{\kappa}_{i,j} = \begin{cases} 2^{i-j}, & \text{if } i + j \text{ even}, \\ 0, & \text{else}, \end{cases}$$

$$\tilde{\kappa}_{i,j} = \begin{cases} \sqrt{2}, & \text{if } i = 0 \text{ and } j \text{ even}, \\ 2, & \text{if } i + j \text{ even}, \\ 0, & \text{else}. \end{cases}$$

$\{U_n\}_{n\in\mathbb{N}_0} \to \{P_n\}_{n\in\mathbb{N}_0}$:

$$\kappa_{i,j} = \frac{2i+1}{2} \frac{\Gamma\left(\frac{j-i}{2} + \frac{1}{2}\right)}{\Gamma\left(\frac{j-i}{2} + 1\right)} \frac{\Gamma\left(\frac{j+i}{2} + 1\right)}{\Gamma\left(\frac{j+i}{2} + \frac{3}{2}\right)},$$

$$\bar{\kappa}_{i,j} = \frac{1}{\sqrt{\pi}} \frac{2^i}{2^j} \frac{\Gamma\left(i + \frac{3}{2}\right)}{\Gamma(i+1)} \frac{\Gamma\left(\frac{j-i}{2} + \frac{1}{2}\right)}{\Gamma\left(\frac{j-i}{2} + 1\right)} \frac{\Gamma\left(\frac{j+i}{2} + 1\right)}{\Gamma\left(\frac{j+i}{2} + \frac{3}{2}\right)},$$

$$\tilde{\kappa}_{i,j} = \frac{\sqrt{2i+1}}{\sqrt{\pi}} \frac{\Gamma\left(\frac{j-i}{2} + \frac{1}{2}\right)}{\Gamma\left(\frac{j-i}{2} + 1\right)} \frac{\Gamma\left(\frac{j+i}{2} + 1\right)}{\Gamma\left(\frac{j+i}{2} + \frac{3}{2}\right)}.$$

A.2 Classical associated functions

A.2.1 Associated Laguerre functions.

Symbol

$$L_n^{(\alpha),m} : \mathbb{R} \to \mathbb{R}, \quad n \in \mathbb{N}_0,\ n \geq |m|,\ \alpha > -1,\ \mu := |m|.$$

Inner product and weight function

$$\langle f, g\rangle = \int_I f(x)g(x)\, w(x)\, \mathrm{d}x, \quad I = [0, \infty), \quad w(x) = x^{\alpha}\mathrm{e}^{-x}.$$

Differential equation

$$\sigma(x)y''(x) + \tau(x)y'(x) + \big(\lambda_n + f_\mu(x)\big)y(x) = 0, \quad y = L_n^{(\alpha),m},$$

$$\sigma(x) = x, \quad \tau(x) = -x + \alpha + 1, \quad \lambda_n = n,$$

$$f_\mu(x) = -\frac{\mu\big(2(x+\alpha)+\mu\big)}{4x}.$$

Differential operator

$$\mathcal{D} = -x\frac{\mathrm{d}^2}{\mathrm{d}x^2} + (x - \alpha - 1)\frac{\mathrm{d}}{\mathrm{d}x} + \frac{\mu\big(2(x+\alpha)+\mu\big)}{4x}.$$

Rodrigues formula

$$\begin{aligned} L_n^{(\alpha),m}(x) &= \left(\frac{\Gamma(n-\mu+1)}{\Gamma(n+1)}\right)^{1/2} x^{\mu/2}\frac{\mathrm{d}^\mu}{\mathrm{d}x^\mu}L_n^{(\alpha)}(x) \\ &= (-1)^\mu \left(\frac{\Gamma(n-\mu+1)}{\Gamma(n+1)}\right)^{1/2} x^{\mu/2} L_{n-\mu}^{(\alpha+\mu)}(x) \\ &= \frac{(-1)^\mu \mathrm{e}^x x^{-\alpha-\mu/2}}{\sqrt{\Gamma(n+1)\Gamma(n-\mu+1)}}\frac{\mathrm{d}^{n-\mu}}{\mathrm{d}x^{n-\mu}}\left(x^{n+\alpha}\mathrm{e}^{-x}\right). \end{aligned}$$

The matrix $\mathbf{G}$

$\{L_n^{(\alpha),m}\}_{n=\mu}^N \to \{L_n^{(\alpha),\hat{m}}\}_{n=\hat{\mu}}^N$:

$$\mathbf{G} = \mathrm{diag}(\mathbf{d}) + \frac{\mu-\hat{\mu}}{4}\left(2 + \frac{\mu-\hat{\mu}}{\alpha+\mu}\right)\left(\mathrm{triu}(\mathbf{u}\,\mathbf{v}^{\mathrm{T}}, 1) + \mathrm{tril}(\mathbf{v}\,\mathbf{u}^{\mathrm{T}}, -1)\right),$$

$$\mathbf{d} = \left(j + \frac{\mu-\hat{\mu}}{4}\left(4 + \frac{\mu-\hat{\mu}}{\alpha+\hat{\mu}}\right)\right)_{j=\hat{\mu}}^{N-(\mu-\hat{\mu})/2},$$

$$\mathbf{u} = \left(\sqrt{\frac{\Gamma(j+\alpha+1)}{\Gamma(j-\hat{\mu}+1)}}\right)_{j=\hat{\mu}}^{N-(\mu-\hat{\mu})/2},$$

$$\mathbf{v} = \left(\sqrt{\frac{\Gamma(j-\hat{\mu}+1)}{\Gamma(j+\alpha+1)}}\right)_{j=\hat{\mu}}^{N-(\mu-\hat{\mu})/2}.$$

A.2.2 Generalized associated Jacobi functions.

Symbol

$$P_n^{(\alpha,\beta),m,m'}:\mathbb{R}\to\mathbb{R},\quad n\in\mathbb{N}_0,\ n\geq n^*,\ \alpha,\beta>-1,$$
$$n^*:=\max\{|m|,|m'|\},\ \mu:=|m'-m|,\ \nu:=|m'+m|.$$

Inner product and weight function

$$\langle f,g\rangle=\int_I f(x)g(x)\,w(x)\,\mathrm{d}x,\quad I=[-1,1],\quad w(x)=(1-x)^\alpha(1+x)^\beta.$$

Differential equation

$$\sigma(x)y''(x)+\tau(x)y'(x)+\big(\lambda_n+f_{\mu,\nu}(x)\big)y(x)=0,\quad y=P_n^{(\alpha,\beta),m,m'},$$
$$\sigma(x)=1-x^2,\quad \tau(x)=-(\alpha+\beta+2)x+\beta-\alpha,\quad \lambda_n=n(n+\alpha+\beta+1),$$
$$f_{\mu,\nu}(x):=-\left(\frac{\mu(2\alpha+\mu)}{2(1-x)}+\frac{\nu(2\beta+\nu)}{2(1+x)}\right).$$

Differential operator

$$\mathcal{D}=-(1-x^2)\frac{\mathrm{d}^2}{\mathrm{d}x^2}+\big((\alpha+\beta+2)x+\alpha-\beta\big)\frac{\mathrm{d}}{\mathrm{d}x}+\frac{\mu(2\alpha+\mu)}{2(1-x)}+\frac{\nu(2\beta+\nu)}{2(1+x)}.$$

Rodrigues formula

$$\begin{aligned}P_n^{(\alpha,\beta),m,m'}(x)&:=C_{n,\mu,\nu}(1-x)^{\mu/2}(1+x)^{\nu/2}P_{n-n^*}^{(\alpha+\mu,\beta+\nu)}(x)\\&=\frac{(-1)^{n-n^*}C_{n,\mu,\nu}}{2^{n-n^*}\Gamma(n-n^*+1)(1-x)^{\alpha+\mu/2}(1+x)^{\beta+\nu/2}}\\&\quad\times\frac{\mathrm{d}^{n-n^*}}{\mathrm{d}x^{n-n^*}}\Big((1-x)^{n-n^*+\alpha+\mu}(1+x)^{n-n^*+\beta+\nu}\Big),\end{aligned}$$

$$\begin{aligned}C_{n,\mu,\nu}=&2^{-n^*}\left(\frac{\Gamma(n+\alpha+1)\Gamma(n+\beta+1)}{\Gamma(n+1)\Gamma(n+\alpha+\beta+1)}\right.\\&\left.\times\frac{\Gamma(n-n^*+1)\Gamma(n+n^*+\alpha+\beta+1)}{\Gamma(n-n^*+\alpha+\mu+1)\Gamma(n-n^*+\beta+\nu+1)}\right)^{1/2}.\end{aligned}$$

The matrix $\mathbf{G}$

$\{P_n^{(\alpha,\beta),m,m'}\}_{n=n^*}^N\to\{P_n^{(\alpha,\beta),\hat m,\hat m'}\}_{n=\hat n^*}^N$:

$$\begin{aligned}\mathbf{G}=\operatorname{diag}(\mathbf{d})&+\frac{(\mu-\hat\mu)(2\alpha+\mu+\hat\mu)}{4(\alpha+\hat\mu)}\left(\operatorname{triu}(\mathbf{u}_1\,\mathbf{v}_1^{\mathrm T},1)+\operatorname{tril}(\mathbf{v}_1\,\mathbf{u}_1^{\mathrm T},-1)\right)\\&+\frac{(\nu-\hat\nu)(2\beta+\nu+\hat\nu)}{4(\beta+\hat\nu)}\left(\operatorname{triu}(\mathbf{u}_2\,\mathbf{v}_2^{\mathrm T},1)+\operatorname{tril}(\mathbf{v}_2\,\mathbf{u}_2^{\mathrm T},-1)\right),\end{aligned}$$

$$\begin{aligned}\mathbf{d}=\Bigg(&j(j+\alpha+\beta+1)\\&+(2j+\alpha+\beta+1)\left(\frac{(\mu-\hat\mu)(2\alpha+\mu+\hat\mu)}{4(\alpha+\hat\mu)}+\frac{(\nu-\hat\nu)(2\beta+\nu+\hat\nu)}{4(\beta+\hat\nu)}\right)\Bigg)_{j=\hat n^*}^N,\end{aligned}$$

$$\mathbf{u}_1 = \left(\left(\frac{(2j+\alpha+\beta+1)\Gamma(j-\hat{n}^*+\alpha+\hat{\mu}+1)\Gamma(j+\hat{n}^*+\alpha+\beta+1)}{\Gamma(j-\hat{n}^*+1)\Gamma(j-\hat{n}^*+\beta+\hat{\nu}+1)} \right)^{1/2} \right)_{j=\hat{n}^*}^{N},$$

$$\mathbf{v}_1 = \left(\left(\frac{(2j+\alpha+\beta+1)\Gamma(j-\hat{n}^*+1)\Gamma(j-\hat{n}^*+\beta+\hat{\nu}+1)}{\Gamma(j-\hat{n}^*+\alpha+\hat{\mu}+1)\Gamma(j+\hat{n}^*+\alpha+\beta+1)} \right)^{1/2} \right)_{j=\hat{n}^*}^{N},$$

$$\mathbf{u}_2 = \left(\left((-1)^j \frac{(2j+\alpha+\beta+1)\Gamma(j-\hat{n}^*+\beta+\hat{\nu}+1)\Gamma(j+\hat{n}^*+\alpha+\beta+1)}{\Gamma(j-\hat{n}^*+1)\Gamma(j-\hat{n}^*+\alpha+\hat{\mu}+1)} \right)^{1/2} \right)_{j=\hat{n}^*}^{N},$$

$$\mathbf{v}_2 = \left(\left((-1)^j \frac{(2j+\alpha+\beta+1)\Gamma(j-\hat{n}^*+1)\Gamma(j-\hat{n}^*+\alpha+\hat{\mu}+1)}{\Gamma(j-\hat{n}^*+\beta+\hat{\nu}+1)\Gamma(j+\hat{n}^*+\alpha+\beta+1)} \right)^{1/2} \right)_{j=\hat{n}^*}^{N}.$$

A.2.3 Associated Jacobi functions.

Symbol

$$P_n^{(\alpha,\beta),m} : \mathbb{R} \to \mathbb{R}, \quad n \in \mathbb{N}_0,\ n \geq |m|,\ \alpha,\beta > -1,\ \mu := |m|.$$

Inner product and weight function

$$\langle f, g\rangle = \int_I f(x)g(x)\, w(x)\, \mathrm{d}x, \quad I = [-1,1], \quad w(x) = (1-x)^\alpha(1+x)^\beta.$$

Differential equation

$$\sigma(x)y''(x) + \tau(x)y'(x) + \big(\lambda_n + f_\mu(x)\big)y(x) = 0, \quad y = P_n^{(\alpha,\beta),m},$$
$$\sigma(x) = 1-x^2, \quad \tau(x) = -(\alpha+\beta+2)x + \beta - \alpha, \quad \lambda_n = n(n+\alpha+\beta+1),$$
$$f_\mu(x) = -\frac{\mu(2\alpha+\mu)}{2(1-x)} - \frac{\mu(2\beta+\mu)}{2(1+x)}.$$

Differential operator

$$\mathcal{D} = -(1-x^2)\frac{\mathrm{d}^2}{\mathrm{d}x^2} + \big((\alpha+\beta+2)x + \alpha - \beta\big)\frac{\mathrm{d}}{\mathrm{d}x} + \frac{\mu(2\alpha+\mu)}{2(1-x)} + \frac{\mu(2\beta+\mu)}{2(1+x)}.$$

Rodrigues formula

$$\begin{aligned} P_n^{(\alpha,\beta),m}(x) &= C_{n,\mu}\big(1-x^2\big)^{\mu/2}\frac{\mathrm{d}^\mu}{\mathrm{d}x^\mu}P_n^{(\alpha,\beta)}(x) \\ &= \frac{(-1)^{n-\mu}}{2^n}\left(\frac{\Gamma(n+\mu+\alpha+\beta+1)}{\Gamma(n-\mu+1)\Gamma(n+1)\Gamma(n+\alpha+\beta+1)}\right)^{1/2} \\ &\quad \times (1-x)^{-\alpha-\mu/2}(1+x)^{-\beta-\mu/2}\frac{\mathrm{d}^{n-\mu}}{\mathrm{d}x^{n-\mu}}\left((1-x)^{n+\alpha}(1+x)^{n+\beta}\right), \end{aligned}$$

$$C_{n,\mu} = \left(\frac{\Gamma(n-\mu+1)\Gamma(n+\alpha+\beta+1)}{\Gamma(n+1)\Gamma(n+\mu+\alpha+\beta+1)}\right)^{1/2}.$$

The matrix $\mathbf{G}$

$\{P_n^{(\alpha,\beta),m}\}_{n=\mu}^N \to \{P_n^{(\alpha,\beta),\hat{m}}\}_{n=\hat{\mu}}^N$:

$$\begin{aligned} \mathbf{G} = \operatorname{diag}(\mathbf{d}) &+ \frac{(\mu-\hat{\mu})(2\alpha+\mu+\hat{\mu})}{4(\alpha+\hat{\mu})}\left(\operatorname{triu}(\mathbf{u}_1\,\mathbf{v}_1^{\mathrm{T}},1) + \operatorname{tril}(\mathbf{v}_1\,\mathbf{u}_1^{\mathrm{T}},-1)\right) \\ &+ \frac{(\mu-\hat{\mu})(2\beta+\mu+\hat{\mu})}{4(\beta+\hat{\mu})}\left(\operatorname{triu}(\mathbf{u}_2\,\mathbf{v}_2^{\mathrm{T}},1) + \operatorname{tril}(\mathbf{v}_2\,\mathbf{u}_2^{\mathrm{T}},-1)\right), \end{aligned}$$

$$\mathbf{d} = \left(j(j+\alpha+\beta+1) + (2j+\alpha+\beta+1)(\mu-\hat{\mu})\left(\frac{2\alpha+\mu+\hat{\mu}}{4(\alpha+\hat{\mu})} + \frac{2\beta+\mu+\hat{\mu}}{4(\beta+\hat{\mu})}\right)\right)_{j=\hat{\mu}}^N,$$

$$\mathbf{u}_1 = \left(\left(\frac{(2j+\alpha+\beta+1)\Gamma(j+\alpha+1)\Gamma(j+\hat{\mu}+\alpha+\beta+1)}{\Gamma(j-\hat{\mu}+1)\Gamma(j+\beta+1)} \right)^{1/2} \right)_{j=\hat{\mu}}^{N},$$

$$\mathbf{v}_1 = \left(\left(\frac{(2j+\alpha+\beta+1)\Gamma(j-\hat{\mu}+1)\Gamma(j+\beta+1)}{\Gamma(j+\alpha+1)\Gamma(j+\hat{\mu}+\alpha+\beta+1)} \right)^{1/2} \right)_{j=\hat{\mu}}^{N},$$

$$\mathbf{u}_2 = \left(\left((-1)^j \frac{(2j+\alpha+\beta+1)\Gamma(j+\beta+1)\Gamma(j+\hat{\mu}+\alpha+\beta+1)}{\Gamma(j-\hat{\mu}+1)\Gamma(j+\alpha+1)} \right)^{1/2} \right)_{j=\hat{\mu}}^{N},$$

$$\mathbf{v}_2 = \left(\left((-1)^j \frac{(2j+\alpha+\beta+1)\Gamma(j-\hat{\mu}+1)\Gamma(j+\alpha+1)}{\Gamma(j+\beta+1)\Gamma(j+\hat{\mu}+\alpha+\beta+1)} \right)^{1/2} \right)_{j=\hat{\mu}}^{N}.$$

A.2.4 Generalized associated Gegenbauer functions.

Symbol

$$C_n^{(\alpha),m,m'}:\mathbb{R}\to\mathbb{R},\quad n\in\mathbb{N}_0,\ n\geq n^*,\ -1/2<\alpha,$$
$$n^*:=\max\{|m|,|m'|\},\ \mu:=|m'-m|,\ \nu:=|m'+m|.$$

Inner product and weight function

$$\langle f,g\rangle=\int_I f(x)g(x)w(x)\,\mathrm{d}x,\quad I=[-1,1],\quad w(x)=(1-x^2)^{\alpha-1/2}.$$

Differential equation

$$\sigma(x)y''(x)+\tau(x)y'(x)+\big(\lambda_n+f_{\mu,\nu}(x)\big)y(x)=0,\quad y=C_n^{(\alpha),m,m'},$$
$$\sigma(x)=1-x^2,\quad \tau(x)=-(2\alpha+1)x,\quad \lambda_n=n(n+2\alpha),$$
$$f_{\mu,\nu}(x)=-\frac{\mu(2\alpha+\mu-1)}{2(1-x)}-\frac{\nu(2\alpha+\nu-1)}{2(1+x)}.$$

Differential operator

$$\mathcal{D}=-(1-x^2)\frac{\mathrm{d}^2}{\mathrm{d}x^2}+(2\alpha+1)x\frac{\mathrm{d}}{\mathrm{d}x}+\frac{\mu(2\alpha+\mu-1)}{2(1-x)}+\frac{\nu(2\alpha+\nu-1)}{2(1+x)}.$$

Rodrigues formula

$$\begin{aligned}
C_n^{(\alpha),m,m'}(x)&=\frac{\Gamma(\alpha+1/2)\Gamma(n+2\alpha)}{\Gamma(n+\alpha+1/2)\Gamma(2\alpha)}P_n^{(\alpha-1/2,\alpha-1/2),m,m'}(x)\\
&=C_{n,\mu,\nu}(1-x)^{\mu/2}(1+x)^{\nu/2}P_{n-n^*}^{(\alpha+\mu-1/2,\alpha+\nu-1/2)}(x),\\
&=\frac{(-1)^{n-n^*}C_{n,\mu,\nu}}{2^{n-n^*}\Gamma(n-n^*+1)(1-x)^{\alpha+\mu/2-1/2}(1+x)^{\alpha+\nu/2-1/2}}\\
&\qquad\times\frac{\mathrm{d}^{n-n^*}}{\mathrm{d}x^{n-n^*}}\Big((1-x)^{n-n^*+\alpha+\mu-1/2}(1+x)^{n-n^*+\alpha+\nu-1/2}\Big),
\end{aligned}$$

$$C_{n,\mu,\nu}=\frac{\Gamma(\alpha+1/2)}{2^{n^*}\Gamma(2\alpha)}\left(\frac{\Gamma(n-n^*+1)\Gamma(n+2\alpha)\Gamma(n+n^*+2\alpha)}{\Gamma(n+1)\Gamma(n-n^*+\alpha+\mu+1/2)\Gamma(n-n^*+\alpha+\nu+1/2)}\right)^{1/2}.$$

The matrix $\mathbf{G}$

$\{C_n^{(\alpha),m,m'}\}_{n=n^*}^N\to\{C_n^{(\alpha),\hat{m},\hat{m}'}\}_{n=\hat{n}^*}^N$:

$$\begin{aligned}
\mathbf{G}=\operatorname{diag}(\mathbf{d})&+\frac{(\mu-\hat{\mu})(2\alpha+\mu+\hat{\mu}-1)}{2\alpha+2\hat{\mu}-1}\left(\operatorname{triu}(\mathbf{u}_1\mathbf{v}_1^{\mathrm{T}},1)+\operatorname{tril}(\mathbf{v}_1\mathbf{u}_1^{\mathrm{T}},-1)\right)\\
&+\frac{(\nu-\hat{\nu})(2\alpha+\nu+\hat{\nu}-1)}{2\alpha+2\hat{\nu}-1}\left(\operatorname{triu}(\mathbf{u}_2\mathbf{v}_2^{\mathrm{T}},1)+\operatorname{tril}(\mathbf{v}_2\mathbf{u}_2^{\mathrm{T}},-1)\right),
\end{aligned}$$

$$\mathbf{d} = \left(j(j+2\alpha) + (j+\alpha) \left(\frac{(\mu-\hat{\mu})(2\alpha+\mu+\hat{\mu}-1)}{2\alpha+2\hat{\mu}-1} + \frac{(\nu-\hat{\nu})(2\alpha+\nu+\hat{\nu}-1)}{2\alpha+2\hat{\nu}-1} \right) \right)_{j=\hat{n}^*}^{N},$$

$$\mathbf{u}_1 = \left(\left(\frac{(j+\alpha)\Gamma(j-\hat{n}^*+\alpha+\hat{\mu}+1/2)\Gamma(j+\hat{n}^*+2\alpha)}{\Gamma(j-\hat{n}^*+1)\Gamma(j-\hat{n}^*+\alpha+\hat{\nu}+1/2)} \right)^{1/2} \right)_{j=\hat{n}^*}^{N},$$

$$\mathbf{v}_1 = \left(\left(\frac{(j+\alpha)\Gamma(j-\hat{n}^*+1)\Gamma(j-\hat{n}^*+\alpha+\hat{\nu}+1/2)}{\Gamma(j-\hat{n}^*+\alpha+\hat{\mu}+1/2)\Gamma(j+\hat{n}^*+2\alpha)} \right)^{1/2} \right)_{j=\hat{n}^*}^{N},$$

$$\mathbf{u}_2 = \left(\left((-1)^j \frac{(j+\alpha)\Gamma(j-\hat{n}^*+\alpha+\hat{\nu}+1/2)\Gamma(j+\hat{n}^*+2\alpha)}{\Gamma(j-\hat{n}^*+1)\Gamma(j-\hat{n}^*+\alpha+\hat{\mu}+1/2)} \right)^{1/2} \right)_{j=\hat{n}^*}^{N},$$

$$\mathbf{v}_2 = \left(\left((-1)^j \frac{(j+\alpha)\Gamma(j-\hat{n}^*+1)\Gamma(j-\hat{n}^*+\alpha+\hat{\mu}+1/2)}{\Gamma(j-\hat{n}^*+\alpha+\hat{\nu}+1/2)\Gamma(j+\hat{n}^*+2\alpha)} \right)^{1/2} \right)_{j=\hat{n}^*}^{N}.$$

A.2.5 Associated Gegenbauer functions.

Symbol

$$C_n^{(\alpha),m} : \mathbb{R} \to \mathbb{R}, \quad n \in \mathbb{N}_0,\ -1/2 < \alpha,\ |m| \le n,\ \mu := |m|.$$

Inner product and weight function

$$\langle f, g \rangle = \int_I f(x) g(x) w(x)\, \mathrm{d}x, \quad I = [-1,1], \quad w(x) = (1-x^2)^{\alpha-1/2}.$$

Differential equation

$$\sigma(x) y''(x) + \tau(x) y'(x) + \big(\lambda_n + f_\mu(x)\big) y(x) = 0, \quad y = C_n^{(\alpha),m},$$

$$\sigma(x) = 1 - x^2, \quad \tau(x) = -(2\alpha+1)x, \quad \lambda_n = n(n+2\alpha),$$

$$f_\mu(x) = -\frac{\mu(2\alpha+\mu-1)}{1-x^2}.$$

Differential operator

$$\mathcal{D} = -(1-x^2)\frac{\mathrm{d}^2}{\mathrm{d}x^2} + (2\alpha+1)x\frac{\mathrm{d}}{\mathrm{d}x} + \frac{\mu(2\alpha+\mu-1)}{1-x^2}.$$

Rodrigues formula

$$\begin{aligned}
C_n^{(\alpha)}(x) &= \frac{\Gamma(\alpha+1/2)\Gamma(n+2\alpha)}{\Gamma(n+\alpha+1/2)\Gamma(2\alpha)} P_n^{(\alpha-1/2,\alpha-1/2),m}(x) \\
&= C_{n,\mu}(1-x^2)^{\mu/2}\frac{\mathrm{d}^\mu}{\mathrm{d}x^\mu} C_n^{(\alpha)}(x) \\
&= \frac{(-1)^{n-\mu}\Gamma(\alpha+\frac{1}{2})}{2^n\Gamma(2\alpha)\Gamma(n+\alpha+1/2)} \left(\frac{\Gamma(n+2\alpha)\Gamma(n+\mu+2\alpha)}{\Gamma(n-\mu+1)\Gamma(n+1)}\right)^{1/2} \\
&\quad \times (1-x^2)^{1/2-\alpha-\mu/2} \frac{\mathrm{d}^{n-\mu}}{\mathrm{d}x^{n-\mu}}\left((1-x^2)^{n+\alpha-1/2}\right),
\end{aligned}$$

$$C_{n,\mu} = \left(\frac{\Gamma(n-\mu+1)\Gamma(n+2\alpha)}{\Gamma(n+1)\Gamma(n+\mu+2\alpha)}\right)^{1/2}.$$

The matrix **G**

$\{C_n^{(\alpha),m}\}_{n=\mu}^N \to \{C_n^{(\alpha),\hat{m}}\}_{n=\hat{\mu}}^N$:

$$\mathbf{G} = \mathrm{diag}(\mathbf{d}) + 2(\mu-\hat{\mu})\frac{2\alpha+\mu+\hat{\mu}-1}{2\alpha+2\hat{\mu}-1}\left(\mathrm{triuc}(\mathbf{u}\,\mathbf{v}^{\mathrm{T}},1) + \mathrm{trilc}(\mathbf{v}\,\mathbf{u}^{\mathrm{T}},-1)\right),$$

$$\mathbf{d} = \left(j(j+2\alpha) + 2(j+\alpha)(\mu-\hat{\mu})\frac{2\alpha+\mu+\hat{\mu}-1}{2\alpha+2\hat{\mu}-1}\right)_{j=\hat{\mu}}^N,$$

$$\mathbf{u} = \left(\left(\frac{(j+\alpha)\Gamma(j+\hat{\mu}+2\alpha)}{\Gamma(j-\hat{\mu}+1)}\right)^{1/2}\right)_{j=\hat{\mu}}^N,$$

$$\mathbf{v} = \left(\left(\frac{(j+\alpha)\Gamma(j-\hat{\mu}+1)}{\Gamma(j+\hat{\mu}+2\alpha)}\right)^{1/2}\right)_{j=\hat{\mu}}^N.$$

A.2.6 Generalized associated Legendre functions.

Symbol

$$P_n^{m,m'} : \mathbb{R} \to \mathbb{R}, \quad n \in \mathbb{N}_0,\ n \geq n^*,$$
$$n^* := \max\{|m|, |m'|\},\ \mu := |m' - m|,\ \nu := |m' + m|.$$

Inner product and weight function

$$\langle f, g \rangle = \int_I f(x) g(x) w(x)\, \mathrm{d}x, \quad I = [-1, 1], \quad w(x) = 1.$$

Differential equation

$$\sigma(x) y''(x) + \tau(x) y'(x) + \big(\lambda_n + f_{\mu,\nu}(x)\big) y(x) = 0, \quad y = P_n^{m,m'},$$
$$\sigma(x) = 1 - x^2, \quad \tau(x) = -2x, \quad \lambda_n = n(n+1),$$
$$f_{\mu,\nu}(x) = -\frac{\mu^2}{2(1-x)} - \frac{\nu^2}{2(1+x)}.$$

Differential operator

$$\mathcal{D} = -(1 - x^2) \frac{\mathrm{d}^2}{\mathrm{d}x^2} + 2x \frac{\mathrm{d}}{\mathrm{d}x} + \frac{\mu^2}{2(1-x)} + \frac{\nu^2}{2(1+x)}.$$

Rodrigues formula

$$\begin{aligned} P_n^{m,m'}(x) &= P_n^{(0,0),m,m'}(x) \\ &= C_{n,\mu,\nu} (1-x)^{\mu/2} (1+x)^{\nu/2} P_{n-n^*}^{(\mu,\nu)}(x) \\ &= \frac{(-1)^{n-n^*} C_{n,\mu,\nu}}{2^{n-n^*} \Gamma(n - n^* + 1) (1-x)^{\mu/2} (1+x)^{\nu/2}} \\ &\quad \times \frac{\mathrm{d}^{n-n^*}}{\mathrm{d}x^{n-n^*}} \Big((1-x)^{n-n^*+\mu} (1+x)^{n-n^*+\nu} \Big). \end{aligned}$$

$$C_{n,\mu,\nu} = 2^{-n^*} \left(\frac{\Gamma(n - n^* + 1) \Gamma(n + n^* + 1)}{\Gamma(n - n^* + \mu + 1) \Gamma(n - n^* + \nu + 1)} \right)^{1/2}.$$

The matrix $\mathbf{G}$

$\{P_n^{m,m'}\}_{n=n^*}^N \to \{P_n^{\hat{m},\hat{m}'}\}_{n=\hat{n}^*}^N$:

$$\begin{aligned} \mathbf{G} = \operatorname{diag}(\mathbf{d}) &+ \frac{\mu^2 - \hat{\mu}^2}{2\hat{\mu}} \big(\operatorname{triu}(\mathbf{u}_1 \mathbf{v}_1^{\mathrm{T}}, 1) + \operatorname{tril}(\mathbf{v}_1 \mathbf{u}_1^{\mathrm{T}}, -1) \big) \\ &+ \frac{\nu^2 - \hat{\nu}^2}{2\hat{\nu}} \big(\operatorname{triu}(\mathbf{u}_2 \mathbf{v}_2^{\mathrm{T}}, 1) + \operatorname{tril}(\mathbf{v}_2 \mathbf{u}_2^{\mathrm{T}}, -1) \big), \end{aligned}$$

$$\mathbf{d} = \left(j(j+1) + (j + 1/2) \left(\frac{\mu^2 - \hat{\mu}^2}{2\hat{\mu}} + \frac{\nu^2 - \hat{\nu}^2}{2\hat{\nu}} \right) \right)_{j=\hat{n}^*}^N,$$

$$\mathbf{u}_1 = \left(\left(\frac{(j + 1/2) \Gamma(j - \hat{n}^* + \hat{\mu} + 1) \Gamma(j + \hat{n}^* + 1)}{\Gamma(j - \hat{n}^* + 1) \Gamma(j - \hat{n}^* + \hat{\nu} + 1)} \right)^{1/2} \right)_{j=\hat{n}^*}^N,$$

$$\mathbf{v}_1 = \left(\left(\frac{(j + 1/2) \Gamma(j - \hat{n}^* + 1) \Gamma(j - \hat{n}^* + \hat{\nu} + 1)}{\Gamma(j - \hat{n}^* + \hat{\mu} + 1) \Gamma(j + \hat{n}^* + 1)} \right)^{1/2} \right)_{j=\hat{n}^*}^N,$$

$$\mathbf{u}_2 = \left(\left((-1)^j \frac{(j+1/2)\Gamma(j-\hat{n}^*+\hat{\nu}+1)\Gamma(j+\hat{n}^*+1)}{\Gamma(j-\hat{n}^*+1)\Gamma(j-\hat{n}^*+\hat{\mu}+1)} \right)^{1/2} \right)_{j=\hat{n}^*}^{N},$$

$$\mathbf{v}_2 = \left(\left((-1)^j \frac{(j+1/2)\Gamma(j-\hat{n}^*+1)\Gamma(j-\hat{n}^*+\hat{\mu}+1)}{\Gamma(j-\hat{n}^*+\hat{\nu}+1)\Gamma(j+\hat{n}^*+1)} \right)^{1/2} \right)_{j=\hat{n}^*}^{N}.$$

A.2.7 Associated Legendre functions.

Symbol

$$P_n^m : \mathbb{R} \to \mathbb{R}, \quad n \in \mathbb{N}_0,\ |m| \le n,\ \mu := |m|.$$

Inner product and weight function

$$\langle f, g \rangle = \int_I f(x)g(x)w(x)\,\mathrm{d}x, \quad I = [-1,1], \quad w(x) = 1.$$

Differential equation

$$\sigma(x)y''(x) + \tau(x)y'(x) + \big(\lambda_n + f_\mu(x)\big)y(x) = 0, \quad y = P_n^m,$$

$$\sigma(x) = 1 - x^2, \quad \tau(x) = -2x, \quad \lambda_n = n(n+1),$$

$$f_\mu(x) = -\frac{\mu^2}{1-x^2}.$$

Differential operator

$$\mathcal{D} = -(1-x^2)\frac{\mathrm{d}^2}{\mathrm{d}x^2} + 2x\frac{\mathrm{d}}{\mathrm{d}x} + \frac{\mu^2}{1-x^2}.$$

Rodrigues formula

$$\begin{aligned} P_n^m(x) &= P_n^{(0,0),m}(x) \\ &= C_{n,\mu}(1-x^2)^{\mu/2}\frac{\mathrm{d}^\mu}{\mathrm{d}x^\mu}P_n(x) \\ &= \frac{(-1)^{n-\mu}}{2^n}\left(\frac{\Gamma(n+\mu+1)}{\Gamma(n-\mu+1)}\right)^{1/2}(1-x^2)^{-\mu/2}\frac{\mathrm{d}^{n-\mu}}{\mathrm{d}x^{n-\mu}}\left((1-x^2)^n\right), \end{aligned}$$

$$C_{n,\mu} = \left(\frac{\Gamma(n-\mu+1)}{\Gamma(n+\mu+1)}\right)^{1/2}.$$

The matrix $\mathbf{G}$

$\{P_n^m\}_{n=\mu}^N \to \{P_n^{\hat{m}}\}_{n=\hat{\mu}}^N$:

$$\mathbf{G} = \mathrm{diag}(\mathbf{d}) + \frac{\mu^2-\hat{\mu}^2}{\hat{\mu}}\left(\mathrm{triuc}(\mathbf{u}\,\mathbf{v}^{\mathrm{T}}, 1) + \mathrm{trilc}(\mathbf{v}\,\mathbf{u}^{\mathrm{T}}, -1)\right),$$

$$\mathbf{d} = \left(j(j+1) + (j+1/2)\frac{\mu^2-\hat{\mu}^2}{\hat{\mu}}\right)_{j=\hat{\mu}}^N,$$

$$\mathbf{u} = \left(\left(\frac{(j+1/2)\Gamma(j+\hat{\mu}+1)}{\Gamma(j-\hat{\mu}+1)}\right)^{1/2}\right)_{j=\hat{\mu}}^N,$$

$$\mathbf{v} = \left(\left(\frac{(j+1/2)\Gamma(j-\hat{\mu}+1)}{\Gamma(j+\hat{\mu}+1)}\right)^{1/2}\right)_{j=\hat{\mu}}^N.$$

A.2.8 Generalized associated Chebyshev functions of first kind.

Symbol

$$T_n^{m,m'} : \mathbb{R} \to \mathbb{R}, \quad n \in \mathbb{N}_0,\ n \geq n^*,$$
$$n^* := \max\{|m|, |m'|\},\ \mu := |m' - m|,\ \nu := |m' + m|.$$

Inner product and weight function

$$\langle f, g\rangle = \int_I f(x)g(x)w(x)\,\mathrm{d}x, \qquad I = [-1,1], \quad w(x) = \frac{1}{\sqrt{1-x^2}}.$$

Differential equation

$$\sigma(x)y''(x) + \tau(x)y'(x) + \big(\lambda_n + f_{\mu,\nu}(x)\big)y(x) = 0, \quad y = T_n^{m,m'},$$
$$\sigma(x) = 1 - x^2, \quad \tau(x) = -x, \quad \lambda_n = n^2,$$
$$f_{\mu,\nu}(x) = -\frac{\mu(\mu-1)}{2(1-x)} - \frac{\nu(\nu-1)}{2(1+x)}.$$

Differential operator

$$\mathcal{D} = -(1-x^2)\frac{\mathrm{d}^2}{\mathrm{d}x^2} + 2x\frac{\mathrm{d}}{\mathrm{d}x} + \frac{\mu(\mu-1)}{2(1-x)} + \frac{\nu(\nu-1)}{2(1+x)}.$$

Rodrigues formula

$$\begin{aligned}
T_n^{m,m'}(x) &= \frac{\Gamma(1/2)\Gamma(n+1)}{\Gamma(n+1/2)} P_n^{(-1/2,-1/2),m,m'}(x) \\
&= C_{n,\mu,\nu}(1-x)^{\mu/2}(1+x)^{\nu/2} P_{n-n^*}^{(\mu-1/2,\nu-1/2)}(x) \\
&= \frac{(-1)^{n-n^*} C_{n,\mu,\nu}}{2^{n-n^*}\Gamma(n-n^*+1)(1-x)^{\mu/2-1/2}(1+x)^{\nu/2-1/2}} \\
&\quad \times \frac{\mathrm{d}^{n-n^*}}{\mathrm{d}x^{n-n^*}}\Big((1-x)^{n-n^*+\mu-1/2}(1+x)^{n-n^*+\nu-1/2}\Big),
\end{aligned}$$

$$C_{n,\mu,\nu} = \frac{\Gamma(1/2)}{2^{n^*}}\left(\frac{\Gamma(n+1)\Gamma(n-n^*+1)\Gamma(n+n^*)}{\Gamma(n)\Gamma(n-n^*+\mu+1/2)\Gamma(n-n^*+\nu+1/2)}\right)^{1/2}.$$

The matrix $\mathbf{G}$

$\{T_n^{m,m'}\}_{n=n^*}^N \to \{T_n^{\hat{m},\hat{m}'}\}_{n=\hat{n}^*}^N$:

$$\begin{aligned}
\mathbf{G} = \mathrm{diag}(\mathbf{d}) &+ \frac{(\mu-\hat{\mu})(\mu+\hat{\mu}-1)}{2\hat{\mu}-1}\left(\mathrm{triu}(\mathbf{u}_1\mathbf{v}_1^{\mathrm{T}}, 1) + \mathrm{tril}(\mathbf{v}_1\mathbf{u}_1^{\mathrm{T}}, -1)\right) \\
&+ \frac{(\nu-\hat{\nu})(\nu+\hat{\nu}-1)}{2\hat{\nu}-1}\left(\mathrm{triu}(\mathbf{u}_2\mathbf{v}_2^{\mathrm{T}}, 1) + \mathrm{tril}(\mathbf{v}_2\mathbf{u}_2^{\mathrm{T}}, -1)\right),
\end{aligned}$$

$$\mathbf{d} = \left(j^2 + j\left(\frac{(\mu-\hat{\mu})(\mu+\hat{\mu}-1)}{2\hat{\mu}-1} + \frac{(\nu-\hat{\nu})(\nu+\hat{\nu}-1)}{2\hat{\nu}-1}\right)\right)_{j=\hat{n}^*}^{N},$$

$$\mathbf{u}_1 = \left(\left(\frac{j\Gamma(j-\hat{n}^*+\hat{\mu}+1/2)\Gamma(j+\hat{n}^*)}{\Gamma(j-\hat{n}^*+1)\Gamma(j-\hat{n}^*+\hat{\nu}+1/2)}\right)^{1/2}\right)_{j=\hat{n}^*}^{N},$$

$$\mathbf{v}_1 = \left(\left(\frac{j\Gamma(j-\hat{n}^*+1)\Gamma(j-\hat{n}^*+\hat{\nu}+1/2)}{\Gamma(j-\hat{n}^*+\alpha+\hat{\mu}+1/2)\Gamma(j+\hat{n}^*)}\right)^{1/2}\right)_{j=\hat{n}^*}^{N},$$

$$\mathbf{u}_2 = \left(\left((-1)^j\frac{j\Gamma(j-\hat{n}^*+\hat{\nu}+1/2)\Gamma(j+\hat{n}^*)}{\Gamma(j-\hat{n}^*+1)\Gamma(j-\hat{n}^*+\hat{\mu}+1/2)}\right)^{1/2}\right)_{j=\hat{n}^*}^{N},$$

$$\mathbf{v}_2 = \left(\left((-1)^j\frac{j\Gamma(j-\hat{n}^*+1)\Gamma(j-\hat{n}^*+\hat{\mu}+1/2)}{\Gamma(j-\hat{n}^*+\hat{\nu}+1/2)\Gamma(j+\hat{n}^*)}\right)^{1/2}\right)_{j=\hat{n}^*}^{N}.$$

A.2.9 Associated Chebyshev functions of first kind.

Symbol

$$T_n^m : \mathbb{R} \to \mathbb{R}, \quad n \in \mathbb{N}_0, \ |m| \le n, \ \mu := |m|.$$

Inner product and weight function

$$\langle f, g\rangle = \int_I f(x)g(x)w(x)\,\mathrm{d}x, \quad I = [-1,1], \quad w(x) = \frac{1}{\sqrt{1-x^2}}.$$

Differential equation

$$\sigma(x)y''(x) + \tau(x)y'(x) + \big(\lambda_n + f_\mu(x)\big)y(x) = 0, \quad y = T_n^m,$$

$$\sigma(x) = 1 - x^2, \quad \tau(x) = -x, \quad \lambda_n = n^2,$$

$$f_\mu(x) = -\frac{\mu(\mu-1)}{1-x^2}.$$

Differential operator

$$\mathcal{D} = -(1-x^2)\frac{\mathrm{d}^2}{\mathrm{d}x^2} + 2x\frac{\mathrm{d}}{\mathrm{d}x} + \frac{\mu(\mu-1)}{1-x^2}.$$

Rodrigues formula

$$\begin{aligned} T_n^m(x) &= \frac{\Gamma(1/2)\Gamma(n+1)}{\Gamma(n+1/2)} P_n^{(-1/2,-1/2),m}(x) \\ &= C_{n,\mu}(1-x^2)^{\mu/2}\frac{\mathrm{d}^\mu}{\mathrm{d}x^\mu}T_n(x) \\ &= \frac{(-1)^{n-\mu}\sqrt{\pi}}{2^n\Gamma(n+1/2)}\left(\frac{n\Gamma(n+\mu)}{\Gamma(n-\mu+1)}\right)^{1/2}(1-x^2)^{1/2-\mu/2}\frac{\mathrm{d}^{n-\mu}}{\mathrm{d}x^{n-\mu}}\left((1-x^2)^{n-1/2}\right), \end{aligned}$$

$$C_{n,\mu} = \left(\frac{\Gamma(n-\mu+1)}{n\Gamma(n+\mu)}\right)^{1/2}$$

The matrix **G**

$\{T_n^m\}_{n=\mu}^N \to \{T_n^{\hat{m}}\}_{n=\hat{\mu}}^N$:

$$\mathbf{G} = \operatorname{diag}(\mathbf{d}) + 2(\mu-\hat{\mu})\frac{\mu+\hat{\mu}-1}{2\hat{\mu}-1}\left(\operatorname{triuc}(\mathbf{u}\,\mathbf{v}^{\mathrm{T}},1) + \operatorname{trilc}(\mathbf{v}\,\mathbf{u}^{\mathrm{T}},-1)\right),$$

$$\mathbf{d} = \left(j^2 + 2j(\mu-\hat{\mu})\frac{\mu+\hat{\mu}-1}{2\hat{\mu}-1}\right)_{j=\hat{\mu}}^N,$$

$$\mathbf{u} = \left(\left(\frac{j\Gamma(j+\hat{\mu})}{\Gamma(j-\hat{\mu}+1)}\right)^{1/2}\right)_{j=\hat{\mu}}^N,$$

$$\mathbf{v} = \left(\left(\frac{j\Gamma(j-\hat{\mu}+1)}{\Gamma(j+\hat{\mu})}\right)^{1/2}\right)_{j=\hat{\mu}}^N.$$

A.2.10 Generalized associated Chebyshev functions of second kind.

Symbol

$$U_n^{m,m'}: \mathbb{R} \to \mathbb{R}, \quad n \in \mathbb{N}_0,\ n \geq n^*,$$
$$n^* := \max\{|m|, |m'|\},\ \mu := |m' - m|,\ \nu := |m' + m|.$$

Inner product and weight function

$$\langle f, g\rangle = \int_I f(x)g(x)w(x)\,\mathrm{d}x, \quad I = [-1,1], \quad w(x) = \sqrt{1-x^2}.$$

Differential equation

$$\sigma(x)y''(x) + \tau(x)y'(x) + \big(\lambda_n + f_{\mu,\nu}(x)\big)y(x) = 0, \quad y = U_n^{m,m'},$$
$$\sigma(x) = 1 - x^2, \quad \tau(x) = -x, \quad \lambda_n = n(n+2),$$
$$f_{\mu,\nu}(x) = -\frac{\mu(\mu+1)}{2(1-x)} - \frac{\nu(\nu+1)}{2(1+x)}.$$

Differential operator

$$\mathcal{D} = -(1-x^2)\frac{\mathrm{d}^2}{\mathrm{d}x^2} + 2x\frac{\mathrm{d}}{\mathrm{d}x} + \frac{\mu(\mu+1)}{2(1-x)} + \frac{\nu(\nu+1)}{2(1+x)}.$$

Rodrigues formula

$$\begin{aligned} U_n^{m,m'}(x) &= \frac{\Gamma(3/2)\Gamma(n+2)}{\Gamma(n+3/2)} P_n^{(1/2,1/2),m,m'}(x) \\ &= C_{n,\mu,\nu}(1-x)^{\mu/2}(1+x)^{\nu/2} P_{n-n^*}^{(\mu+1/2,\nu+1/2)}(x) \\ &= \frac{(-1)^{n-n^*} C_{n,\mu,\nu}}{2^{n-n^*}\Gamma(n-n^*+1)(1-x)^{\mu/2+1/2}(1+x)^{\nu/2+1/2}} \\ &\quad \times \frac{\mathrm{d}^{n-n^*}}{\mathrm{d}x^{n-n^*}}\left((1-x)^{n-n^*+\mu+1/2}(1+x)^{n-n^*+\nu+1/2}\right), \end{aligned}$$

$$C_{n,\mu,\nu} = \frac{\Gamma(3/2)}{2^{n^*}}\left(\frac{\Gamma(n+2)\Gamma(n-n^*+1)\Gamma(n+n^*+2)}{\Gamma(n+1)\Gamma(n-n^*+\mu+3/2)\Gamma(n-n^*+\nu+3/2)}\right)^{1/2}.$$

The matrix $\mathbf{G}$

$\{U_n^{m,m'}\}_{n=n^*}^N \to \{U_n^{\hat{m},\hat{m}'}\}_{n=\hat{n}^*}^N$:

$$\begin{aligned} \mathbf{G} = \operatorname{diag}(\mathbf{d}) &+ \frac{(\mu-\hat{\mu})(\mu+\hat{\mu}+1)}{2\hat{\mu}+1}\left(\operatorname{triu}(\mathbf{u}_1\,\mathbf{v}_1^{\mathrm{T}}, 1) + \operatorname{tril}(\mathbf{v}_1\,\mathbf{u}_1^{\mathrm{T}}, -1)\right) \\ &+ \frac{(\nu-\hat{\nu})(\nu+\hat{\nu}+1)}{2\hat{\nu}+1}\left(\operatorname{triu}(\mathbf{u}_2\,\mathbf{v}_2^{\mathrm{T}}, 1) + \operatorname{tril}(\mathbf{v}_2\,\mathbf{u}_2^{\mathrm{T}}, -1)\right), \end{aligned}$$

$$\mathbf{d} = \left(j(j+2\alpha) + (j+1)\left(\frac{(\mu-\hat{\mu})(\mu+\hat{\mu}+1)}{2\hat{\mu}+1} + \frac{(\nu-\hat{\nu})(\nu+\hat{\nu}+1)}{2\hat{\nu}+1} \right) \right)_{j=\hat{n}^*}^{N},$$

$$\mathbf{u}_1 = \left(\left(\frac{(j+1)\Gamma(j-\hat{n}^*+\hat{\mu}+3/2)\Gamma(j+\hat{n}^*+2)}{\Gamma(j-\hat{n}^*+1)\Gamma(j-\hat{n}^*+\hat{\nu}+3/2)} \right)^{1/2} \right)_{j=\hat{n}^*}^{N},$$

$$\mathbf{v}_1 = \left(\left(\frac{(j+1)\Gamma(j-\hat{n}^*+1)\Gamma(j-\hat{n}^*+\hat{\nu}+3/2)}{\Gamma(j-\hat{n}^*+\hat{\mu}+3/2)\Gamma(j+\hat{n}^*+2)} \right)^{1/2} \right)_{j=\hat{n}^*}^{N},$$

$$\mathbf{u}_2 = \left(\left((-1)^j \frac{(j+1)\Gamma(j-\hat{n}^*+\hat{\nu}+3/2)\Gamma(j+\hat{n}^*+2)}{\Gamma(j-\hat{n}^*+1)\Gamma(j-\hat{n}^*+\hat{\mu}+3/2)} \right)^{1/2} \right)_{j=\hat{n}^*}^{N},$$

$$\mathbf{v}_2 = \left(\left((-1)^j \frac{(j+1)\Gamma(j-\hat{n}^*+1)\Gamma(j-\hat{n}^*+\hat{\mu}+3/2)}{\Gamma(j-\hat{n}^*+\hat{\nu}+3/2)\Gamma(j+\hat{n}^*+2)} \right)^{1/2} \right)_{j=\hat{n}^*}^{N}.$$

A.2.11 Associated Chebyshev functions of second kind.

Symbol

$$U_n^m : \mathbb{R} \to \mathbb{R}, \quad n \in \mathbb{N}_0, \ |m| \le n, \ \mu := |m|.$$

Inner product and weight function

$$\langle f, g \rangle = \int_I f(x) g(x) w(x) \, \mathrm{d}x, \quad I = [-1, 1], \quad w(x) = \sqrt{1 - x^2}.$$

Differential equation

$$\sigma(x) y''(x) + \tau(x) y'(x) + \big(\lambda_n + f_\mu(x)\big) y(x) = 0, \quad y = U_n^m,$$
$$\sigma(x) = 1 - x^2, \quad \tau(x) = -3x, \quad \lambda_n = n(n+2),$$
$$f_\mu(x) = -\frac{\mu(\mu+1)}{1 - x^2}.$$

Differential operator

$$\mathcal{D} = -(1 - x^2) \frac{\mathrm{d}^2}{\mathrm{d}x^2} + 2x \frac{\mathrm{d}}{\mathrm{d}x} + \frac{\mu(\mu+1)}{1 - x^2}.$$

Rodrigues formula

$$\begin{aligned} U_n^m(x) &= \frac{\Gamma(3/2)\Gamma(n+2)}{\Gamma(n+3/2)} P_n^{(1/2,1/2),m}(x) \\ &= C_{n,\mu} \big(1 - x^2\big)^{\mu/2} \frac{\mathrm{d}^\mu}{\mathrm{d}x^\mu} U_n(x) \\ &= \frac{(-1)^{n-\mu}\Gamma(3/2)}{2^n \Gamma(n+3/2)} \left(\frac{(n+1)\Gamma(n+\mu+2)}{\Gamma(n-\mu+1)} \right)^{1/2} \\ &\quad \times (1 - x^2)^{-1/2-\mu/2} \frac{\mathrm{d}^{n-\mu}}{\mathrm{d}x^{n-\mu}} \left((1 - x^2)^{n+1/2} \right), \end{aligned}$$

$$C_{n,\mu} = \left(\frac{(n+1)\Gamma(n-\mu+1)}{\Gamma(n+\mu+2)} \right)^{1/2}.$$

The matrix $\mathbf{G}$

$\{U_n^m\}_{n=\mu}^N \to \{U_n^{\hat{m}}\}_{n=\hat{\mu}}^N$:

$$\mathbf{G} = \mathrm{diag}(\mathbf{d}) + 2(\mu - \hat{\mu}) \frac{\mu + \hat{\mu} + 1}{2\hat{\mu} + 1} \left(\mathrm{triuc}(\mathbf{u}\,\mathbf{v}^\mathrm{T}, 1) + \mathrm{trilc}(\mathbf{v}\,\mathbf{u}^\mathrm{T}, -1) \right),$$

$$\mathbf{d} = \left(j(j+2) + 2(j+1)(\mu - \hat{\mu}) \frac{\mu + \hat{\mu} + 1}{2\hat{\mu} + 1} \right)_{j=\hat{\mu}}^N,$$

$$\mathbf{u} = \left(\left(\frac{(j+1)\Gamma(j+\hat{\mu}+2)}{\Gamma(j-\hat{\mu}+1)} \right)^{1/2} \right)_{j=\hat{\mu}}^N,$$

$$\mathbf{v} = \left(\left(\frac{(j+1)\Gamma(j-\hat{\mu}+1)}{\Gamma(j+\hat{\mu}+2)} \right)^{1/2} \right)_{j=\hat{\mu}}^N.$$

Bibliography

[1] Alpert, B. K., and Rokhlin, V. A fast algorithm for the evaluation of Legendre expansions. *SIAM J. Sci. Stat. Comput. 12* (1991), 158 – 179.

[2] Andrews, G. E., Askey, R., and Roy, R. *Special Functions.* Cambridge University Press, Cambridge, UK, 2000.

[3] Arfken, G. B. *Mathematical methods for physicists*, 3rd ed ed. Academic Press, Orlando, 1985.

[4] Askey, R. *Orthogonal Polynomials and Special functions.* SIAM, Philadelphia, PA, USA, 1975.

[5] Björk, Å., and Dahlquist, G. *Numerical Methods.* Prentice Hall, Englewood Cliffs, NJ, USA, 1974.

[6] Bochner, S. Über Sturm-Liouvillesche Polynomsysteme. *Math. Z. 29*, 1 (1929), 730 – 736.

[7] Boyd, J. P. *Chebyshev and Fourier Spectral Methods*, second ed. Dover Press, New York, NY, USA, 2000.

[8] Bunch, J. R., Nielsen, C. P., and Sorensen, D. C. Rank-one modification of the symmetric eigenproblem. *Numer. Math. 31* (1978), 31 – 48.

[9] Chandrasekaran, S., and Gu, M. A divide-and-conquer algorithm for the eigendecomposition of symmetric block-diagonal plus semiseparable matrices. *Numer. Math. 96* (2004), 723 – 731.

[10] Chihara, T. *An Introduction to Orthogonal Polynomials.* Gordon and Breach, New York, NY, USA, 1978.

[11] Chirikjian, G. S., and Kyatkin, A. *Engineering Applications of Noncommutative Harmonic Analysis: with Emphasis on Rotation and Motion Groups.* CRC Press, Boca Raton, FL, USA, 2001.

[12] Cipra, B. A. The Best of the 20th Century: Editors Name Top 10 Algorithms. *SIAM News 33* (2000).

[13] Cooley, J. W., and Tukey, J. W. An algorithm for machine calculation of complex Fourier series. *Math. Comput. 19* (1965), 297 – 301.

[14] Cuppen, J. J. M. A divide and conquer method for the symmetric tridiagonal eigenproblem. *Numerische Mathematik 36* (1981), 177 – 195.

[15] Dongarra, J. J., and Sorensen, D. C. A fully parallel algorithm for the symmetric eigenvalue problem. *SIAM J. Sci. Stat. Comput. 8*, 2 (1987), 139 – 154.

[16] Driscoll, J. R., and Healy, D. Computing Fourier transforms and convolutions on the 2–sphere. *Adv. in Appl. Math. 15* (1994), 202 – 250.

[17] Driscoll, J. R., Healy, D., and Rockmore, D. Fast discrete polynomial transforms with applications to data analysis for distance transitive graphs. *SIAM J. Comput. 26* (1996), 1066 – 1099.

[18] Dutt, A., Gu, M., and Rokhlin, V. Fast algorithms for polynomial interpolation, integration and differentiation. Tech. Rep. YALEU/DCS/RR-977, Yale University, Department of Computer Science, 1993.

[19] Dutt, A., Gu, M., and Rokhlin, V. Fast algorithms for polynomial interpolation, integration and differentiation. *SIAM J. Numer. Anal. 33* (1996), 1689 – 1711.

[20] Eidelman, Y., Gohberg, I., and Olshevsky, V. Eigenstructure of order-one-quasiseparable matrices. Three-term and two-term recurrence relations. *Linear Algebra Appl. 405* (2005), 1 – 40.

[21] Erdős, P., and Turán, P. On Interpolation I. Quadrature and mean convergence in the Lagrange interpolation. *Ann. of Math. (2) 38* (1937), 142 – 155.

[22] FRIGO, M., AND JOHNSON, S. G. FFTW, C subroutine library. http://www.fftw.org.

[23] GAUTSCHI, W. *Orthogonal Polynomials – Computation and Approximation.* Numerical Mathematics and Scientific Computation. Oxford University Press, Oxford, UK, 2004.

[24] GOLDBERG, D. What Every Computer Scientist Should Know about Floating Point Arithmetic. *ACM Computing Surveys 23* (1991), 5 – 48.

[25] GOLUB, G. H. Some Modified Matrix Eigenvalue Problems. *SIAM Review 15*, 2 (1973), 318 – 334.

[26] GOLUB, G. H., AND VAN LOAN, C. F. *Matrix Computations*, second ed. The Johns Hopkins University Press, Baltimore, MD, USA, 1993.

[27] GRADSTEIN, I., AND RYSHIK, I. *Tables of Series, Products, and Integrals*, vol. 2. Verlag Harri Deutsch, Thun, Frankfurt am Main, Germany, 1981.

[28] GREENGARD, L., AND ROKHLIN, V. A fast algorithm for particle simulations. *J. Comput. Phys. 73* (1987), 325 – 348.

[29] GU, M., AND EISENSTAT, S. C. A stable and efficient algorithm for the rank-one modification of the symmetric eigenproblem. *SIAM J. Matrix Anal. Appl. 15*, 4 (1994), 1266 – 1276.

[30] GU, M., STANLEY, AND EISENSTAT, C. A stable and fast algorithm for updating the singular value decomposition. Tech. Rep. YALE/DCS/TR966, Yale University, 1993.

[31] HEALY, D., KOSTELEC, P., MOORE, S., AND ROCKMORE, D. FFTs for the 2-sphere - improvements and variations. *J. Fourier Anal. Appl. 9* (2003), 341 – 385.

[32] HEALY, D. M., KOSTELEC, P. J., AND ROCKMORE, D. Towards Safe and Effective High-Order Legendre Transforms with Applications to FFTs for the 2-sphere. *Adv. Comput. Math. 21* (2004), 59 – 105.

[33] HIELSCHER, R., PRESTIN, J., AND VOLLRATH, A. Fast summation of functions on the rotation group. *Mathematical Geosciences 42* (2010), 773–794.

[34] HIGHAM, N. J. *Accuracy and Stability of Numerical Algorithms.* SIAM, Philadelphia, PA, USA, 1996.

[35] HORN, R. A., AND JOHNSON, C. R. *Matrix Analysis.* Cambridge University Press, Cambridge, 1985.

[36] HUA, L. K. *Harmonic Analysis of Functions of Several Complex Variables in the Classical Domains.* Translations of Mathematical Monographs. American Mathematical Society, Providence, RI, USA, 1963.

[37] KAHAN, W. Rank-1 Perturbed Diagonal's Eigensystem. *unpublished manuscript* (1989).

[38] KEINER, J. Gegenbauer Polynomials and Semiseparable Matrices. *Electron. Trans. Numer. Anal. 30* (2008), 26 – 53.

[39] KEINER, J. Computing with Expansions in Gegenbauer Polynomials. *SIAM J. Sci. Comput. 31* (2009), 2151 – 2171.

[40] KEINER, J., KUNIS, S., AND POTTS, D. NFFT 3.0, C subroutine library. http://www.tu-chemnitz.de/~potts/nfft.

[41] KEINER, J., KUNIS, S., AND POTTS, D. Fast summation of radial functions on the sphere. *Computing 78* (2006), 1 – 15.

[42] KEINER, J., KUNIS, S., AND POTTS, D. Efficient reconstruction of functions on the sphere from scattered data. *J. Fourier Anal. Appl. 13* (2007), 435 – 458.

[43] KEINER, J., KUNIS, S., AND POTTS, D. Using NFFT3 - a software library for various nonequispaced fast Fourier transforms. *ACM Trans. Math. Software 36* (2009),

Article 19, 1 – 30.

[44] KEINER, J., AND POTTS, D. Fast evaluation of quadrature formulae on the sphere. *Math. Comput. 77* (2008), 397 – 419.

[45] KEINER, J., AND PRESTIN, J. A fast algorithm for spherical basis approximation. In *Frontiers in Interpolation and Approximation* (2006), N. K. Govil, H. N. Mhaskar, R. N. Mohapatra, Z. Nashed, and J. Szabados, Eds., Pure and Applied Mathematics, Taylor & Francis Books, Boca Raton, Florida.

[46] KEINER, J., AND WATERHOUSE, B. J. Fast principal components analysis method for finance problems with unequal time steps. In *Monte Carlo and Quasi-Monte Carlo Methods 2008*, P. L' Ecuyer and A. B. Owen, Eds. Springer Berlin Heidelberg, 2009, pp. 455–465.

[47] KNOPP, K. *Theory and Applications of Infinite Series.* Blackie and Son, Ltd., London, UK, 1963.

[48] KOORNWINDER, T. The addition formula for Jacobi polynomials, I, Summary of results. *Indag. Math. 34* (1972), 188 – 191.

[49] KOORNWINDER, T. The addition formula for Jacobi polynomials and spherical harmonics. *SIAM J. Appl. Math. 25* (1973), 236 – 246.

[50] KOORNWINDER, T. Jacobi polynomials, II. An analytic proof of the product formula. *SIAM J. Math. Anal. 5* (1974), 125 – 137.

[51] KOORNWINDER, T. Jacobi polynomials, III. An analytic proof of the addition formula. *SIAM J. Math. Anal. 6* (1975), 533 – 543.

[52] KOSTELEC, P. J., AND ROCKMORE, D. N. FFTs on the rotation group. *J. Fourier Anal. Appl. 14* (2008), 145 – 179.

[53] KOVACS, J. A., CHACÓN, P., CONG, Y., METWALLY, E., AND WRIGGERS, W. Fast rotational matching of rigid bodies by fast Fourier transform acceleration of five degrees of freedom. *Acta Crystallogr. Sect. D 59*, 8 (2003), 1371 – 1376.

[54] KUNIS, S. A note on stability results for scattered data interpolation on Euclidean spheres. *Adv. Comput. Math.* (accepted).

[55] KUNIS, S., AND POTTS, D. Fast spherical Fourier algorithms. *J. Comput. Appl. Math. 161* (2003), 75 – 98.

[56] LEBEDEV, N. N. *Special Functions and Their Applications.* Dover Publications, Inc., New York, NY, USA, 1972.

[57] LI, R. Solving secular equations stably and efficiently. Tech. rep., University of California at Berkeley, Berkeley, CA, USA, 1993.

[58] LUKE, Y. L. *The Special Functions and Their Approximations*, vol. 1. Academic Press, New York, NY, USA, 1969.

[59] MARONA, P., AND DA ROCHA, Z. Connection coefficients between orthogonal polynomials and the canonical sequence. *Numer. Algorithms 47* (2008), 291 – 314.

[60] MASTRONARDI, N., CAMP, E. V., AND BAREL, M. V. Divide and conquer algorithms for computing the eigendecomposition of symmetric diagonal-plus-semiseparable matrices. *Numer. Algorithms 39* (2005), 379 – 398.

[61] MOHLENKAMP, M. J. A fast transform for spherical harmonics. *J. Fourier Anal. Appl. 5* (1999), 159 – 184.

[62] MÜLLER, C. *Spherical Harmonics.* Springer, Aachen, 1966.

[63] NEVAI, P. G. *Orthogonal Polynomials.* American Mathematical Society, Providence, RI, USA, 1980.

[64] NIKIFOROV, A. F., AND UVAROV, V. B. *Special Functions of Mathematical Physics.* Birkhäuser, Basel, Switzerland, 1988.

[65] OLVER, F. W. J. *Asymptotics and Special Functions.* Academic Press, New York,

NY, USA, 1974.

[66] Potts, D., Prestin, J., and Vollrath, A. A fast algorithm for nonequispaced Fourier transforms on the rotation group. *Numer. Algorithms* (2009). accepted.

[67] Potts, D., Steidl, G., and Tasche, M. Fast and stable algorithms for discrete spherical Fourier transforms. *Linear Algebra Appl. 275/276* (1998), 433 – 450.

[68] Potts, D., Steidl, G., and Tasche, M. Fast Fourier transforms for nonequispaced data: A tutorial. In *Modern Sampling Theory: Mathematics and Applications* (Boston, MA, USA, 2001), J. J. Benedetto and P. J. S. G. Ferreira, Eds., Birkhäuser, pp. 247 – 270.

[69] Ritchie, D. W., and Kemp, G. J. L. Protein docking using spherical polar Fourier correlations. *PROTEINS: Struct. Funct. Genet. 39* (2000), 178 – 194.

[70] Rokhlin, V., and Tygert, M. Fast algorithms for spherical harmonic Expansions. *SIAM J. Sci. Comput. 27* (2006), 1903 – 1928.

[71] Sorensen, D. C., and Tang, P. T. P. On the orthogonality of eigenvectors computed by divide-and-conquer techniques. *SIAM J. Numer. Anal. 28*, 6 (1991), 1752 – 1775.

[72] Suda, R., and Takami, M. A fast spherical harmonics transform algorithm. *Math. Comput. 71* (2002), 703 – 715.

[73] Szegő, G. *Orthogonal Polynomials*, 4th ed. Amer. Math. Soc., Providence, RI, USA, 1975.

[74] Temme, N. M. *Special Functions.* John Wiley & Sons, Inc., New York, NY, USA, 1996.

[75] Tropp, E. A. Dimensional interpolation and addition formulae for Jacobi polynomial and hypergeometric function. Tech. Rep. 99-71, University of Texas at Austin, 1999.

[76] Tygert, M. Fast algorithms for spherical harmonic expansions II. *J. Comput. Phys. 227* (2008), 4260 – 4279.

[77] Vandebril, R., Barel, M. V., Golub, G., and Mastronardi, N. A bibliography on semiseparable matrices. *Calcolo 42* (2005), 249 – 270.

[78] Vandebril, R., van Barel, M., and Mastronardi, N. *Matrix Computations and Semiseparable Matrices Volume 1 – Linear Systems.* The Johns Hopkins University Press, Baltimore, MD, USA, 2008.

[79] Vandebril, R., van Barel, M., and Mastronardi, N. *Matrix Computations and Semiseparable Matrices Volume 2 – Eigenvalue and Singular Value Methods.* The Johns Hopkins University Press, Baltimore, MD, USA, 2008.

[80] Varshalovich, D., Moskalev, A., and Khersonski, V. *Quantum Theory of Angular Momentum.* World Scientific Publishing, Singapore, 1988.

[81] v.d. Boogaart, K. G., Hielscher, R., Prestin, J., and Schaeben, H. Kernel-based methods for inversion of the radon transform on SO(3) and their applications to texture analysis. *J. Comput. Appl. Math. 199* (2007), 122 – 140.

[82] Wilkinson, J. *The Algebraic Eigenvalue Problem.* Clarendon Press, Oxford, UK, 1965.